QMC   701299 3

TURN

METEOROLOGICAL GLOSSARY

Met. O. 842
A.P.897

Meteorological Office

Meteorological Glossary

Compiled by D. H. McIntosh, M.A., D.Sc.

London: Her Majesty's Stationery Office: 1972

U.D.C. 551.5(038)

First published 1916
Fifth edition 1972

© *Crown copyright 1972*

Printed and published by
HER MAJESTY'S STATIONERY OFFICE

To be purchased from
49 High Holborn, London WC1V 6HB
13a Castle Street, Edinburgh EH2 3AR
109 St Mary Street, Cardiff CF1 1JW
Brazennose Street, Manchester M60 8AS
50 Fairfax Street, Bristol BS1 3DE
258 Broad Street, Birmingham B1 2HE
80 Chichester Street, Belfast BT1 4JY
or through booksellers

Price £2·75 net

SBN 11 400208 8*

PREFACE TO THE FIFTH EDITION

When, in 1967, the *Meteorological glossary* came under consideration for reprinting, it was decided to ask Dr McIntosh to undertake a revised edition, with co-operation from within the Meteorological Office. The opportunity has been taken in this edition, to delete some terms which are considered no longer appropriate, and to include various new entries and revisions which stem from recent advances and practice.

Units of the Système International have been adopted in this edition. In some cases, however, the traditional British or metric units are also included because of existing World Meteorological Organization recommendations and for the convenience of user interests during the period before complete national and international adoption of SI units.

Meteorological Office, 1970.

PREFACE TO THE FOURTH EDITION

In 1916, during the directorship of Sir Napier Shaw, the Meteorological Office published two pocket-size companion volumes, the 'Weather map' to explain how weather maps were prepared and used by the forecasters, and the 'Meteorological glossary' to explain the technical meteorological terms then employed. With the advance of the science and the elaboration of its techniques the publications have been in continuous demand, many times reprinted and on several occasions completely revised. The second edition was in 1930, the third in 1938–39, and soon after World War II it was obvious that radical revision was necessary once again. In 1956 the fourth edition of the 'Weather map' was issued but, for the first time in forty years, it was not found possible to prepare simultaneously a new edition of the 'Meteorological glossary', which had become very much out-of-date and in need of a complete remodelling. For earlier editions the task had been shared amongst the professional staff of the Office and the result was an interesting and up-to-date volume containing much useful information, although the freedom allowed to the many contributors had led to a unique volume of quite uneven character with articles varying from brief dictionary definitions to encyclopaedic essays. For the fourth edition the number of new entries was to be larger than ever before and it was decided that the need was now for a more systematic reference work containing a brief definition of all the terms in ordinary use rather than for a compilation of miscellaneous articles giving information which would more properly be looked for in one of the many modern textbooks. For this purpose a single author, assisted if need be by expert referees, would, it seemed, be advantageous and the Office was fortunate in finding in Dr D. H. McIntosh of the University of Edinburgh a physicist and meteorologist of wide experience willing to undertake the major task. Before being passed for printing every article has been read critically by more than one member of the scientific staff of the Office and Dr. McIntosh has shown a remarkable readiness to compromise. In this way it is hoped that the excellence of the author's original draft has been fully retained while providing a work which will adequately meet the needs of the official service. It would, however, be too much to hope that no further improvement will be possible and the Office will be pleased to receive from any source criticisms and suggestions calculated to increase the value of the work, not only for the professional meteorologist, but for interested people everywhere.

Meteorological Office, 1962.

ACKNOWLEDGEMENTS

The Meteorological Office is indebted to the following for permission to reproduce many of the illustrations appearing in this publication:

	Plates
G. V. Black, Esq.	2
H. A. D. Cameron, Esq.	3
M. W. Dybeck, Esq.	24
J. Paton, Esq.	1, 18
Rev. Fr. C. Rey, S.J.	25
J. Steljes, Esq.	19, 21
J. E. Tinkler, Esq.	20
The Royal Meteorological Society (Clarke-Cave Collection)	5, 6, 8, 9, 10, 11, 12, 13, 15, 16
The United States Navy	26

The remaining Plates, Nos., 4, 7, 14, 17, 22, 23, 27 and 28 are all Crown copyright.

A

ablation: The disappearance of snow and ice by melting and evaporation. The chief meteorological factor which controls the rate of ablation is air temperature; subsidiary factors are humidity, wind speed, direct solar radiation and rainfall. The rate of ablation of a snow-field is also affected by such non-meteorological factors as size, slope and aspect of snow-field, depth and age of snow, and nature of underlying surface.

abroholos: A violent squall on the coast of Brazil, occurring mainly between May and August.

absolute extremes: See EXTREMES.

absolute humidity: An alternative for VAPOUR CONCENTRATION.

absolute instability: See STABILITY.

absolute instrument: An instrument with which measurements may be made in units of mass, length, and time (or in units of a known and direct relationship to these) and against which other non-absolute instruments may be calibrated. See also STANDARD.

absolute stability: See STABILITY.

absolute temperature: See TEMPERATURE SCALES.

absolute vorticity: See VORTICITY.

absolute zero (of temperature): Temperature of $-273 \cdot 15°C$, the zero on the kelvin (absolute) scale. See TEMPERATURE SCALES.

absorption: Removal of radiation from an incident solar or terrestrial beam, with conversion to another form of energy—electrical, chemical or heat.
 The absorption of radiation by the gases of the atmosphere is highly selective in terms of wavelengths and may depend also on pressure and temperature. Numerical expression is given by the law, variously known as Beer's, Bouguer's, or Lambert's law, applicable to monochromatic radiation:

$$I = I_0 e^{-\alpha m}$$

where I_0 is the intensity of incident radiation, I the intensity after passing through mass m of absorbing substance and α the absorption coefficient. An alternative expression is

$$I = I_0 e^{-\alpha x}$$

where x is the path length through the absorbing substance.
 The effectiveness of a gas as an absorber of solar or terrestrial radiation depends

on the width and strength (absorption coefficient) of the absorption lines and bands, the concentration of the gas, and the wavelength positions of the bands relative to the maximum of the Planck curve (see RADIATION) at solar or terrestrial temperature, respectively. The relative energies involved in atmospheric absorption processes are represented in Figure 1, in which the Planck curves appropriate to solar and terrestrial radiation temperatures (6000 and 250 K, respectively) are shown with equal areas to represent over-all balance of the two fluxes. Fine structure of the absorption bands is omitted. The main constituents of the atmosphere, N_2 and O_2, are almost completely transparent except in the far ultra-violet: minor constituents such as O_3, CO_2, H_2O, N_2O have intense absorption bands, mainly in the infra-red and longer wavelengths. The product of (a) and (c) in Figure 1 represents the energy absorbed by the stratosphere, and the product of (a) and ((b)−(c)) that absorbed by the troposphere.

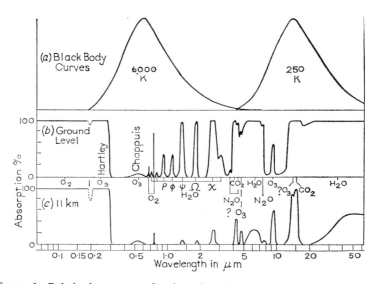

FIGURE 1—Relative importance of various absorptions in the atmosphere
(a) Black-body emission for 6000 and 250 K.
(b) Atmospheric absorption spectrum for a solar beam reaching ground level.
(c) Atmospheric absorption spectrum for a solar beam reaching the temperate tropopause.

ρ, φ, ψ, Ω and χ are the near infra-red bands of water vapour. In (b) and (c) no fine structure is shown although all the infra-red bands have in fact a very complex structure. Most of the bands shown have a broad doublet structure, but even this is smoothed out. It has been assumed that the atmosphere contains 3 mm of ozone at s.t.p., all above 11 km; that it contains 2 g/cm² of water vapour above ground level and 10^{-3} g/cm² above 11 km; that carbon dioxide and nitrous oxide are mixed in equal proportions with the atmosphere at all heights. The small window (dotted) at 0·2μm and the two ozone bands marked (?) have not yet been observed at the two levels concerned.

(GOODY, R. M. and ROBINSON, G. D.; Radiation in the troposphere and lower stratosphere. *Q J R Met Soc*, London, 77, 1951, p. 153.)

Since water has an appreciable absorption coefficient in solar radiation wavelengths greater than about one micrometre, thick clouds are able to absorb over 20 per cent of incident solar radiation. Towards terrestrial radiation, clouds and fog behave as almost perfect black bodies. The surface of the earth is very variable in its absorption of solar radiation (see ALBEDO) but absorbs nearly all incident terrestrial radiation.

acceleration: Rate of change of velocity with respect to time. Like velocity, acceleration has both magnitude and direction; thus uniform circular motion, involving change of direction without change of speed, implies acceleration ('centripetal acceleration'). The dimensions are $L\,T^{-2}$.

The 'absolute acceleration' of the air ($\dot{\mathbf{V}}_a$), acceleration measured relative to axes fixed in space, is equal to the force acting per unit mass of air (Newton's second law of motion). The important forces per unit mass in air motion are the PRESSURE GRADIENT FORCE $\left(-\dfrac{1}{\rho}\nabla p\right)$, the Newtonian force of GRAVITY directed towards the earth's centre ($\mathbf{g}_a$), and the 'frictional' forces arising from eddy VISCOSITY or molecular viscosity (**F**). Thus we have the 'equation of absolute motion':

$$\dot{\mathbf{V}}_a = -\frac{1}{\rho}\nabla p + \mathbf{g}_a + \mathbf{F}.$$

The air's 'relative acceleration' ($\dot{\mathbf{V}}$), i.e. its acceleration measured with respect to axes fixed in the earth, may be shown to be related to $\dot{\mathbf{V}}_a$ and to two other accelerations, namely the CENTRIPETAL ACCELERATION of the coinciding point of the earth ($\dot{\mathbf{V}}_e$) and the CORIOLIS ACCELERATION ($-2\Omega \wedge \mathbf{V}$, where Ω is the earth's angular velocity and $\wedge$ denotes the vector cross product).

The relationship is
$$\dot{\mathbf{V}} = \dot{\mathbf{V}}_a - 2\Omega \wedge \mathbf{V} - \dot{\mathbf{V}}_e.$$

Substitution for $\dot{\mathbf{V}}_a$ in the first equation gives

$$\dot{\mathbf{V}} = -\frac{1}{\rho}\nabla p + \mathbf{g}_a - \dot{\mathbf{V}}_e - 2\Omega \wedge \mathbf{V} + \mathbf{F}.$$

Since also the force of gravity along the local vertical (**g**) is given by $\mathbf{g} = \mathbf{g}_a - \dot{\mathbf{V}}_e$, we have, by substitution, the 'equation of relative motion'

$$\dot{\mathbf{V}} = -\frac{1}{\rho}\nabla p - 2\Omega \wedge \mathbf{V} + \mathbf{g} + \mathbf{F}.$$

The relative acceleration of the air ($\dot{\mathbf{V}}$) is of fundamental importance in dynamical meteorology because it is directly related to the development of pressure systems. While, in theory, $\dot{\mathbf{V}}$ may be obtained from the equation of relative motion, this is not possible in practice because it is a small residual of much larger and imperfectly known terms.

The equation of relative motion has its most practical application in 'steady' motion ($\dot{\mathbf{V}} = 0$) in which all the forces acting on the air are balanced. In particular, when the flow is horizontal and frictionless, the forces **g** and **F** may be neglected and the equation becomes that for GEOSTROPHIC flow.

Common situations in which the air is subject to acceleration, with accompanying AGEOSTROPHIC component of wind, are (a) curved flow (see GRADIENT WIND), (b) change of pressure gradient in direction of flow (see CONFLUENCE), (c) local change of pressure gradient with time (see ISALLOBARIC WIND).

The degree of acceleration experienced in flight is usually expressed in the unit g, this being the acceleration produced by gravity at the earth's surface, namely 9·8 m/s².

acclimatization: The process of adjustment of an animal, normally with implied physiological effects, to a marked change of physical environment, such as to a large change of altitude or a change to extreme conditions of temperature and/or humidity.

accretion: In meteorology, this usually refers to the growth of an ice particle by collision with water drops. The term is also used in the more general sense of growth of water drops, or ice particles, by collision. See PRECIPITATION, ICE ACCRETION.

accumulated temperature: The integrated excess or deficiency of temperature measured with reference to a fixed datum over an extended period of time. If on a given day the temperature is above the datum value for n hours and the mean temperature during that period exceeds the datum line by m degrees, the accumulated temperature for the day above the datum is nm degree-hours or $nm/24$ degree-days. By summing the daily entries arrived at in this way, the accumulated temperature above or below the datum value may be evaluated for periods such as a week, a month, a season or a year.

In practice, daily values of accumulated temperature are derived not from hourly values but by a method involving the use of daily maximum (X) and minimum (N) temperatures: empirical formulae relating to X, N and the datum value (D) are used when D lies between X and N.

The datum value which has been used in relation to agriculture is 42°F (6°C); this is widely used as the critical temperature above which the growth of vegetation in a European climate is initiated and maintained. For the study of heating problems a datum value of 60°F is used by British engineers. Meteorological Office *Professional Note* No. 125* contains average monthly and yearly values of accumulated temperature with respect to datum values of 42°, 50°, 60° and 70°F at each of 49 stations in the British Isles for the period 1921–50. The data were derived by an empirical method which uses the average value and standard deviation of monthly mean temperature.

accuracy: In physical measurement, the closeness with which an observation of a quantity, or the mean of a series of observations, is considered to approach the unknown true value of the quantity. See also ERROR.

acoustic sounding: Investigation of the properties of the atmosphere by the propagation and reception of sound waves.

acre-inch: An obsolescent term for the volume of water (weighing about 100 tons ($\approx 10^5$ kg)) which would cover one acre to the depth of one inch (1 acre = 4046·86m². 1 inch = 25·4 mm).

actinic rays: Radiation which effects chemical changes, as in photography. The term is also loosely used to signify ULTRA-VIOLET RADIATION.

actinometer: An early name for an instrument which measures solar radiation, usually at normal incidence, as in the Linke–Fuessner and Michelson actinometers (see PYRHELIOMETER). The corresponding term for a recording instrument is 'actinograph'. The name 'actinometer' is also applied to an instrument which measures the intensity of ACTINIC RAYS. See also PYRANOMETER and PYRHELIOMETER.

actinon: Gas, of atomic mass 219 and atomic number 86, which is a radioactive isotope of RADON. It occurs in minute concentration in the atmosphere and plays a small part in the IONIZATION of the air at low levels.

---

* SHELLARD, H. C.; Averages of accumulated temperature and standard deviation of monthly mean temperature over Britain, 1921–50. *Prof Notes Met Off*, London, **8**, No. 125, 1959.

adiabatic: An adiabatic process (thermodynamic) is one in which heat does not enter or leave the system. (Greek, *a* not, and *diabaino* pass through.)

Because the atmosphere is compressible and pressure varies with height adiabatic processes play a fundamental role in meteorology. Thus, if a parcel of air rises it expands against its lower environmental pressure; the work done by the parcel in so expanding is at the expense of its internal energy and its temperature falls, despite the fact that no heat leaves the parcel. Conversely, the internal energy of a falling parcel is increased and its temperature raised, as a result of the work done on the air in compressing it.

Observation shows that such processes determine, to a large extent, the vertical temperature distribution within the troposphere. It also supports the view that, to a first approximation, it is justifiable to treat the vertically moving, individual masses of air of indefinite size (termed 'parcels') as CLOSED SYSTEMS which move through the environment without unduly disturbing it or exchanging heat with it. Various non-adiabatic processes such as condensation, evaporation, radiation, and turbulent mixing also operate to produce temperature changes in the free atmosphere but their effects are generally negligible in comparison with those caused by appreciable vertical motion.

Such adiabatic or 'dynamical' temperature changes proceed at a definite rate. For dry (unsaturated) air the change in temperature per unit height change (i.e. the LAPSE rate) is given by the equivalent expressions:

$$\frac{\gamma - 1}{\gamma} \frac{g}{R} \text{ or } \frac{g}{c_p}$$

where R is the gas constant for air (287 J/kg degK), g the acceleration of gravity (9·8 m/s^2) and γ the ratio of specific heats of dry air at constant pressure (c_p) and constant volume (c_v) respectively (1·4), from which the 'dry adiabatic lapse rate' (DALR) is about 0·98 degC per 100 m or, with sufficient accuracy, 1 degC per 100 m (5·4 degF per 1000 ft).

For a saturated rising parcel the fall of temperature is checked by the latent heat liberated. The 'saturated adiabatic lapse rate' (SALR) is therefore less than that for unsaturated air by an amount which varies with temperature and pressure: at lower levels in temperate latitudes the SALR is about half that of the DALR. Since widespread vertical motion occurs in the TROPOSPHERE, the average lapse rate in this region lies between the DALR and SALR.

Two extreme types of process involving ascent of saturated air may be visualized: (i) a reversible, adiabatic ascent at the SALR, in which all products of condensation—cloud, rain, hail or snow—are retained within the ascending air, partake of the temperature changes of the air, and are available for evaporation at the appropriate stages on subsequent descent of the air, which is also at the SALR; (ii) an irreversible and, strictly, non-adiabatic process, in which all products of condensation are removed during ascent and in which subsequent descent of the air is at the DALR. The latter process corresponds much more closely to what happens in the atmosphere than does the former and is termed a pseudo-adiabatic process. Because of the loss of the heat content of the precipitated water, cooling on ascent in a pseudo-adiabatic process is at a rate slightly in excess of the SALR, but the difference between the rates is negligible; thus, saturated adiabatics and pseudo-adiabatics (lines on an AEROLOGICAL DIAGRAM representing the respective lapse rates) are, for practical purposes, identical. The important distinction between the pseudo-adiabatic and reversible processes lies in the different rates of temperature change undergone by the air on subsequent descent. See also ENTROPY, ISENTROPIC, ADIABATIC EQUATIONS.

adiabatic atmosphere: A hypothetical atmosphere characterized by the dry adiabatic lapse rate throughout. It is also termed 'neutral atmosphere' or 'convective atmosphere'. See ADIABATIC.

adiabatic diagram: An alternative for AEROLOGICAL DIAGRAM, or THERMODYNAMIC DIAGRAM.

adiabatic equations: The three (equivalent) relationships between the variables of state, p (pressure), v (specific volume), and T (temperature), of a PERFECT GAS in an ADIABATIC process.

They are:

(i) $pv^\gamma =$ constant, (ii) $Tv^{\gamma-1} =$ constant, (iii) $Tp^{-(\gamma-1)/\gamma} =$ constant, where γ is the ratio of the specific heats, c_p/c_v. The third relationship is the one most often used in meteorology. For dry air at ordinary temperatures, $\gamma = 1.40$ and

$$\frac{\gamma - 1}{\gamma} = \frac{c_p - c_v}{c_p} = \frac{R}{c_p} = 0.286.$$

The same value may be used as an approximation for moist air. See SPECIFIC HEAT.

adiabatic region: That region of the atmosphere (the TROPOSPHERE) where the temperature LAPSE rate is determined mainly by vertical ADIABATIC motion of the air; it is also termed the 'convective region'.

adsorption: The penetration of a substance, e.g. gas or thin film of liquid, into the surface layer of a solid with which it is in contact.

advection: The process of transfer (of an air-mass property) by virtue of motion. In particular cases, attention may be confined to either the horizontal or vertical components of the motion. The term is, however, often used to signify horizontal transfer only.

advection fog: Fog formed by the passage of relatively warm, moist and stable air over a cool surface. It is associated mainly with cool sea areas, particularly in spring and summer, and may affect adjacent coasts. It may occur also over land in winter, particularly when the surface is frozen or snow-covered—sometimes then, however, in conjunction with RADIATION FOG.

advective change of temperature: That contribution to local temperature change which is caused by either (or both) horizontal or vertical ADVECTION of air. The horizontal component of change, which is generally the more effective in the troposphere, is proportional to the horizontal temperature gradient at the level concerned and to the wind speed in the direction of this gradient; the vertical component of change is proportional to the vertical wind velocity and to the static stability of the air and depends also on whether or not the air is saturated.

aerobiology: The study of the part played by the earth's atmosphere in the movement of living animal and plant organisms.

aerodrome meteorological minima: Limiting meteorological conditions prescribed for the purpose of determining the usability of an aerodrome either for take-off or for landing of aircraft.

aerodynamic roughness, smoothness: A physical boundary is 'aerodynamically rough' when fluid flow is turbulent down to the boundary itself. Over such a boundary the velocity profile and surface drag are independent of the fluid viscosity (v) but depend on a ROUGHNESS LENGTH (z_0) which is related to the height and spacing of the roughness elements of the surface. A surface is 'aerodynamically smooth' if there exists a layer adjacent to it in which the flow is laminar and in which the velocity profile and surface drag are related to the fluid viscosity. A surface which is aerodynamically smooth at low speed of flow may become aerodynamically rough at a higher speed. A surface may thus be described as rough or smooth only in terms of the associated flow; alternatively, the flow itself may be described as aerodynamically rough or smooth.

In meteorology, nearly all surfaces are aerodynamically rough for any significant wind speed.

aerodynamics: The study of the forces and reactions arising from the motion of bodies, more particularly the parts of an aircraft, through the air. It is from such forces or reactions that the lifting power of an aircraft is obtained.

aerogram: An AEROLOGICAL DIAGRAM, due to A. Refsdal, in which the abscissa is $\log T$ and the ordinate $T \log p$.

aerological diagram: A graphical representation of the observations of pressure, temperature and humidity, made in a vertical sounding of the atmosphere.

The reference lines which facilitate the plotting of a sounding and its assessment after plotting are isobars, isotherms, dry adiabatics, saturated (pseudo-) adiabatics and saturation moisture lines. Each of the diagrams in common use—the tephigram, emagram and Stüve diagram—has its own particular advantage. Each has, exactly or very nearly, the property of being a true thermodynamic ('equivalent') diagram in that equal area represents equal energy at any point of the diagram; possession of this property simplifies energy and height (geopotential) calculations.

The tephigram (T–φ gram)—see Figure 2—has rectangular Cartesian coordinates in which the abscissa is temperature (T) and the ordinate ENTROPY (φ), (entropy is now normally designated S, not φ) i.e. $\log \theta$, where θ is the dry-bulb potential temperature; the dry adiabatics ($\theta =$ constant) are therefore straight lines perpendicular to the isotherms. The basic co-ordinates in the emagram are T and $\log p$ ($p =$ pressure); in a 'rectangular' model the T and $\log p$ axes are perpendicular to each other; in the 'oblique' emagram these axes meet at an angle of 45°, thus making the dry adiabatics (slightly curved) and isotherms meet at an angle of about 90°. This latter is a decided advantage in assessing the static STABILITY of an ascent curve since the normal range of lapse rate of temperature in the troposphere is between the isothermal and dry adiabatic rates. In the Stüve diagram rectangular co-ordinates of T and p^κ ($\kappa = 0{\cdot}286$) are used; the dry adiabatics are then straight lines (as are the isobars and isotherms) but the property of strict equivalence of energy and area is sacrificed.

aerology: A word denoting the study of the atmosphere, but generally used in the sense of a study limited to the atmosphere above the surface layers.

aeronomy: A term sometimes used to denote that branch of atmospheric physics which is concerned with those regions, upwards of about 50 km, where DISSOCIATION and IONIZATION are fundamental properties.

aerosol: In meteorology, an aggregation of minute particles (solid or liquid) suspended in the atmosphere. See NUCLEUS.

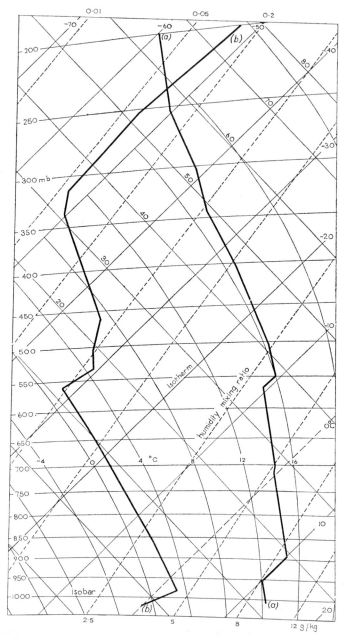

FIGURE 2—Aerological diagram (tephigram) illustrating pressure–temperature plots of two ascents made at Crawley (51° 05′ N, 0° 13′ W).
Curve (a): ascent made in maritime tropical air, 2330 GMT, 11 December 1961.
Curve (b): ascent made in Arctic air, 2330 GMT, 5 December 1961.

afterglow: See ALPINE GLOW.

ageostrophic wind: The VECTOR difference between the actual wind and the GEOSTROPHIC WIND (see Figure 3); it is also called the 'geostrophic departure' and 'geostrophic deviation'.

Actual wind = geostrophic wind + ageostrophic wind.

The ageostrophic wind is of fundamental importance in that it is necessarily associated with CONVERGENCE or DIVERGENCE and vertical motion in the atmosphere. The ISALLOBARIC WIND is a particular example of ageostrophic wind; an ageostrophic component is also present in the SURFACE WIND and GRADIENT WIND.

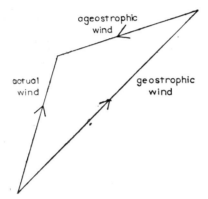

FIGURE 3—Ageostrophic wind.

aggregation: The process of growth of snowflakes or ice crystals by collision and adherence.

agroclimatology: The study of those aspects of climate which are relevant to the problems of agriculture. Such study involves types of data, e.g. earth temperature and accumulated temperature, which are often not considered in more general CLIMATOLOGY.

agrometeorology: Meteorology relevant to problems of agriculture. It is concerned, in particular, with the surface layers of the atmosphere and with the conditions in the top layer of the earth's surface which are associated with variations of the meteorological elements.

air: The mixture of gases which form the earth's ATMOSPHERE. In the absence of dust and water vapour, the composition of the air up to about 20 km is taken to be as shown in Table I; the percentage composition by volume and weight is given, i.e. FRACTIONAL VOLUME ABUNDANCE and MIXING RATIO.

Of the gases shown in Table I only carbon dioxide and ozone have appreciable local variations of concentration. There are also, mainly in the low atmosphere, minute variable quantities of such gases as radon, actinon, thoron, sulphur dioxide, hydrogen chloride, methane and nitrous oxide.

Ordinary (moist) air may be regarded as a mixture of dry air and water vapour. The concentration of water vapour in surface air varies from a small fraction of one per cent to over three per cent; in general the concentration decreases with increasing altitude.

TABLE I—*Composition of dry air*

	Molecular weight ($^{12}C = 12·000$)	Proportional composition By volume per cent	By weight per cent
Dry air	28·966	100·0	100·0
Nitrogen	28·013	78·09	75·54
Oxygen...	31·999	20·95	23·14
Argon	39·948	0·93	1·27
Carbon dioxide	44·010	0·03	0·05
Neon	20·183	$1·8 \times 10^{-3}$	$1·2 \times 10^{-3}$
Helium	4·003	$5·2 \times 10^{-4}$	$7·2 \times 10^{-5}$
Krypton	83·800	$1·0 \times 10^{-4}$	$3·0 \times 10^{-4}$
Hydrogen	2·016	$5·0 \times 10^{-5}$	$4·0 \times 10^{-6}$
Xenon	131·300	$8·0 \times 10^{-6}$	$3·6 \times 10^{-5}$
Ozone	47·998	$1·0 \times 10^{-6}$	$1·7 \times 10^{-6}$

air discharge: A term sometimes used to denote a lightning flash from cloud to air. See LIGHTNING.

airglow: General term for the radiation which is emitted continuously by the upper atmosphere. The day, twilight and night emissions are termed DAYGLOW, TWILIGHT-GLOW and NIGHTGLOW, respectively, the last being the most extensively studied.

airlight: The increase in apparent brightness of a distant object viewed in daylight, owing to scattering of light towards the observer by particles held in suspension in the atmosphere, and by air molecules, between the observer and object. Airlight, and therefore object brightness, increase with object distance owing to increase in the number of scattering agents. A critical point is reached, limiting daylight VISIBILITY, at which the brightness of a suitable object is just indistinguishable from its background. See also KOSCHMIEDER'S LAW, CONTRAST THRESHOLD OF THE EYE.

air mass: A body of air in which horizontal gradients of temperature and humidity are relatively slight and which is separated from an adjacent body of air by a more or less sharply defined transition zone (FRONT) in which these gradients are relatively large.

The horizontal dimensions of air masses are normally hundreds or even thousands of kilometres. The term is, however, also used in relation to phenomena of much smaller scale, e.g. the sea-breeze.

Homogeneity in a body of air is produced by prolonged contact, in a 'source region', with an underlying surface of uniform temperature and humidity. The main source regions are those in which occur the permanent or semi-permanent anticyclones, with rather indeterminate boundaries, which are a prominent feature of the GENERAL CIRCULATION—the subtropical, polar, and winter continental anticyclones. Slow transformation of the air-mass properties acquired at the source region is effected on subsequent movement of the air from the region, mainly through its contact with a different surface, but to an appreciable extent also by radiation and large-scale vertical motion. The synoptic meteorology of middle latitudes, in particular, is dominated by considerations of the air-mass properties originally acquired and the manner of their recent modification—whether warming or cooling, becoming more or less moist or more or less stable.

Air masses are classified into groups designated as 'polar' (P) or 'tropical' (T), maritime (m) or continental (c), defining the basic temperature and humidity characteristics, respectively; more generally, a twofold classification in terms of both elements, e.g. *mT* or *cP*, is used. Further divisions are sometimes made into Arctic or Antarctic (*A*) air and into classes of more local significance, e.g. Mediterranean air.

air-mass analysis: Synoptic identification of AIR MASSES and location of boundaries (FRONTS) between adjacent air masses.

The identifying of an air mass implies the finding of its source region and would appear to entail the retrospective tracking of the air back to such a region. In practice, such a procedure is very seldom required because of the continuity provided by a series of synoptic charts. An air mass moving from its source region usually has in early stages a well-defined front at its junction with the adjacent air mass. Further identification of the limits of the air masses is largely made in terms of the movement of this front which has a high degree of continuity with time.

The locating of fronts on a surface synoptic chart rests in part on identifying the line or zone of maximum horizontal gradient of air temperature and humidity, and in part on locating the physical and dynamical effects frequently associated with a junction of two air masses—precipitation, and discontinuities of pressure tendency and wind velocity. Cloud type, visibility, and lapse rate of temperature are further parameters in which there is normally a discontinuity at an air-mass boundary.

The locating of fronts on surface synoptic charts is made more difficult by the fact that, at land stations, such elements as temperature and, to a lesser extent, dew-point are not 'conservative', i.e. they are readily changed by such processes as radiational warming and cooling, often to very different degrees at different places. Wind velocity and, more particularly, visibility are other surface elements which at land stations may not be 'representative' of the air mass as a whole. Such lack of representativeness is much less true of observed elements in the upper atmosphere. Temperature and humidity of individual elements of air in the free atmosphere are, however, by no means conservative in vertical motion, and recourse is sometimes made, in identifying air masses, to such derived parameters as POTENTIAL TEMPERATURE and WET-BULB POTENTIAL TEMPERATURE which are conservative, or quasi-conservative, for certain specific processes to which the air may be subjected.

air-mass climatology: Description of climate in terms of the frequencies and properties of the different types of AIR MASS which affect a specified region in a specified period.

air-mass thunderstorm: A THUNDERSTORM which is formed by convection within an air mass, usually by heating of the lower layers. By implication, it is one in whose formation neither a front nor large-scale dynamical lifting of the air mass plays an important part.

air-meter: An instrument for measuring the flow of air. It consists of a light 'windmill' in which inclined vanes are carried on the spokes of a wheel arranged to rotate about a horizontal axis. A system of counters is provided to show the number of rotations of the wheel. Calibration is effected in terms of a speed unit so that the instrument acts as a convenient portable ANEMOMETER. Both 'sensitive' (low-speed) and 'high-speed' air-meters are used.

air pocket: An obsolescent term for a region of descending air in which an aircraft experiences a proportionate decrease of lift.

Air pockets are usually experienced in association with convective-type storms ('downdraughts') and, in strong and squally winds, on the leeward side of hills, buildings and other obstructions. The turbulence produced by an obstacle to wind flow extends to a height which increases with temperature lapse rate up to three or four times that of the obstacle.

Aitken nucleus: See NUCLEUS.

albedo: A measure of the reflecting power of a surface, being that fraction of the incident RADIATION (total or monochromatic) which is reflected by a surface.

Typical values of total albedo (per cent) of various surfaces are: forest 5 to 10; wet earth 10; rock 10 to 15; dry earth 10 to 25; sand 20 to 30; grass 25; old snow 55; fresh snow 80. The albedo of a water surface varies from about 5 per cent at high solar elevation to 70 per cent at low solar elevation. The albedo of clouds is difficult to measure but is known to depend on cloud type and thickness; estimates of an average value in time and space vary from 50 to 65 per cent.

The albedo of the earth–atmosphere system as a whole ('planetary albedo') is estimated to be about 0·4 (i.e. 40 per cent). This signifies that about four-tenths of the incident solar radiation is returned to space, without change of wavelength, by reflection from clouds and the earth's surface and by back scattering from air molecules and dust. A similar value, with real variations up to about 5 per cent, is inferred from photometric comparisons of the earth-lit and sun-lit segments of the moon.

albedometer: An instrument for measuring the ALBEDO of a surface.

Aleutian low: A depression, centred near the Aleutian Islands in the North Pacific, which is a conspicuous feature of the northern hemisphere surface mean pressure chart in winter. The depression has an average central pressure below 1000 mb in January, and represents the aggregate of the many depressions which affect this region in winter.

alidade: An instrument for measuring the angular elevation of an object, e.g. a cloud feature or a searchlight spot. The object is sighted by the observer along a rod whose angular position (0°–90°) with respect to the horizontal is obtained by reading from an engraved scale of degrees. Both fixed and portable (or 'hand') alidades are used.

alpha (or α) particle: A particle emitted spontaneously from the nuclei of certain radioactive elements. It is identical with a helium nucleus, comprising two neutrons and two protons, and therefore carries a positive charge of two units.

α particles are of such low penetrative power (only a few centimetres in air) that the particles emitted by radioactive materials in the earth's crust are insignificant in forming IONS. α particles are, however, also emitted from the radioactive gases, mainly RADON, ACTINON and THORON at low atmospheric levels, and are responsible for a significant part of the ionization of the air at these levels, over land. See also BETA PARTICLE, GAMMA RADIATION.

alpine glow: A series of phenomena seen in mountainous regions about sunrise and sunset.

Two principal phases are generally recognized:

(i) The true alpine glow. At sunset this phase begins when the sun is 2° above the horizon; snow-covered mountains in the east are seen to assume a series of tints from yellow to pink, and finally purple. As this phase is due mostly to direct illumination by the sun it terminates when the mountain tops pass into the SHADOW OF THE EARTH. The alpine glow is most striking when there are clouds in the western sky and the illumination of the mountains is intermittent.

(ii) The afterglow. This begins when the sun is well below the horizon, 3° or 4°. The lighting is faint and diffuse with no sharp boundary and occurs only when the PURPLE LIGHT is manifest in the sky.

alter shield: A form of RAIN-GAUGE shield consisting of a funnel-shaped construction of spaced slats surrounding the gauge. The purpose of the shield, which is not used by the Meteorological Office, is to improve the representativeness of the gauge catch, especially snow, in windy conditions. See also EXPOSURE.

alti-electrograph: A balloon-borne device for obtaining records of the vertical component of the electric field inside thunderstorms. The direction and magnitude of the current flow between two vertically separated points is recorded on a disc of pole-finding paper. A chemical reaction at the electrodes pressing on the paper makes a trace, the width of which varies with current.

altimeter: An instrument for determining the altitude (generally of an aircraft) with respect to a datum level. The two main types are (i) a radio altimeter and (ii) a pressure altimeter.

The pressure altimeter is an ANEROID BAROMETER which is calibrated directly in height units on the basis of the 'altimeter equation':

$$z = \frac{R\bar{T}_v}{g} \log_e \frac{p_0}{p_1} \approx 67 \cdot 4 \; \bar{T}_v \log_{10} \frac{p_0}{p_1} \text{ (metres)}$$

$$\approx 221 \cdot 1 \; \bar{T}_v \log_{10} \frac{p_0}{p_1} \text{ (feet)}$$

where z is the height, R the specific gas constant for dry air, $\bar{T}_v$ the mean virtual air temperature in the air column at the bottom and top of which the pressures are p_0 and p_1, respectively and g the gravitational acceleration.

The assumptions made in the instrument graduation are that p_0 has the value appropriate to MSL pressure of the ICAO STANDARD ATMOSPHERE, and that the variation of mean virtual temperature with height corresponds to that in the standard atmosphere.

Corrections may be required to take account of the fact that actual conditions differ from those of the standard atmosphere. The first concerns p_0 and involves an ALTIMETER SETTING; the required correction is the same at all altitudes. Thus an ICAO altimeter, for example, shows the height interval in the ICAO standard atmosphere between the altimeter-setting pressure and the ambient pressure. The second correction involves a positive (negative) correction when actual $\bar{T}_v$ is higher (lower) than that of the standard atmosphere; this correction is negligible at low altitudes. See also D-VALUE.

altimeter setting: The altimeter setting, designated QNH in the aircraft Q-CODE, is defined as that value of pressure, for a particular aerodrome and time, which, when set on the sub-scale of a standard ALTIMETER (based, for example, on the ICAO STANDARD ATMOSPHERE), will cause the altimeter to read the height of the aerodrome when the aircraft is at rest on the aerodrome. For procedure for obtaining QNH see *Observer's handbook.**

---

*London, Meteorological Office. Observer's handbook, 3rd edn. London, HMSO, 1969, p. 104.

altitude: The angular distance of an object above the horizon: synonymous with angle of ELEVATION.

In meteorology, altitude generally signifies height above mean sea level (geometric metres or feet); in dynamical meteorology, however, height is usually expressed in GEOPOTENTIAL metres or feet. In aviation, altitude signifies geometric height of an aircraft above mean sea level.

altocumulus (Ac): One of the CLOUD GENERA (Latin, *altum* height, and *cumulus* heap).

'White or grey, or both white and grey, patch, sheet or layer of cloud, generally with shading, composed of laminae, rounded masses, rolls, etc., which are sometimes partly fibrous or diffuse and which may or may not be merged; most of the regularly arranged small elements usually have an apparent width of between one and five degrees.'* See Plates 8 and 16. See also CLOUD CLASSIFICATION.

altostratus (As): One of the CLOUD GENERA (Latin, *altum* height, and *stratus* spread out).

'Greyish or bluish cloud sheet or layer of striated, fibrous or uniform appearance, totally or partly covering the sky, and having parts thin enough to reveal the sun at least vaguely, as through ground glass. Altostratus does not show halo phenomena.'* See Plate 9. See also CLOUD CLASSIFICATION.

ambient pressure, temperature etc.: The pressure or temperature etc. in that part of the atmosphere which immediately surrounds a specified physical entity such as a cloud or thermal.

amorphous clouds: A term used in respect of a more or less continuous layer of low clouds without regular features and generally associated with rain, as for example NIMBOSTRATUS.

amplitude: For a true wave variation, the magnitude of the maximum departure of the quantity concerned from its mean value. For a more complex type of oscillation, the amplitude is usually taken as being half the mean difference between maxima and minima. See also WAVE MOTION.

anabatic wind: A local wind which blows up a slope heated by sunshine. It is a feature which is much less common than its converse, the KATABATIC WIND.

anafront: As defined by T. Bergeron, a FRONT (warm or cold) along which the warm air is ascending relative to the cold air. Frontal activity is generally well marked with such fronts.

analogue: In synoptic meteorology, a past synoptic situation which resembles the current situation over an appreciable area.

Analogues are usually selected from the same time of year as the current situation. The sequence of weather which followed an analogue is sometimes used as the basis of a weather forecast, in both short-range and long-range forecasting. In the former case the analogues refer to the surface (and upper air) synoptic charts which are drawn as a routine. In the latter case they may refer to charts which show, for example, the distribution of the values of monthly temperature anomalies (departures from average) at stations distributed over a large area.

---

* Geneva, World Meteorological Organization. International cloud atlas, Vol. 1. Geneva, WMO, 1956, p. 11.

analysis: In synoptic meteorology, the co-ordination by means of isopleths or representative symbols of the elements plotted on a surface or upper air chart ('surface analysis' and 'upper air analysis', respectively), generally for the purpose of making a weather forecast.

Surface analysis comprises both 'isobaric analysis' (the drawing of isobars) and 'air-mass analysis' (the identification of air masses and drawing of fronts); other types of isopleth, notably ISALLOBARS, may be drawn on the surface chart as aids in analysis. In upper air (isobaric) analysis contours are drawn, sometimes also streamlines, isopleths of temperature and isopleths of wind direction and speed.

anemogram: The record of an ANEMOGRAPH.

anemograph: An instrument for recording the speed, and sometimes also the direction, of the wind. Each of the basic types of anemometer may be used as the wind-speed sensor of the anemograph, and, combined with a suitable WIND VANE, as the wind-speed and direction sensors.

For normal purposes the cup-contact, cup-generator and pressure-tube anemographs are most commonly used. The cup-contact anemometer can record mean wind speed over a given time interval by displaying its contacts as marks on a chart moving with time, the marks being readily counted by means of the scale markings on the chart. The cup-generator anemometer can record instantaneous wind speed by feeding its output into a moving-coil pen-recording meter.

The pressure-tube anemograph, now obsolescent, makes use of the difference of pressure set up between two pipes, one of which is kept facing the wind by the action of a wind vane, while the other is connected to a system of suction holes on a vertical tube. The difference of pressure so produced is arranged to raise a float carrying a pen, the height of which above the zero position is made proportional to wind speed by suitable design of the float. The instrument takes a certain time to respond to changes of wind speed, but gusts and lulls with periods down to a few seconds are indicated accurately, though they may not be distinguishable on the record. Either 'direct recording' at a point underneath the instrument head, or 'remote recording' at a considerable distance from the instrument, may be arranged.

anemometer: An instrument for determining the speed of the wind. One or other of three properties of the wind is used in such instruments: (i) its kinetic energy, which causes rotation as in cup anemometers and anemometers of the windmill type (AIR-METER); (ii) its pressure, as in the pressure-tube ANEMOGRAPH and pressure-plate anemometer; (iii) its cooling power, as in the hot-wire anemometer.

Cup anemometers consist of three or four cups, conical or hemispherical in shape, mounted symmetrically about a vertical axis. In the cup-contact anemometer, the closing of electrical contacts produces an intermittently audible note the rate of recurrence of which is proportional to the wind speed. In the cup-generator anemometer, the rotating cups are made to generate a voltage which registers on a dial calibrated in knots or miles per hour. In the cup-counter or run-of-wind anemometer, the integrated flow of the air in miles is registered on a counter.

In the pressure-plate anemometer, the deflexion of a flat plate placed in the wind is measured; its use is confined mainly to atmospheric turbulence measurement. In the hot-wire anemometer, the current required to maintain constant the electrical resistance (and so the temperature) of a fine platinum wire which is exposed to the wind may be used as a measure of the wind speed; alternatively, a large resistance may be placed in series with the wire, the current kept constant and the varying potential drop across the wire used as a measure of the wind. This instrument is used when rapidity of response to wind fluctuations is important.

See also EXPOSURE.

aneroid barometer: The aneroid ('without liquid') barometer was invented by Lucien Vidie in about 1843. In its simplest form it consists of a shallow capsule of thin, corrugated metal which is exhausted of air. The faces are kept apart by the stiffness of the metal or by a separate spring. Compensation for temperature is provided by a bimetallic link or (over a limited range of pressure) by a small quantity of residual air in the capsule. In some instruments several capsules are employed. The relative movements of the faces due to changes of atmospheric pressure are conveyed and magnified through a train of levers to a chain which actuates a pointer on a dial.

An aneroid is light, portable and convenient. Since, however, it may be subject to errors introduced by imperfect elasticity of the metal etc., and to changes of zero (both generally small in modern instruments), it requires occasional checking against a mercury barometer and in the past has normally been used only when the use of a mercury barometer was impracticable. Recently, however, aneroid barometers which are as accurate and reliable as a Kew-pattern barometer have been developed.

angels, radar: Radar reflections obtained on certain occasions of apparently clear air. One type, comprising clouds of point-target echoes, has been definitely associated with birds. It is probable that other types are associated with abrupt changes of refractive index of the air.

ångström unit: A measure of wavelength, equal to 10^{-10}m, denoted by Å. The nearest standard SI UNITS are the nanometre, 10^{-9}m, and the picometre, 10^{-12}m.

angular momentum: The angular momentum (or moment of momentum) per unit mass of a body rotating about a fixed axis is the product of the linear velocity of the body and the perpendicular distance of the body from the axis of rotation. The dimensions are L^2T^{-1}.

A point P of the earth in latitude φ has a west–east (zonal) velocity of $\Omega\, a \cos \varphi$ where Ω is the earth's angular velocity and a its radius (see Figure 4). Thus, at this latitude the absolute angular momentum of air of relative zonal velocity u (and not too far removed from the surface of the earth)

$$= (u + \Omega\, a \cos \varphi)\, a \cos \varphi$$
$$= u\, a \cos \varphi + \Omega\, a^2 \cos^2 \varphi.$$

The first of these terms is the 'relative angular momentum' of the air, reckoned positive for positive u. The second term is the angular momentum of the coinciding point of the earth.

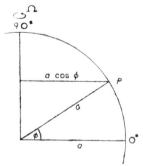

FIGURE 4—Angular momentum.

The transfer of angular momentum effected by the atmosphere and the manner in which the entire earth–atmosphere system conserves absolute angular momentum over a long period of time are of fundamental significance in the explanation of the GENERAL CIRCULATION of the atmosphere.

angular velocity: The angular velocity of a moving line is the time rate of change of the angle between the line and a fixed line in a plane containing two successive positions of the moving line. Angular velocity is represented by a vector normal to this plane. A suitable convention is adopted as to which direction of rotation is considered positive. The angular velocity of a moving point about a fixed point is the angular velocity of the line joining the two points. The angular velocity of a moving point about a fixed axis is the rate of change of the angle between a plane drawn through the axis and the moving point, and a fixed plane passing through the axis. The angular velocity of a solid body about an axis is the angular velocity of any point of the solid body about that axis. The dimensions are T^{-1}.

Angular velocity is a vector quantity which is normally measured either in revolutions per unit time or in radians per unit time. Since there are 2π radians per revolution, ω (radians per second) is related to N (revolutions per minute) by the expression $\omega = \pi N/30$.

angular velocity of the earth: The ANGULAR VELOCITY of the earth (Ω) may be represented by a vector parallel to the axis of the earth and directed northwards. This vector may be resolved into components $\Omega \cos \varphi$ about the line directed towards local north (rotation in a vertical plane) and $\Omega \sin \varphi$ about the local vertical (rotation in a horizontal plane)—see Figure 5. The total vector is of magnitude 2π rad/sidereal day $= 7\cdot292 \times 10^{-5}$ rad/s.

$\Omega \sin \varphi$ is the only component which is significant in large-scale motion, which is almost entirely horizontal. The rotation is counter-clockwise in the northern hemisphere, clockwise in the southern hemisphere; it is thus cyclonic in both cases.

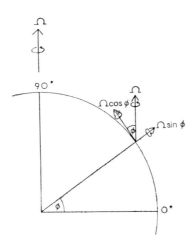

FIGURE 5—Angular velocity of the earth.

anomalous audibility: Audibility of sound waves over a region which is separated from the source of the waves by a 'zone of silence'. See AUDIBILITY.

anomalous radio propagation: Under normal atmospheric conditions, radio waves up to a few metres in length, transmitted in beams (for example, radar) in a direction close to the horizon, attain ranges on the earth's surface which approximate only to the distance of the optical horizon, being limited by the curvature of the earth. In abnormal atmospheric conditions, determined by the abnormal variation with height of the REFRACTIVE INDEX of the low atmosphere, the waves may be subject to 'anomalous propagation' in which they suffer such downward refraction as to extend their normal range by many times, exceptionally from, say, 50 km to some hundreds of kilometres.

This effect, also known as 'superrefraction', is associated with the existence of a so-called 'radio duct' in conditions of a large inversion of temperature, or large decrease of humidity with height, or both, extending through an atmospheric layer near the surface of the earth. Such conditions would be expected, for example, in air moving out from a warm, dry land over a cool sea. The depth of the duct required for marked abnormality of propagation increases with wavelength; it is, for example, some 15 metres for a wavelength of 3 cm, and about 200 metres for a wavelength of one metre. The frequency of the phenomenon is, therefore, much greater at the short wavelengths.

anomalous sound propagation: The propagation of sound waves along a path, from source to receiver, other than close to the earth's surface, as the result of which ANOMALOUS AUDIBILITY occurs. See AUDIBILITY.

anomaly: The departure of an element from its long-period average value for the place concerned. The space distribution of such anomalies at a specified time is known as an 'anomaly pattern'.

The term is also used in other senses; for example, a place that is relatively warm for its latitude (as western Norway in winter) is sometimes said to have a positive temperature anomaly.

Antarctic air: An AIR MASS, originating over the Antarctic continent, which is cold and dry in all seasons. It is sometimes designated continental polar (*cP*) air.

Antarctic Circle: The parallel of latitude 66° 33′ S, south of which lies the 'Antarctic zone', or 'southern polar zone'.

Antarctic front: A FRONT which develops and persists around the Antarctic continent in about latitudes 60°–65°S, and divides ANTARCTIC AIR from the maritime POLAR AIR to the north. The front is often found to exist round a large part of the hemisphere.

anthelion: A colourless mock sun (PARHELION) appearing at the point of the sky opposite to and at the same altitude as the sun. The phenomenon is rare. Rather more frequently, oblique arcs crossing at that point are reported. The phenomenon is no doubt caused by the reflection of light from ice crystals, but the exact explanation is in doubt.

anticorona: An alternative for GLORY.

anti-crepuscular rays: See CREPUSCULAR RAYS.

anticyclogenesis: The initiation of anticyclonic circulation, or its strengthening around an existing ANTICYCLONE.

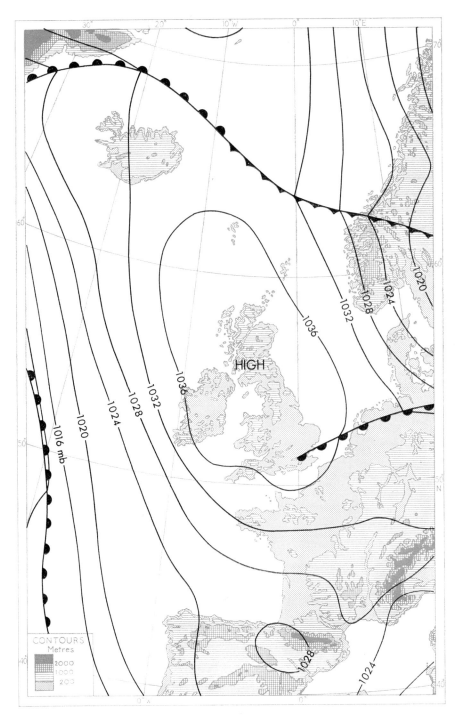

FIGURE 6—Anticyclone over the British Isles, 06 GMT, 20 December 1961.

anticyclolysis: The disappearance or weakening of anticyclonic circulation around an existing ANTICYCLONE.

anticyclone: That atmospheric pressure distribution in which there is a high central pressure relative to the surroundings. It is characterized on a synoptic chart by a system of closed isobars, generally approximately circular or oval in form, enclosing the central high pressure (see Figure 6). The term 'anticyclone' was selected (by F. Galton in 1861) as implying the possession of characteristics opposite to those found in the cyclone or depression. Thus, the circulation about the centre of an anticyclone is clockwise in the northern hemisphere (counter-clockwise in the southern), and the weather is generally quiet and settled.

Two contrasting types of anticyclone, 'warm' and 'cold', are recognized. The former has a warm troposphere (though sometimes cold in a shallow layer near the earth's surface), high tropopause and cold stratosphere; the anticyclonic circulation is deep and the feature slow moving. The preferred region is the subtropical belt at 30°–40° latitude, for example the Azores and Bermuda regions, but with frequent extensions across west Europe; typical warm anticyclones also occur in higher latitudes in association with BLOCKING. In contrast, the cold anticyclone has a relatively cold baroclinic troposphere, low tropopause, warm stratosphere and shallow circulation. It often forms in the cold air behind a depression and moves fairly rapidly in a direction between south and east (northern hemisphere), sometimes then slowly transforming to a warm type. Persistent radiational cooling in winter over high-latitude continents produces semi-permanent cold anticyclones, or 'continental anticyclones', as in Siberia and North America.

The dynamical structure of the anticyclone is one of horizontal convergence at high levels, horizontal divergence at low levels and slowly subsiding air throughout a large part of the troposphere. The SUBSIDENCE proceeds at a maximum rate in early stages of formation of the anticyclone and is greatest in mid-troposphere. The subsidence results, through dynamical warming of the air, in a decrease in relative humidity and an increase of static STABILITY of the air, often with the formation of an 'anticyclonic inversion' of temperature. These processes often result in fine, cloudless weather. There are, however, important exceptions: in summer, sea and coastal fog and, in winter, widespread stratocumulus ('ANTICYCLONIC GLOOM'), forming in moist air at the base of the inversion, or radiation fog, forming in conditions of little cloud, are commonly associated with an anticyclone.

anticyclonic gloom: Conditions of poor illumination occurring with an overcast sky of stratus or stratocumulus beneath an inversion in an ANTICYCLONE. It is particularly associated with the accumulation, in and below the cloud, of the pollution products of industrial areas.

antisolar point: That point, below or above the horizon, towards which the extension of the line from the sun to an observer's eye is directed.

anti-trades: The anti-trades, or countertrades, are upper winds which in some low-latitude areas prevail above the TRADE WINDS and which, more or less opposite in direction to the trade winds, are responsible for poleward transport of air aloft. The transition to anti-trades, marked by a temperature inversion, occurs at some 1–1·5 km in high trade-wind latitudes in eastern oceanic regions and at increasingly high levels equatorwards and to the west.

antitriptic wind: A class of winds in which the effect of FRICTION or eddy VISCOSITY predominates over the CORIOLIS ACCELERATION and the CENTRIPETAL ACCELERATION effects. The air movement is from high to low pressure, a quasi-steady state being established by the effect of friction or eddy viscosity. Such winds occur mainly in low latitudes where the Coriolis acceleration is very small, or in small-scale systems (e.g. sea-breeze) in higher latitudes.

antitwilight (or antitwilight arch): Alternatives for COUNTERGLOW.

anvil cloud: Cloud having at its top a projecting point or wedge like an anvil. The form is usually assumed by the tops of fully developed CUMULONIMBUS clouds as in Plate 15. The anvil is considered to indicate in nearly all cases the presence of ice crystals or snowflakes; it quickly assumes a fibrous or nebulous appearance and is often stretched out from the main cloud columns by changes of wind with height. The term 'anvil cirrus' is applied to an anvil which becomes separated from the main cloud.

aphelion: That point of the orbit of a planet or comet which is farthest from the sun. Aphelion for the earth occurs on about 1 July; the sun–earth distance is then 1·5 per cent greater than the yearly mean distance.

apogee: That point of the orbit of a satellite, natural or artificial, which is farthest from the earth.

apparent form of the sky: The somewhat 'flattened' appearance of the sky presented to most observers. Among the effects of this appearance is the tendency to overestimate the elevation angles of objects in the sky.

Appleton layer: A layer of the IONOSPHERE at some 300 km height, now usually termed the F_2-layer. See also F-LAYER.

applied meteorology: The application of meteorological knowledge in a wide variety of activities, for example, industry, transport, hydrology and agriculture, for the purpose of using meteorological conditions to the best advantage. Forecasting, or climatology, or the application of the results of meteorological research in specific problems may be involved. For certain operations, notably in agriculture, the application of meteorological knowledge may include the altering of meteorological conditions on a small scale, as in the construction of SHELTER-BELTS.

APT: Abbreviation for AUTOMATIC PICTURE TRANSMISSION.

aqueous vapour: An alternative for WATER VAPOUR.

Arago's point: A 'neutral' point of the sky at which the normally observed polarization of the light from a clear sky disappears; that discovered by Arago in the early 19th century is some 20° above the ANTISOLAR POINT. Other neutral points were subsequently discovered by Babinet and Brewster about 20° above and below the sun, respectively. Day-to-day and systematic diurnal and seasonal variations of position of these points, probably associated with haze variations, have been discovered. See also POLARIZATION.

arcs of contact: Upper and lower 'arcs of contact' to the 22° HALO occasionally form, the lower arc being very rare. At high solar elevations the arcs may appear concave towards the sun. At low solar elevations the higher arc is convex towards the sun. The points of contact may have the appearance of 'mock suns' (PARHELIA) and may display brilliant colour.

Arctic air: An AIR MASS originating in the snow- and ice-covered Arctic and travelling almost directly south. In the British Isles this air mass is confined mainly to winter and spring and is associated with a northerly wind, snow showers (especially on north-facing coasts and hills), low temperatures, exceptional visibility and steep lapse rate.

Arctic Circle: The parallel of latitude 66° 33′ N, north of which lies the 'Arctic zone', or 'northern polar zone'.

Arctic front: A FRONT which separates ARCTIC AIR to the north from maritime polar air or continental polar air to the south. A section of the Arctic front is often found in the area from south Greenland to north of Norway in winter and spring.

Arctic sea smoke: If, when cold air moves over warm water, the vapour pressure at the water surface exceeds the saturation vapour pressure at the air temperature, then evaporation from the water surface proceeds at a higher rate than can be accommodated by the air. The excess water vapour over that required to saturate the air condenses and, in the unstable conditions present in the layer near the surface, the condensed water is carried continuously upwards to evaporate into the drier air above. 'Steam' or 'smoke' thus appears to rise off the water surface. If an inversion exists near the water surface, fog may be confined below the inversion and become dense.

The phenomenon occurs, for example, over inlets of the sea in high latitudes; over newly formed openings in pack ice; over lakes and streams on calm, clear nights; over damp ground heated by bright sunshine in cool conditions. Alternative names are 'frost smoke', 'sea smoke', 'steam fog', 'warm-water fog', 'water smoke' and 'the barber'. See Plate 3.

arcus (arc): A supplementary cloud feature (Latin, *arcus* arch).

'A dense, horizontal roll with more or less tattered edges, situated on the lower front part of certain clouds and having, when extensive, the appearance of a dark menacing arch.

This supplementary feature occurs with CUMULONIMBUS and, less often, with CUMULUS.'* See also CLOUD CLASSIFICATION.

argon: The most abundant of the INERT GASES, comprising 0.93×10^{-2} and 1.27×10^{-2} part per part of dry air by volume and weight, respectively. Its molecular weight is 39·948. Its inertness and relatively high density render it a suitable tracer (in terms, for example, of the argon: nitrogen ratio) of the degree of the GRAVITATIONAL SEPARATION of the atmospheric constituents.

arid climate: A climate in which the rainfall is insufficient to support vegetation is termed arid. Köppen and Geiger† in their *Klimakarte der Erde* used the following formulae for the limits of rainfall for arid and semi-arid climates:

Rainfall mainly in cold season	$R = 2t$
Rainfall evenly distributed throughout year	$R = 2t + 14$
Rainfall mainly in hot season	$R = 2t + 28$

where t is the mean annual temperature in degrees Celsius. If the annual rainfall (cm) is less than R and greater than $R/2$ the climate is steppe or semi-arid; if it is less than $R/2$ the climate is desert or arid. See also DESERT, STEPPE.

---

* Geneva, World Meteorological Organization. International cloud atlas, Vol. 1. Geneva, WMO, 1956, p. 17.
† KÖPPEN, W. and GEIGER, R.; Klimakarte der Erde. Gotha, Justus Perthes, 1928.

arithmetic mean: If $x_1, x_2, \ldots x_n$ are n measurements of the same kind, then

$$\frac{1}{n} \sum_{j=1}^{n} x_j$$

is the arithmetic mean, generally denoted by $\bar{x}$. See also MEAN.

artificial-ice nucleus: A freezing or sublimation NUCLEUS generated for the purpose of CLOUD SEEDING.

artificial precipitation: See CLOUD SEEDING.

ascendent: See GRADIENT.

A-scope indicator: See RADAR METEOROLOGY.

ash: The incombustible solid material released when a substance is burned: a small proportion escapes to the atmosphere to contribute to ATMOSPHERIC POLLUTION.

The proportion of total ash that emerges into the atmosphere, and also the average size of emerging particles, depend on the velocity of the flue gases and are, in the absence of a grit arrester, much greater for industrial than for domestic chimneys. Emerging particle sizes range from 0·2 cm downwards, all except the smallest particles being deposited near the source. Measurement of such ash and other deposited material is made by the DEPOSIT GAUGE.

aspect: The aspect of sloping ground is the geographic direction in which the line of greatest downslope points. The 'aspect angle' is, in the northern hemisphere, the angle between this line and geographic south, usually reckoned positive eastwards, negative westwards.

aspirated psychrometer: A PSYCHROMETER in which a high rate of ventilation is provided artificially as in the ASSMANN PSYCHROMETER or WHIRLING PSYCHROMETER.

Assmann psychrometer: A PSYCHROMETER in which a definite rate of ventilation is secured by drawing the air over the thermometer bulbs by means of a fan driven by a motor and in which the thermometers are mounted in a polished metal frame as a protection against solar radiation.

atmosphere: The gaseous envelope which is held to the earth by gravitational attraction and which, in large measure, rotates with it. The internal motions of the atmosphere caused by solar radiational heating constitute together with their physical effects, the main concern of meteorology. The term is also used of the gaseous envelopes of planets and stars.

The composition of the earth's atmosphere is discussed under AIR. Of special importance in meteorology are the time and space variations of water vapour. A distinction is drawn, in terms of atmospheric composition, between the homosphere, a region extending from the surface to about 80 km in which the permanent gaseous constituents are well mixed, and the heterosphere, the region above 80 km in which the processes of dissociation of oxygen and diffusion effect a change in composition with height.

The atmosphere's outermost fringe has been termed the 'exosphere', at the lower limit of which (about 700 km) the escape to space of neutral particles was calculated to begin. S. Chapman has, however, suggested that since the particles at such levels are mainly ionized and so are controlled not by diffusion but by the earth's magnetic field, the atmosphere may be considered to extend to a height of several or even many earth radii, where the density falls to that of interplanetary gas.

Division of the atmosphere into the regions TROPOSPHERE, STRATOSPHERE, MESOSPHERE, and THERMOSPHERE, according to characteristic temperature lapse rates, is illustrated in Figure 7; the approximate pressure and density variations with height are also shown. The terms OZONOSPHERE, CHEMOSPHERE and IONOSPHERE are also used of certain regions in which specified processes occur. See also STANDARD ATMOSPHERE.

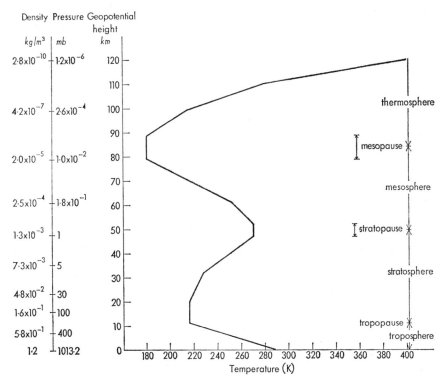

FIGURE 7—Average temperature structure of the atmosphere from 0 to 120 km. Data are taken from the *U.S. Standard Atmosphere, 1962**, which depicts idealized middle-latitude year-round mean conditions. From 0 to 32 km this corresponds with the ICAO STANDARD ATMOSPHERE† as adopted by WMO.

atmospheric boil: An alternative for SHIMMER.

atmospheric chemistry: A term which is generally taken to comprise the study of the chemical composition of the air (mainly that near the earth's surface), of aerosols, and of rainfall. Less commonly, it is used in a wider sense so as to include the processes of PHOTOCHEMISTRY which are important at levels above about 40 km.

Monthly samples of air and rainfall, collected at a network of stations which was augmented and made world-wide during the International Geophysical Year, are

---

* Washington, United States Committee on Extension to the Standard Atmosphere. U.S. Standard atmosphere, 1962. Washington, 1962.
† Montreal International Civil Aviation Organization. Manual of the ICAO standard atmosphere, 2nd edn. Montreal, 1964.

chemically analysed for the presence and concentration of the inorganic substances Na, K, Mg, Ca, NH_3, Cl, S, NO_3 and CO_3. The pH value of the sample (the logarithm, to base 10, of the inverse of the concentration of hydrogen in the solution) has also been determined.

Widely different results, difficult to interpret, have been obtained with the air sampling measurements, which have been concerned mainly with the contained particulate matter. More consistent results, though with unexplained space and time variations, have been obtained with the rainfall sampling measurements. In particular, a deficiency of Cl relative to Na and other ions which occur in sea water has been found.

In addition to the above, surface air has been sampled for carbon dioxide and ozone. Also, various measurements have been made of the nature and amount of the radioactive material from nuclear explosions which is contained in the troposphere and stratosphere, or is deposited as dust on the earth's surface, or is washed out of the atmosphere by precipitation.

atmospheric electricity: The various electrical phenomena which occur naturally in the lower atmosphere: the IONOSPHERE is conventionally excluded, except in so far as it reacts on the electrical properties of the lower atmosphere. While the THUNDERSTORM is the most familiar manifestation of atmospheric electricity, substantial electrical effects exist also in fine-weather regions and have been continuously measured at various places since the latter part of the 19th century. It is now generally held that the two classes of phenomena are closely linked.

The fine-weather electric field recorded in surface POTENTIAL GRADIENT measurements is directed downwards to earth and has a mean value of about 100 volts per metre. Such a field implies a negative charge on the earth's surface of about 1 coulomb per 1000 km^2 (3 e.s.u./m^2) in fine-weather regions.

Balloon measurements show that in such regions the potential gradient decreases with height, rapidly at first, then more slowly; a normal positive SPACE CHARGE is implied in the low atmosphere, with a surplus of positive over negative ions. Integration of the measured electric field with height shows that relative to the earth's surface, a positive potential of about 4×10^5 volts exists at a height of about 15 km, with little further addition to this value with further height increase owing to the high conductivity of the air at such and higher levels.

Various ionizing radiations render the air electrically conducting, though only feebly so near the earth's surface—see ION. A so-called 'conduction current' flows in fine-weather regions under the action of the field which prevails there (positive ions down to negatively charged earth, negative ions up); this is partially offset by a 'convection current' which transfers positive space charge bodily upwards from the low atmosphere. The measured air–earth current averages about 2×10^{-12} A/m^2. Its space and time variations are considerably smaller than those of the potential gradient; in particular, decrease of the electric field with increasing height is compensated by increasing conductivity, so maintaining an almost constant current, in accordance with Ohm's law. After about 1952 changes occurred in atmospheric conductivity, especially at low levels, because of the deposition of RADIOACTIVE FALLOUT from nuclear explosions.

Calculation shows that the air–earth current would within a few minutes destroy the electric field which gives rise to it and so would itself die out but for the action of a continuous compensating mechanism. The origin of the required 'supply current', for long a matter of doubt and controversy, is now thought to lie in rainstorms and thunderclouds, as was originally suggested by C. T. R. Wilson. Various mechanisms operate in the 'disturbed weather' regions:

  (i) Large electrical charges accumulate at the base of thunderclouds; when such clouds are low, the electric field between cloud and ground often exceeds the value required for brush discharge, especially from elevated or

pointed conductors. Since thundercloud bases carry a predominant negative charge, positive ions are mainly discharged from earth, leaving there a net negative charge.
  (ii) Air-to-ground lightning flashes convey predominantly negative charge to ground.
  (iii) Precipitation carries a net positive charge to ground.
  (iv) Aircraft measurements above thunderclouds indicate that negative ions flow downwards from the high atmosphere to the positively charged thundercloud tops, leaving a net positive charge on the high atmosphere.

The sum of effects (i) and (ii) exceeds effect (iii), resulting in the conveying of a net negative charge to ground in disturbed regions; this charge and the positive charge conveyed upwards above thunderstorms quickly spread over the conducting earth and high atmosphere, respectively, and give rise to the downward-directed field 'leakage current' observed in fine-weather regions. Over-all (space and time) balance exists between the supply and leakage currents. A major piece of supporting evidence for this mechanism is the existence of a systematic diurnal variation of the fine-weather field in phase with integrated thunderstorm activity over the world (maximum about 18 GMT, minimum about 04 GMT), observed in those regions (e.g. oceanic) where complications of diurnal variation of low-level conductivity are absent.

atmospheric optics: The optical phenomena of interest to meteorologists include, among many other examples, the BLUE OF THE SKY, SUNRISE AND SUNSET COLOURS, the production of RAINBOWS, CORONAE, HALOES, MIRAGES, PARHELIA, the fading of daylight, and the TWINKLING of stars. See, for example, Humphreys,* and Minnaert.†

atmospheric pollution: Contamination of the atmosphere by gases and solids produced in the burning of natural and artificial fuels, in chemical and some other industrial processes, and in nuclear explosions; the term may be considered also to include contamination produced by such processes as accumulation of cosmic dust, raising by wind of surface dust, eruption of volcanoes, decay of vegetation, evaporation of sea salt spray, and natural radioactivity.

Coal burning is the main source of pollution in Great Britain and other industrialized countries, the main pollutants so formed being SULPHUR DIOXIDE, SMOKE, and ASH, with smaller quantities of hydrogen chloride, hydrogen fluoride and silicon tetrafluoride. Sulphur dioxide is also produced by burning heavy fuel oils. Petroleum burning, producing mainly hydrocarbons which react with sunlight to form ozone, is an increasing source of pollution. Of increasing importance also, on both a local and world-wide scale, are the radioactive products of nuclear reactions.

Atmospheric pollution measurements have been made in Great Britain since 1914 at an expanding network of stations. Standardization of method has since been achieved, the measurements being made mainly by local authorities, government departments and nationalized and private industries. The standard measurements are of sulphur dioxide, smoke and deposited pollution (see DEPOSIT GAUGE). Measurement of radioactive pollution is mainly the responsibility of the Atomic Energy Authority.

Systematic variations of smoke and sulphur dioxide concentration are found on several time scales. Typical results for cities in Great Britain are:

  (i) A daily variation, with primary minimum in early hours of morning,

---

  * HUMPHREYS, W. J.; Physics of the air, 3rd edn. New York, McGraw–Hill Book Co., 1940.
  † MINNAERT, M.; Light and colour in the open air. London, G. Bell and Sons Ltd., 1959.

secondary minimum in mid-afternoon, and total daily range up to about 5 : 1 in winter months.
(ii) Week-end decrease of some 20 to 40 per cent.
(iii) Winter to summer decrease of some 2 : 1 or 3 : 1.
(iv) A SECULAR TREND, irregular but negative at most places.

The smoke and sulphur dioxide output variations at source, combined with the systematic variations of such meteorological factors as temperature lapse rate and turbulence near the ground, wind velocity, and rainfall explain the nature of the daily and annual variations of concentration; the weekly variation and secular trend are explained by output variation alone.

The relatively large particles which make up most of the ash which reaches the atmosphere soon fall to ground and so are little affected by meteorological conditions. The time variations of suspended or deposited ash are thus controlled almost entirely by output variations.

Superimposed on the systematic variations of smoke and sulphur dioxide concentration are large day-to-day changes, associated essentially with changes in the important meteorological factors; the change from winter minimum to winter SMOG concentration represents, for example, an increase of 1 : 20 or more.

Mathematical treatment by O. G. Sutton of the distribution of airborne gaseous pollution has shown (i) that maximum concentration is about four times greater, and occurs at a point about five times farther from source, in large-inversion conditions compared with large-lapse conditions, and (ii) that great benefit, in terms of maximum concentration at ground level from a given source, is derived in all lapse-rate conditions by increase of chimney height.

Concern for the physiological and financial damage wrought by atmospheric pollution has, since the first half of the 19th century, led to local and national legislation aimed at its mitigation. Among the methods adopted, so far on a limited scale only, are the more efficient use of solid fuel by appropriate design of fire or boiler and by skilled or automatic stoking, the replacement of coal by smokeless fuel, the use of grit arresters, the removal by washing of sulphur dioxide from flue gases, the use of high chimneys, and the control of pollution emission in cases of prolonged unfavourable meteorological conditions.

atmospheric pressure: See PRESSURE.

atmospherics: Electrical impulses of natural origin, mainly originating in LIGHTNING discharges, which cause crashing or grinding noises in a wireless receiver; the phenomenon is also termed 'static'. The impulses which originate within a radius of some thousands of kilometres reach a particular receiver on a multitude of paths suffering reflections from the lower ionosphere and the earth's surface, and constitute a varying background noise-level in the receiver. Since the impulses decrease in strength with increasing distance from their place of origin, the noise-level is greatest in low latitudes where thunderstorms are most frequent.

While the phenomenon constitutes a serious difficulty in long-distance wireless transmission, it is used to advantage in geophysics in various ways: first, by suitable arrangement of apparatus, in the locating of thunderstorms; second, by examination of the wave-forms, in the study of the nature of lightning discharges; third, in the determination of diurnal and seasonal variations of thunderstorm activity; fourth, in the detection of sudden enhancements of atmospherics and thus of associated solar flares (see SEA).

The term 'sferics' is the accepted contraction of the word 'atmospherics' for meteorological purposes. See also SFERICS FIX.

atmospheric tides: Effects directly analogous to the simple gravitational tides familiar in the oceans are produced in the atmosphere by the action of the moon, the elements chiefly involved being pressure and wind. The tides are most conveniently identified at the earth's surface from the pressure variations, which are positive or negative with respect to the mean pressure level according as the air is heaped up over, or drawn away from, a locality. Two pressure maxima (one at the longitude corresponding to that 'under' the tide-producing body and the other at the antipodal point) and two pressure minima (at longitudes 90° from the pressure maxima) occur at any given epoch. At any given place the effect is such as to produce a semi-diurnal oscillation. When the pressure data are arranged in lunar time, there emerges a systematic effect of 'correct' phase and very small amplitude (0·09 mb near the equator, decreasing towards either pole).

Although the gravitational tidal action of the sun is smaller than that of the moon by the factor 2·4, the amplitude of the atmospheric oscillation which is governed by solar time is much the greater; the related effect on surface pressure is clearly visible on low-latitude barograms—see DIURNAL VARIATION. The explanation is that the solar-controlled oscillations are caused by a combination of thermal and gravitational action, the former being the more important; conventionally both agencies are considered in discussions of atmospheric tides.

The main solar tidal components are those of 24-hour and 12-hour periods. They are termed the S_1 and S_2 components respectively, the former being much the more variable in phase and amplitude. The mean amplitude of S_1 is about half that of S_2.

In middle latitudes the amplitude of the solar tidal wind is a small fraction of 1 m/s at the surface and it changes little throughout the troposphere. Between the low and high stratosphere it increases from less than 1 m/s to about 7 m/s and there is a further large increase to some 15–30 m/s at 80–90 km; at these and higher levels the tidal component is a very significant, and sometimes dominant, part of the total wind.

The RESONANCE hypothesis that the large magnitude of S_2 requires for its explanation the existence of an atmospheric FREE PERIOD very close to 12 hours is not supported by more recent theory according to which sufficient explanation of the magnitude of S_2 lies in the semi-diurnal nature of the temperature wave which affects the whole of the atmosphere. The smaller value of S_1 is explained by the structure of the atmosphere, which tends to suppress this oscillation.

atmospheric window: A term applied to that region of the ABSORPTION spectrum of water vapour which extends from about 8·5 to 11 micrometres. Ground radiation in this range of wavelengths is, in contrast to ground radiation of other wavelengths, little absorbed by water vapour and, in the absence of cloud, escapes to space. See TERRESTRIAL RADIATION, GREENHOUSE EFFECT.

attached thermometer: A thermometer attached with its bulb within the metal tube surrounding a mercury BAROMETER. A reading of this thermometer is required in order that an appropriate 'temperature correction' may be applied to the barometer reading if the thermometer has a reading other than the STANDARD TEMPERATURE of the barometer.

In some marine barometers the attached thermometer is incorporated in the GOLD SLIDE.

attachment: In meteorological literature, often used with particular reference to the disappearance of free ELECTRONS by their attachment to neutral oxygen atoms or molecules, thus forming negative IONS. The rate of the process is expressed by an 'attachment coefficient' with dimensions L^3T^{-1}.

attenuation: In geophysics, the depletion of electromagnetic energy (e.g. solar radiation, radio waves, radar waves) which is effected by the earth's atmosphere and its constituents. The rate of attenuation is represented by the 'attenuation coefficient' σ, which includes the effects both of ABSORPTION and of SCATTERING of the radiation, and is defined for monochromatic radiation by the equation, analogous to Beer's law which applies to absorption only,

$$I = I_0 e^{-\sigma x}$$

where I_0 is the intensity of radiation emitted at the source (or incident on the top of the atmosphere in the case of solar radiation) and I is the intensity after path length x through the absorbing and scattering medium.

The extent to which the above law may be applied over the whole spectrum, or a substantial part of the spectrum, depends on the degree of variation of attenuation coefficient with wavelength.

See also EXTINCTION COEFFICIENT, TURBIDITY.

audibility: The audibility of a sound in the atmosphere may be measured by the distance from its source at which it remains just audible.

The extent to which the loudness of sound decreases with distance from source at a rate other than the 'theoretical' inverse square of distance law is determined by the structure of the atmosphere. The factors mainly involved are the vertical distribution of temperature and wind and probably also the degree of atmospheric turbulence. It is, in turn, possible to infer much about the height variations of temperature and wind in the atmosphere from observations of the space distribution of audibility and time of arrival of sound waves on occasions of natural or artificial explosions.

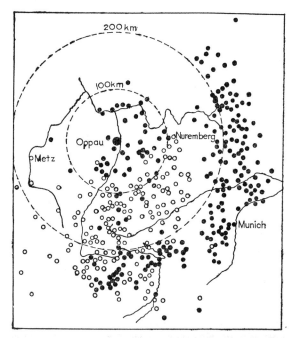

FIGURE 8—Audibility of the explosion on 21 September 1921 at Oppau, Germany.
The circles represent observers who did not hear the explosion, while dots represent those who did.
(GOODY, R. M.; The physics of the stratosphere. Cambridge Monographs on Physics, Cambridge University Press, 1954).

PLATE 1 Auroral corona: Perthshire.

PLATE 2 Auroral rays, Inverness-shire.

PLATE 3 Arctic sea smoke.

PLATE 4 Banner cloud over the Matterhorn.

The REFRACTION of sound waves in a calm atmosphere is governed by the lapse rate of temperature. The waves are bent upwards in conditions of strong lapse, and audibility is correspondingly low; conversely, it is high during a surface inversion because of downward bending of the waves sufficient for them to return to ground. Audibility is greater downwind than upwind, especially when there is a rapid increase of wind with height. A turbulent atmosphere is considered to limit audibility because of the associated mass exchange in the vertical and the accompanying dissipation of the sound energy. Audibility is generally greater, sometimes by a factor of 10 or 20, by night than by day because the conditions required for great audibility are much more characteristic of night than of day. For the same reason great audibility is a prominent feature of polar latitudes.

The 'anomalous audibility' associated with the so-called ANOMALOUS SOUND PROPAGATION is produced not by low-level meteorological conditions but by those at high levels. This feature, observed with large explosions, is characterized by a 'zone of silence' beginning at a distance from the source greater than that reached by the surface-propagated wave, with an outer zone of audibility at still greater distances (see Figure 8). Typical distances from the source are 100 and 200 km, respectively; the zone boundaries may be approximate circles with the source as centre or may depart appreciably from this. The cause of the outer zone is the refraction of sound waves in the high atmosphere to an extent which is sufficient to bend them back to earth. Calculations made from observations of the distribution of audibility and the time of arrival of sound waves on occasions of large explosions show that the major cause of the refraction is a region of high atmospheric temperature at a height of some 50 km, and that the effect of the refraction caused by high-level winds is mainly to make the shapes of the zone boundaries less circular.

aureole: A luminous bluish-white area, of small angular radius, immediately surrounding the sun or moon and bounded by a brownish-red ring. See CORONA.

The term 'aureole' is also used by some authors for the bright area, with no definite boundary, seen round the sun when the latter is not veiled by thin cloud.

aurora: This term (Latin for 'dawn') is applied to the phenomenon in which visible light is emitted by the high atmosphere.

In the 'auroral zones' of maximum frequency, some 20° from the geomagnetic poles, aurora is visible on almost every clear night; the northern auroral zone lies just north of Norway, south of Iceland and Greenland, over northern Canada and north of Siberia. The estimated mean annual frequency of nights of visible aurora, but for the intervention of cloud, twilight and moonlight, is about 150 in northern Scotland and about 10 in southern England; the corresponding frequencies of overhead displays are about 10 and 1, respectively. The terms 'aurora borealis' or 'northern lights' and 'aurora australis' or 'southern lights' apply, respectively, to the northern and southern hemispheres.

In Great Britain, aurora is usually seen near the northern horizon as a 'glow', or as a quiet 'arc', a grey-white feature with relatively sharp lower border. In a great display, during which the aurora may extend so far equatorwards as to be visible in the tropics, other auroral forms appear, with much movement and colour, of which the most characteristic are yellow–green and red. Greenish 'rays' may cover most of the sky polewards of the magnetic zenith, ending in an arc which is usually folded and sometimes with red lower border, the display then resembling a moving 'drapery'. If the display passes overhead, the parallel rays moving along the lines of force appear, by perspective, to converge at the magnetic zenith, thus producing a 'corona'. The later stages of a great display are usually marked by 'flaming', in which light surges upwards from the horizon. The display ends with a polewards recession of the auroral forms, the rays often then degenerating into diffuse surfaces of white light. Numerous reports of audible aurora and of aurora reaching almost to the ground in great displays are generally discredited.

The distribution of aurora in time and space has been determined by visual observation and by simultaneous photography from a number of stations. More recently, 'all-sky' automatic cameras and radio-echo equipment have also been used. The observations have shown that aurora is most frequent (in places equatorwards of the auroral zone) towards midnight and near the equinoxes, that auroral processes occur also during the day, and that auroral light originates at heights varying between 70 and 1000 km, with a marked peak frequency at about 100 km.

In spectral measurement of auroral light, molecular and atomic nitrogen and oxygen (atmospheric) and atomic hydrogen lines are the chief of those identified; the hydrogen line is subject to Doppler shift (see DOPPLER EFFECT) towards shorter wavelengths, implying the entry of high-speed protons into the atmosphere. The characteristic yellow–green and red colours are atomic-oxygen emission lines at 5577 Å and 6300 Å, respectively. The latter is emitted at higher levels; when red coloration appears at low levels, the emission is in a band of molecular nitrogen.

Aurora, like geomagnetic disturbance with which it is very closely associated, is caused by the entry into the high atmosphere of a stream of charged solar particles which are deflected by the earth's permanent magnetic field and precipitate over limited regions of the atmosphere. The various auroral forms arise from the primary and secondary collision processes of the solar particles with atmospheric gases and free electrons, and probably also in part from electric discharge processes which result from the generation of powerful electric fields in the high atmosphere.

See Plates 1 and 2. See also GEOMAGNETISM.

austausch: A German term signifying the mixing and exchange implicit in atmospheric TURBULENCE. The term 'austausch coefficient' is an alternative for EXCHANGE COEFFICIENT.

autobarotropy: The (idealized) atmospheric state in which surfaces of constant pressure and density (or specific volume) remain always in coincidence. See BAROTROPIC.

autoconvective lapse rate: That temperature LAPSE rate which defines a state of constant atmospheric density with height; for lapse rates in excess of this the density increases with height. The lapse rate is given by the expression g/R, where g is gravitational acceleration and R the gas constant for air, and has the value 3·42 degC/100 m, i.e. about $3\frac{1}{2}$ times that of the dry adiabatic lapse rate. Lapse rates well in excess of the autoconvective lapse rate have been observed near the earth's surface (approximately the lowest 15 metres) during intense INSOLATION.

The implication contained in the term that this lapse rate represents critical conditions for the 'automatic' establishment of convection is erroneous. Because ascending air cools on expansion, the critical condition is not one of increasing density with height, as in an incompressible fluid, but a lapse rate in excess of the adiabatic lapse rate. See ADIABATIC.

autocorrelation: Autocorrelation, also termed 'serial correlation' and 'lag correlation', signifies CORRELATION within a series of observations often spaced at equal intervals of time. The measure of internal correlation—the 'autocorrelation coefficient' or 'serial correlation coefficient'—is often employed as a measure of the degree of PERSISTENCE within geophysical TIME SERIES; such coefficients, evaluated for various interval lags, may also be combined in the CORRELOGRAM as a means of testing for PERIODICITY.

The autocorrelation coefficient for a lag of L intervals of time is given by

$$r_L = \frac{\sum\limits_{i=1}^{i=N-L}(x_i x_{i+L})}{(N-L)\sigma_x^2}$$

where x_i, x_{i+L} are departures from the mean of the series of N observations and σ_x is the STANDARD DEVIATION of the series.

The autocorrelation coefficient for successive daily values of pressure and temperature at places in the British Isles is about 0·8.

automatic data-processing: A systematic sequence of operations on data, performed largely automatically, the object of which is to extract or revise information. HOLLERITH SYSTEMS and electronic digital computer systems are examples of automatic data processing equipment.

A computer system may be regarded as comprising four units—programmer, operator, software and hardware. The programmer devises the instructions for the computer's operation, the programmed instructions being collectively known as software. The operator is responsible for the operation of the programmes; the mechanical and electronic components which he operates are termed the hardware.

The hardware of a large modern computer generally consists of a central processor (arithmetic unit and control unit), main and backing stores, and various peripheral devices for input and output. The main store is a fast access store and often consists of ferrite cores; the backing store is an additional slow-access store, often consisting of magnetic drums or discs, which may be needed for more complex operations. The control unit directs the sequence and timing of operations, interpreting the coded instructions from the main store and stimulating the appropriate circuits to execute the instructions.

When a programmer is presented with a problem he first breaks it down in the form of a 'flow diagram'. This shows pictorially the sequence of operations to be followed by the computer in solving the problem. The sequence of instructions is then coded in precise detail; accuracy and completeness of the instructions are tested in trial runs on the computer and any necessary amendments are made.

Three categories of programme may be distinguished. First, a 'Director' type of programme, supplied with the computer, is read into the machine at the start of each working day. Its main functions are to control the progress of all other programmes which may be operating simultaneously and to allocate storage space and the input and output devices required. The second category of programme covers an operation common to many problems handled by a particular type of computer; a procedure of this kind, made permanently available for insertion in larger programmes as required, is called a library sub-routine. Finally there is the programme devised for the problem in hand, perhaps incorporating one or more library sub-routines.

A programme may be written in 'machine language', i.e. one appropriate to a particular type of computer, or in a 'high-level language'. A machine-language programme, though detailed and rather difficult to devise, is economical in computer time; it is therefore appropriate to a repetitive operation such as numerical weather forecasting. A high-level language programme is more easily written and may be appropriate to more than one type of computer but is less efficient in its use of computer time. Widely used high-level languages include ALGOL and FORTRAN in science and COBOL in business.

The operator provides the computer with the data and programmed instructions. The input material may be on magnetic tape, paper tape, or punched or magnetic cards; other devices such as character-recognition units may be used. Output from a computer may be on tape or cards, or in the form of figures and letters from a line printer, or as lines or plotted characters on a chart, or as display on a cathode-ray tube.

automatic picture transmission (APT): As used in meteorology, this term denotes the automatic transmission of photographs of the earth and clouds by weather SATELLITES.

A scanned image is transmitted from the satellite to a ground station where the image is rebuilt on a facsimile machine. A ground station usually obtains three pictures, covering about 6 million square kilometres of the earth's surface, on each transit of the satellite.

automatic weather station: A STATION, situated generally in an isolated location, at which observations of meteorological elements are made, and from which the observations are transmitted, by automatic methods which require no permanent staff for their operation.

autumn: See SEASONS.

available potential energy: As defined and used by E. N. Lorenz, this term denotes that (small) part of the total POTENTIAL ENERGY of the atmosphere available for conversion to KINETIC ENERGY under adiabatic flow. It is the difference between the total potential energy of the atmosphere at a given time and the hypothetical minimum total potential energy that would result after adiabatic redistribution of the mass of the atmosphere such that the density stratification was everywhere horizontal. The available potential energy is some ten times greater than the total kinetic energy but is only about one per cent of the total potential energy.

The concept is strictly applicable only to the atmosphere as a whole and is significant in studies of the GENERAL CIRCULATION.

avalanche wind: An avalanche often causes, in advance and at the sides of the descending mass of snow or ice, a very high wind capable of causing destruction at some distance from the avalanche itself. It is known as the 'avalanche wind' or 'avalanche blast'.

average: An alternative for ARITHMETIC MEAN.

aviation forecast: A meteorological forecast issued for the purpose of aviation, generally covering conditions over a specific route and at a terminal or terminals.

The main elements included are the weather, clouds (type, base, thickness), visibility, upper winds, surface winds at terminals, height of the 0°C isotherm, icing and turbulence risks.

Avogadro's law: At normal temperature and pressure the weight of any gas in grammes which is numerically equal to its MOLECULAR WEIGHT occupies 22·4 litres. The law implies that equal volumes of gases, under the same conditions of temperature and pressure, contain the same number of molecules.

Application of this law allows of ready conversion from mass concentration or density (e.g. mg/m^3) of a particular gas in a mixture of gases to FRACTIONAL VOLUME ABUNDANCE (e.g. parts per million) of the gas, or vice versa. Thus, for CO_2 (molecular weight $12 + 32 = 44$) the relationship is 1 mg/m^3 = 0·51 p.p.m.

Avogadro's number: The number (N) of molecules per MOLE of a gas, equal to $6·02252 \pm 0·00028 \times 10^{23}$. See also AVOGADRO'S LAW.

azimuth: The azimuth of an object is the horizontal angle between the observer's MERIDIAN and the line joining observer and object. It is normally measured in degrees (0°–360°) clockwise from true (geographic) north.

Azores anticyclone: Part of the subtropical high-pressure belt of the northern hemisphere. On mean surface-pressure charts the anticyclone in summer has a central pressure of about 1027 mb at about 35° N, with the axis of a ridge extending across northern France and northern Germany; in winter the central pressure is about 1024 mb at about 30° N, with the axis of the ridge then lying across southern Spain.

B

Babinet's point: See ARAGO'S POINT.

Babinet's principle: The DIFFRACTION produced by a drop, for example a drop in a water cloud, is the same as that produced by a small aperture in a screen.

back-bent occlusion: A FRONT which is formed in the rear quadrant of a depression and which is attributed to the air-mass contrast originally developed at the OCCLUSION. As the low-pressure centre moves along the occlusion (usually eastwards) the line of air-mass contrast moves south-eastwards behind the centre as a back-bent occlusion. Secondary cold fronts also appear in the rear quadrant of depressions and do not necessarily arise from back-bent occlusions.

backing: The changing of the wind in a counter-clockwise direction, in either hemisphere.

baguio: A local name by which the TROPICAL CYCLONES experienced in the Philippine Islands are known. A number of the cyclones or typhoons of the western Pacific cross the Philippines; in addition, there is a class of cyclone which is especially associated with these islands, occurring from July to November.

balance equation: An equation expressing the balance between the horizontal non-divergent motion and the corresponding pressure field. The equation may be written

$$f\nabla^2\psi + \nabla\psi \cdot \nabla f + 2\left[\frac{\partial^2\psi}{\partial x^2}\frac{\partial^2\psi}{\partial y^2} - \left(\frac{\partial^2\psi}{\partial x\partial y}\right)^2\right] = g\nabla^2 z$$

where ψ is the STREAM FUNCTION, f the CORIOLIS PARAMETER, g the gravitational acceleration, z the height of the pressure surface, and ∇^2 the LAPLACIAN operator.

Since the flow in the middle troposphere is usually close to a state of non-divergence, the wind at such levels is better derived from a STREAM FUNCTION than from an assumption of geostrophic motion. Solution of the balance equation gives the required stream function.

The horizontal component of this equation is obtained by taking the DIVERGENCE of the EQUATIONS OF MOTION and omitting the lower-order terms involving divergence. It has proved very important in its application to NUMERICAL WEATHER FORECASTS since it makes it possible to use PRIMITIVE EQUATIONS without introducing spurious high-frequency disturbances of large amplitude.

ballistics: The science of gunnery. Range tables for ordnance are constructed on the assumption that certain meteorological conditions—those of a hypothetical 'standard ballistic atmosphere' and no wind—are applicable. In actual firings, variations from these conditions will occur and corrections must be applied when laying the gun, or the shell will not hit the desired target. The application of meteorology to gunnery came into great importance during the First World War and a specialized branch of the subject grew out of the requirements of artillery units for upper air data. Methods for computing EQUIVALENT CONSTANT WIND and BALLISTIC TEMPERATURE are now in common use at army meteorological stations.

ballistic temperature: A temperature computed for the purposes of BALLISTICS from a knowledge of the actual temperature distribution in the atmosphere. The temperature is such that if it were a surface temperature and the lapse rate had the value assigned in the 'standard ballistic atmosphere', the effect on the motion of a shell would be the same as that of the observed temperature distribution. A ballistic density is similarly defined.

ballistic wind: An average wind computed for the purposes of BALLISTICS. It is that wind, constant in speed and direction, which would produce the same displacement of a shell in flight as would the actual winds, varying with height, met by the shell.

ball lightning: A rare form of LIGHTNING in which a persistent and moving luminous white or coloured sphere is seen; the explanation of this form of lightning is controversial as was formerly its existence.

Reports of the sphere dimensions vary from a few centimetres to about a metre but are most commonly from 10 to 20 cm. Duration varies from a few seconds to several minutes. Many reported cases follow a brilliant lightning flash and may be physiological in nature (after-image); other reported cases have, however, occurred without a preceding flash. Sometimes more than one sphere is seen by an observer, or a sphere is reported in the same locality by various observers. The speed of travel is generally about a walking pace. Spheres have been reported to vanish harmlessly, to bounce from the ground or an obstruction, or to pass into or out of rooms leaving, in some cases, sign of their passage as, for example, a hole in a window pane.

balloon sounding: Exploration of the earth's atmosphere by means of a balloon inflated with a gas lighter than air (hydrogen or helium). Such a balloon may be followed by theodolite (PILOT BALLOON) or, if a reflector is attached, by radar (RADAR WIND) to obtain upper air winds. Automatic instruments may be attached to obtain readings of temperature and relative humidity at various pressure levels (RADIOSONDE). The maximum altitude attained by a balloon depends on its FREE LIFT. Radiosonde balloons generally reach from 18 to 30 km.

Very large balloons carrying considerable equipment loads are used on occasion for geophysical research, e.g. in cosmic rays. Others carry a special device so that they may remain for a considerable period at a predetermined level ('constant-level balloon'). Others are used for launching small rockets—see ROCKOON. Early research balloons were manned, with corresponding limitation in height attained. Among the early ascents made for meteorological purposes were those by J. Welsh in 1852 and by J. Glaisher between 1862 and 1866. Captive balloons are sometimes used to obtain meteorological data in the lower atmosphere.

banner cloud: A stationary cloud attached to, and extending downwind from, an isolated mountain peak, the cloud appearance being one of an extended flag. The requirements are an isolated sharp peak, strong wind and relatively moist air. Among well-known examples of the cloud are those associated with Mount Everest and the Matterhorn.

The physical explanation normally advanced for the cloud formation is the lifting of air in a lee eddy from a lower level than on the windward side. An aerodynamic pressure reduction in the lee of the peak may contribute to the cloud formation in a manner analogous to the formation of aircraft wing-tip CONDENSATION TRAILS. See Plate 4.

bar: A unit of atmospheric pressure equal to the pressure of 29·530 in or 750·062 mm of mercury under the standard conditions of temperature 0°C (density of mercury 13 595·1 kg/m^3) and gravitational acceleration 9·80665 m/s^2.

1 bar = 10^3 MILLIBARS = 10^6 DYNES/cm^2 = 10^5 NEWTONS/m^2.
See STANDARD DENSITY, STANDARD GRAVITY, SI UNITS.

barat: A strong, north-westerly squall on the north coast of the island of Celebes, most frequent from December to February.

barber: A little-used alternative for ARCTIC SEA SMOKE.

baroclinic: A baroclinic atmosphere is one in which surfaces of pressure and density (or specific volume) intersect at some level or levels. The atmosphere is always, to some extent, baroclinic. Strong baroclinicity implies the presence of large horizontal temperature gradients and thus of strong THERMAL WINDS.

baroclinic instability: A type of dynamic instability, associated with a strongly BAROCLINIC region of the atmosphere, which is considered to be responsible for at least part of the development of wave disturbances within the strong westerly wind flow which frequently occurs in middle and high latitudes. Growth of the disturbances is characterized by ascent of the warmer, and descent of the colder, air masses, representing a decrease of potential energy and an associated release of kinetic energy. Theory indicates that the degree of instability of the disturbances depends, among other factors, on their wavelength.

baroclinic wave: A wave depression which forms in a strongly BAROCLINIC region of the atmosphere. BAROCLINIC INSTABILITY is important in the intensification of such waves.

barogram: The record made by a BAROGRAPH.

barograph: A recording BAROMETER. That in common use is essentially an ANEROID BAROMETER which is arranged to give continuous recording. Those used by the Meteorological Office are known as the 'open scale' and the 'small pattern', which differ in the type of aneroid element and lever mechanism used. In ship barographs, use is made of an increase of lag coefficient and anti-vibration mounting.

Two non-portable types of barograph, used at some observatories, are the 'float-barograph', in which the movements of the mercury in a barometer are communicated to a recording pen by means of a float, and the 'photobarograph', in which the position of the mercury meniscus is recorded photographically.

barometer: An instrument for measuring atmospheric pressure. The mercury barometer, for example the FORTIN BAROMETER or the KEW-PATTERN BAROMETER, is satisfactory for normal use but is difficult to transport. In these instruments the height of the mercury column in a glass tube, one end closed and uppermost, above the level of mercury in an open vessel (cistern) in which the open end of the tube is immersed, is used as a measure of atmospheric pressure. Barometer 'corrections' must be applied to take account of departures of mercury and scale temperature, and of local GRAVITY, from the 'standard' values assumed in the calibration of the instrument.

A non-mercury type of barometer in common use is the ANEROID BAROMETER.
See also STANDARD DENSITY, STANDARD GRAVITY, STANDARD TEMPERATURE.

barometric characteristic: In synoptic meteorology, an observation of the shape of the BAROGRAPH trace during the three hours prior to the observation. Such observations are plotted on synoptic charts in internationally agreed symbols which simulate the general form of the barograph trace.

barometric tendency: In synoptic meteorology, an observation of the barometric change during a specified period (usually three hours) prior to the observation. Such observations, plotted on synoptic charts, are the basis on which ISALLOBARS are drawn.

barothermograph: An instrument which records temperature and pressure simultaneously, two pens being used to record separate traces of temperature and pressure.

The term has also been loosely used in connection with the obsolete Dines METEOROGRAPH in which temperature was recorded as a function of pressure; in this instrument the recording pen was made to move in one direction by a change of pressure, and in a direction almost at right angles by a change of temperature.

barotropic: A barotropic atmosphere is that hypothetical atmosphere in which surfaces of pressure and density (or specific volume) coincide at all levels. The concept of barotropy, though idealized, gives a useful first approximation in some types of atmospheric problem. The contrasting atmospheric state is the BAROCLINIC.

barotropic wave: A wave-like disturbance in a BAROTROPIC atmosphere. Since such an atmosphere is likely to be approximately non-divergent, the concept of barotropic waves has been applied at about 600 mb (so-called 'level of non-divergence'). Forecasting of the movement of such waves is based on the conservation of absolute VORTICITY by individual air particles. The same principle is applied in respect of LONG WAVES (or Rossby waves) which are essentially barotropic waves in airflow of relatively uniform speed.

bathythermograph: An instrument which provides a continuous trace of sea-temperature variations with depth.

A liquid-in-glass thermometer moves a metal point in one direction across a smoked glass slide while the slide itself is made to move, by metal bellows which are sensitive to pressure (depth) variations, in a direction at right angles to this. The instrument is lowered from a ship to a selected level and is then recovered. The instrument is often used from ships which are under way but then seldom to depths below about 300 m.

beaded lightning: An alternative for PEARL-NECKLACE LIGHTNING.

bearing: The true (geographic) bearing of an object is synonymous with its AZIMUTH. Magnetic bearing is the corresponding angle measured clockwise from magnetic north. Approximate bearings may be named as the compass points, N, NE, etc.

Beaufort notation: A code of letters indicating the state of the weather, past or present. The code was originally introduced by Admiral Beaufort for use at sea but is equally convenient for use on land. Additions have been made to the original schedule. A phenomenon of moderate intensity is indicated by the corresponding small letter: if of slight intensity the suffix $_0$ is added, if of greater intensity a capital letter is used. Continuity is indicated by repetition of the letter, intermittency by prefixing the letter i, and showers by prefixing the letter p. Thus, for example, ir$_0$

denotes intermittent slight rain, and pS heavy snow shower. See *Observer's handbook*.*

Beaufort scale: Wind force is estimated on a numerical scale ranging from 0, calm, to 12, hurricane, first adopted by Admiral Beaufort. The specification of the steps of the scale originally given had reference to a man-of-war of the period 1800–50 and therefore now possesses little more than historic interest.

Details of the Beaufort scale are contained in Table II. The velocity equivalents were originally based on the empirical relationship between estimated number and measured velocity, $V = 1.87 \sqrt{B^3}$, where V is in miles per hour, and B is the corresponding Beaufort number. The pressure equivalents were derived from the formula $p = 0.003 V^2$, where p is in pounds per square foot and V is in miles per hour.

Beaumont period: A period of 48 consecutive hours during which the dry-bulb temperature in the screen has been 10°C or above and the relative humidity has been 75 per cent or above on at least 46 of the 48 hourly observations; the period is used as a criterion for the issue of a POTATO BLIGHT WARNING.

Beer's law: See ABSORPTION.

Bénard cell: When a fluid is carefully heated from below or cooled from above in the laboratory, a cellular pattern of CONVECTION may be established in which the motion in the centres of the cells (upward or downward) is opposed to that on the periphery. Such cells are known as Bénard cells. There is some evidence that such cells develop at times in layered cloud after dusk by radiative cooling at the cloud top.

Bergeron (–Findeisen) theory: That theory which attributes the initiation of precipitation from a cloud to the presence of ice crystals among predominantly supercooled water droplets. See PRECIPITATION.

berg wind: A local name for the offshore FÖHN-type wind in South Africa.

Bermuda anticyclone: That cell of the semi-permanent subtropical high-pressure belt which is frequently centred near Bermuda in the western part of the North Atlantic Ocean.

Bernoulli's theorem: In an inviscid fluid in steady motion the sum per unit mass of the kinetic energy ($v^2/2$), the potential energy possessed by virtue of being in a pressure field (p/ρ), and the gravitational potential energy (gz) is constant;

$$v^2/2 + p/\rho + gz = \text{constant}.$$

Here v is fluid velocity, p pressure, ρ density, g gravitational acceleration, and z height above a selected reference level.

beta (or β) particle: A swiftly moving ELECTRON emitted spontaneously by certain radioactive elements. Beta particles have moderate penetrative power amounting to some yards in air near ground level; their emission by natural radioactive materials in the ground contributes significantly (about one-fifth) to the total IONIZATION of air at low levels over land. See also ALPHA PARTICLE, GAMMA RADIATION.

---

* London, Meteorological Office. Observer's handbook, 3rd edn. London, HMSO, 1969, pp. 68–70.

bi-directional vane (or **bivane**): An instrument comprising two sensitive vanes of similar characteristics, free to turn about vertical and horizontal axes, respectively. The instrument is used, with direct pen or with electrical recording, to indicate simultaneous variations of the horizontal and vertical components of the wind in low-level turbulence measurements. See *Geophysical Memoirs* No. 65.*

billow clouds: Parallel rolls of cloud, separated by relatively narrow, clear spaces. The phenomenon has been explained in various ways: as waves formed at a surface of discontinuity; as convection cells initiated by shearing motion at an almost horizontal boundary between two airstreams; by cellular motion initiated by the development of static instability in a shallow layer of air.

The term is often loosely applied to clouds of the variety UNDULATUS.

bimetallic thermograph: A THERMOGRAPH in which the sensitive element is a curved strip formed by welding together two metals differing in their coefficients of expansion. The changes of curvature undergone by the strip with changes of temperature are used to actuate a recording pen through a lever mechanism.

binomial distribution: When the PROBABILITY of occurrence of an event in a single trial is p, then the number of occurrences in samples each consisting of n independent trials accords with the binomial distribution of MEAN DEVIATION np and STANDARD DEVIATION $\sqrt{(np(1-p))}$.

Thus the probability $p(x)$ that an event of probability p occurs x times in n independent trials is

$$p(x) = \frac{n!}{x!(n-x)!} \cdot p^x(1-p)^{n-x}.$$

The binomial distribution finds a useful application in cases of two contrasting alternatives, e.g. rain or no rain or testing of two forecasting systems, provided that the result of each trial is independent of the results of earlier trials. See FREQUENCY DISTRIBUTION, NORMAL FREQUENCY DISTRIBUTION, POISSON DISTRIBUTION.

bioclimatology: The study of climate in relation to life and health.

biosphere: That part of the earth's envelope, comprising the seas, lower atmosphere and surface layer of the earth's crust, in which living organisms exist in their natural state.

bise: A cold, dry wind which blows in the winter in the mountainous regions of southern France from the north, north-east or north-west. The cold, north-west wind which occurs in Languedoc, near the Mediterranean coast, differs from the MISTRAL in that it is accompanied by heavy clouds and has been given the name '*bise noire*'.

Bishop's ring: A dull reddish-brown ring which is seen round the sun in a clear sky. In the middle of the day the inner radius of the ring is about 10°, the outer 20°; when the sun is low the ring is larger. Bishop's ring was first seen after the great eruption of Krakatoa in 1883 and remained visible till the spring of 1886. It was also seen after the eruptions of Soufrière in St Vincent and Mt Pelée in Martinique in 1902, after the north Siberian meteorite in 1908, and at the time of nearest approach to the earth of Halley's comet on 18 and 19 May 1910. The phenomenon is attributed to DIFFRACTION associated with fine dust in the high atmosphere.

---

* BEST, A. C.; Transfer of heat and momentum in the lowest layers of the atmosphere. *Geophys Mem, London*, **7**, No. 65, 1935.

TABLE II—*Beaufort scale*:

Force	Description	Specifications for use on land	Specifications for use at sea
0	Calm	Calm, smoke rises vertically.	Sea like a mirror.
1	Light air	Direction of wind shown by smoke drift, but not by wind vanes.	Ripples with the appearance of scales are formed, but without foam crests.
2	Light breeze	Wind felt on face; leaves rustle; ordinary vane moved by wind.	Small wavelets, still short but more pronounced. Crests have a glassy appearance and do not break.
3	Gentle breeze	Leaves and small twigs in constant motion; wind extends light flag.	Large wavelets. Crests begin to break. Foam of glassy appearance. Perhaps scattered white horses.
4	Moderate breeze	Raises dust and loose paper; small branches moved.	Small waves, becoming longer; fairly frequent white horses.
5	Fresh breeze	Small trees in leaf begin to sway; crested wavelets form on inland waters.	Moderate waves, taking a more pronounced long form; many white horses are formed. Chance of some spray.
6	Strong breeze	Large branches in motion; whistling heard in telegraph wires; umbrellas used with difficulty.	Large waves begin to form; the white foam crests are more extensive everywhere. Probably some spray.
7	Near gale	Whole trees in motion; inconvenience felt when walking against wind.	Sea heaps up and white foam from breaking waves begins to be blown in streaks along the direction of the wind.
8	Gale	Breaks twigs off trees; generally impedes progress.	Moderately high waves of greater length; edges of crests begin to break into the spindrift. The foam is blown in well-marked streaks along the direction of the wind.
9	Strong gale	Slight structural damage occurs (chimney pots and slates removed).	High waves. Dense streaks of foam along the direction of the wind. Crests of waves begin to topple, tumble and roll over. Spray may affect visibility.
10	Storm	Seldom experienced inland; trees uprooted; considerable structural damage occurs.	Very high waves with long overhanging crests. The resulting foam, in great patches, is blown in dense white streaks along the direction of the wind. On the whole the surface of the sea takes a white appearance. The 'tumbling' of the sea becomes heavy and shock-like. Visibility affected.
11	Violent storm	Very rarely experienced; accompanied by widespread damage.	Exceptionally high waves (small and medium-sized ships might be for a time lost to view behind the waves). The sea is completely covered with long white patches of foam lying along the direction of the wind. Everywhere the edges of the wave crests are blown into froth. Visibility affected.
12	Hurricane	—	The air is filled with foam and spray. Sea completely white with driving spray; visibility very seriously affected.

Specifications and equivalent speeds

Force	Mean pressure (at standard density)* mb	Equivalent speed at 10 m above ground					
		Knots		Miles per hour		Metres per second	
		Mean	Limits	Mean	Limits	Mean	Limits
0	0	0	<1	0	<1	0·0	0·0–0·2
1	0·01	2	1–3	2	1–3	0·8	0·3–1·5
2	0·04	5	4–6	5	4–7	2·4	1·6–3·3
3	0·13	9	7–10	10	8–12	4·3	3·4–5·4
4	0·32	13	11–16	15	13–18	6·7	5·5–7·9
5	0·62	19	17–21	21	19–24	9·3	8·0–10·7
6	1·1	24	22–27	28	25–31	12·3	10·8–13·8
7	1·7	30	28–33	35	32–38	15·5	13·9–17·1
8	2·6	37	34–40	42	39–46	18·9	17·2–20·7
9	3·7	44	41–47	50	47–54	22·6	20·8–24·4
10	5·0	52	48–55	59	55–63	26·4	24·5–28·4
11	6·7	60	56–63	68	64–72	30·5	28·5–32·6
12	≥7·7	—	≥64	—	≥73	—	≥32·7

* The pressure due to the wind on any object exposed to it arises from the impact of the air on the windward side and suction on the leeward side; the mean pressure depends on the shape and size of the object. The values given apply to a disc of approximate area 0·1 to 10 m^2. See STANDARD DENSITY.

black-body radiation: See RADIATION.

black-bulb thermometer: A mercurial maximum thermometer with blackened bulb, mounted in an evacuated outer glass sheath and exposed horizontally to the sun's rays for the purpose of ascertaining the maximum temperature 'in the sun'. On account of the difficulty of obtaining comparable results with different instruments and of interpreting the indications of an individual instrument, black-bulb thermometers are not now recommended as a means of measuring solar radiation. Such measurements are carried out with a PYRANOMETER or PYRHELIOMETER.

black frost: A condition in which the temperature of the ground cools to a sub-freezing temperature but does not reach a value so low as the FROST-POINT of the adjacent air. There is then no deposit of HOAR-FROST on the ground or on terrestrial objects which thus remain 'black' in appearance. The phenomenon is associated with relatively dry air.

black ice: A popular alternative for GLAZE, often used with reference to its occurrence on road surfaces. A thin sheet of ice, relatively dark in appearance, may be formed when light rain or drizzle falls on a road surface which is at a temperature below 0°C. It may also be formed when supercooled fog droplets are intercepted by bridges, trees, etc.

blizzard: A term originally applied to the intensely cold north-westerly gales accompanied by fine, drifting snow which may set in with the passage of a depression across the United States in winter. The term has come to be applied to any high wind accompanied by great cold and drifting or falling snow.

blocking: The term applied in middle latitude synoptic meteorology to the situation in which there is interruption of the normal eastward movement of depressions, troughs, anticyclones and ridges for at least a few days.

A blocking situation is dominated by an anticyclone whose circulation extends to the high troposphere. The ZONAL CIRCULATION to the west is transformed into MERIDIONAL CIRCULATIONS branching polewards and equatorwards. For Europe and the North Atlantic the longitude most favoured for blocking is about 10°W. Over this region the percentage frequency of days of blocking situation has a maximum in April and a minimum in August.

blood-rain: Rain of a red colour which leaves a red stain on the ground. The coloration is due to dust particles contained by the drops; the particles are carried from a sandy region by upper air winds, sometimes for long distances. The phenomenon has been observed frequently in Italy, for example, but has also been known to occur in the British Isles, notably on 1 July 1968.

blowing dust or sand: 'Dust or sand, raised by the wind to moderate heights above the ground. The horizontal visibility at eye level is sensibly reduced'.*

blue moon, sun: A rare phenomenon in which more intense particle SCATTERING of red light than of blue light makes the directly viewed luminary appear blue (or green).

A conspicuous event of this kind occurred in the British Isles on 26 September 1950 and was due to scattering by smoke particles which originated in earlier forest fires in Alberta, Canada. The smoke was shown by aircraft reconnaissance to be

---

* Geneva, World Meteorological Organization. International cloud atlas, Vol. 1. Geneva, WMO, 1956, p. 71.

between about 10 and 13 kilometres. Application of scattering theory to the differential extinction of light measured in different parts of the visible spectrum indicated that the predominant radius of the scattering particles was about $0\cdot5\mu$m.

blue of the sky: The blue of the sky is caused by the SCATTERING of sunlight by the individual molecules of the air—so-called RAYLEIGH SCATTERING. For such small particles, Rayleigh showed that the proportion of scattered light is greater for shorter than for longer wavelengths (inversely proportional to the fourth power of the wavelength). Thus the light which reaches the observer after scattering is rich in the short blue and violet waves (to the latter of which the eye is the less sensitive); while the light reaching the observer directly is deficient in the short waves—hence the yellow or red disc of the sun.

The light scattered by dust particles suspended in the air normally has no wavelength dependence—but see BLUE MOON, SUN. Such particles therefore introduce a white tinge to the blue sky to a degree which increases with the dust concentration. Thus, for example, the sky is of a deeper blue in polar than in anticyclonic air, and also as seen from a mountain top compared with from a low level.

boiling-point: That temperature at which, under conditions of equilibrium between a plane surface of a liquid and its overlying vapour, the existing (saturation) VAPOUR PRESSURE is equal to the external pressure. The variation of the boiling-point of water with external pressure is given under HYPSOMETER.

bolide: The name applied to a METEOR so large as to cause an explosion on being destroyed in the atmosphere.

hologram: The record of a BOLOMETER.

bolometer: An instrument for the determination of the intensity of RADIATION, employing a blackened conductor whose change of resistance with temperature gives a measure of the quantity required. The instrument is often employed in the investigation of the distribution of energy in the spectrum, especially in the infra-red region: it is then called a 'spectrobolometer'.

Boltzmann's constant: The universal constant (k) given by the ratio R^*/N where R^* is the universal GAS CONSTANT and N is AVOGADRO'S NUMBER.

k has the value $1\cdot3805 \times 10^{-23}$ J/degK and appears in equations involving the expression of energy in terms of temperature.

bora: A cold, often very dry, north-easterly wind which blows, sometimes in violent gusts, down from the mountains on the eastern coast of the Adriatic. Strongest and most frequent in winter and on the northern part of the coast, it occurs when pressure is high over central Europe and the Balkans and low over the Mediterranean. If associated with a depression over the Adriatic it is accompanied by heavy cloud and rain or snow. The term is also applied to cold, squally, downslope winds in other parts of the world.

boreal climate: In W. Köppen's classification, a CLIMATE characterized by a snowy winter and warm summer, with a large annual range of temperature, such as obtains over the European, Asian and American continents between about latitudes 60° and 40° N.

Bouguer's halo: A rare type of HALO, white in colour and centred on the ANTISOLAR POINT with an inner radius of about 35° (also called ULLOA'S CIRCLE).

Bouguer's law: See ABSORPTION.

boundary layer: That layer of a fluid adjacent to a physical boundary in which the fluid motion is much affected by the boundary and has a mean velocity less than the free-stream value. In the 'surface boundary layer' of the atmosphere, of depth up to about 100 metres, the motion is controlled predominantly by the presence of the earth's surface, while within an overlying layer ('planetary boundary layer') with top at about 600 metres, effects on air motion by the boundary remain significant. See also FRICTION LAYER.

Bourdon tube: A curved tube of elliptical cross-section, in which changes of volume cause changes of curvature; these are used to actuate a pointer or recording pen. It may be used as a barometer or pressure gauge, being evacuated and responding to variations of external pressure; or for example, it may be connected to a steel thermometer bulb, when it responds to variations in the volume of the mercury which fills the bulb, the connecting capillary tubing and the Bourdon tube.

Bowen ratio: The ratio (R) of the amount of sensible heat to that of latent heat (see HEAT) lost by a surface to the atmosphere by the processes of conduction and turbulence.

Even over water surfaces, very variable values of R are found (including negative values, which signify sensible heat transfer from atmosphere to surface). An average value of about $+0.1$ is considered to hold for oceans, implying that 90 per cent of the heat energy received by oceans is used in evaporation.

Boyden index: An INSTABILITY INDEX (I) given by

$$I = Z - T_{700} - 200$$

where Z is the 1000–700-mb thickness (decametres) and T_{700} is the 700-mb temperature (°C).

Thunderstorms become increasingly likely the further I increases above a threshold value of 94. Allowance for the movement of the isopleths of I over an area (for example by the 700-mb wind) enables some account to be taken of advective changes.

Boyle's law: At constant temperature, the volume of a given mass of gas (or the SPECIFIC VOLUME) is inversely proportional to the pressure on the gas.

brave west winds: A nautical expression denoting the prevailing westerly winds of temperate latitudes. The region of strong westerly winds of the southern hemisphere (latitudes 40° S to 50° S) is termed the ROARING FORTIES.

breakaway depression: A term applied to a WARM-FRONT WAVE or WARM-OCCLUSION DEPRESSION because of the tendency of such secondary depressions to move eastwards away from the parent depression after formation.

breeze: A wind of moderate strength. (See BEAUFORT SCALE.) The word is generally applied to winds, caused by convection, which occur regularly during the day or night; they include LAND- AND SEA-BREEZES, MOUNTAIN BREEZE, VALLEY BREEZE.

Brewster's point: See ARAGO'S POINT.

briefing, meteorological: Oral explanation by a meteorologist of existing and expected meteorological conditions.

bright bands: In meteorological literature, usually refers to quasi-horizontal regions of enhanced reflection obtained in radar scanning of a precipitation belt in the vertical plane. The most conspicuous of the bright bands is that associated with the melting of snowflakes (see MELTING BAND). Other bright bands which occasionally appear at higher levels are thought to mark regions of appreciable changes of size or shape of solid precipitation elements. They have been ascribed, for example, to 'precipitation streaks', the shapes of which are changed by vertical wind shear as they are moved through the radar beam by the wind. See also RADAR METEOROLOGY.

British Rainfall Organization: This term was first used after the death of G. J. Symons in 1900 to describe the voluntary organization of rainfall observers which he had built up since about 1860. In 1919 the work of the Organization was transferred to the Meteorological Office. The term continued in use on the title-page of *British Rainfall* and elsewhere for several decades but is now (since about 1958) no longer used.

British Standard Time (BST): A standard of time introduced experimentally for three years in the British Isles on 18 February, 1968, with the intention that it should operate throughout the year, replacing the Greenwich Mean Time–BRITISH SUMMER TIME alternations. These alternations recommenced when Greenwich Mean Time was reintroduced on 31 October 1971.

British Standard Time is, like British Summer Time, one hour in advance of Greenwich Mean Time.

British Summer Time (BST): The standard of time previously in common use in the British Isles during a period in summer which was defined by the Summer Time Act, 1925. British Summer Time is one hour in advance of Greenwich Mean Time; 9 GMT is thus the same as 10 BST.

Summer time was first introduced in 1916. The periods during which British Summer Time was in operation between 1916 and 1961, and also those periods between 1941 and 1945 when 'double summer time' (clock time two hours ahead of GMT) was in operation are listed, for example, in the Fourth Edition (1963) of this book. From 1962 to 1967 the dates of the beginning and ending of British Summer Time were as follows:

1962	25 March–28 October	1963	31 March–27 October
1964	22 March–24 October	1965	21 March–24 October
1966	20 March–23 October	1967	19 March–29 October

In 1968 British Summer Time was supplanted by BRITISH STANDARD TIME.

British thermal unit: A unit of energy, defined as the heat required to raise the temperature of one pound of water by 1 degF.

$$1 \text{ Btu} = 252 \cdot 1 \text{ CALORIES}$$
$$= 1055 \text{ JOULES}.$$

Brocken spectre: When an observer stands on a hill partially enveloped in mist and in such a position that his shadow is thrown on to the mist he may get the illusion that the shadow is a person seen dimly through the mist. The illusion that this person or 'spectre' is at a considerable distance is accompanied by the illusion that he is gigantic. The Brocken is a mountain in Germany. See GLORY.

Brückner cycle: The name usually given to a feature observed by Brückner (1890) who by drastic smoothing was able to discern a series of 25 'cycles' covering the years A.D. 1020 to 1890 in which comparatively cool, rainy periods alternated with warmer, drier periods. The individual 'cycles', which were of very small amplitude, varied in duration from 20 to 50 years with an average length of 34·8 years.

Brunt–Väisälä frequency: The frequency $N/2\pi$ where N is defined by the equation

$$N^2 = \frac{g}{\theta}\frac{\partial \theta}{\partial z}$$

where $\partial\theta/\partial z$ is the vertical gradient of potential temperature and g the gravitational acceleration.

With certain simplifying assumptions $N/2\pi$ is the natural frequency of small vertical oscillations of a parcel of air about its equilibrium position in a stable atmosphere.

brush discharge: Discharge of electricity from sharp points on a conductor. See ST ELMO'S FIRE.

Buchan spells: A total of nine periods during the year which were advanced by Alexander Buchan (1867), on the basis of some 50 years' observations, as constituting fairly reliable periods of unseasonal cold (six cases) or warmth (three cases) in south-east Scotland. The periods were: cold, 7–14 February, 11–14 April, 9–14 May, 29 June–4 July, 6–11 August and 6–13 November; warm, 12–15 July, 12–15 August, 3–14 December.

While evidence has since been advanced in support of similar spells in other parts of Europe, the reality of the spells as statistically significant features has also since been disputed. The reliability of the spells is certainly too low to permit of their direct application in long-range weather forecasting.

bumpiness: A condition of the atmosphere characterized by rapid variations, including generally alterations of sign, of the vertical component of velocity experienced by an aircraft in flight.

Bumpiness is associated generally with either (or both) convection currents in an unstable atmosphere or a flow of air across surface irregularities. It is much more common and intense over the land than over the sea. It is generally most marked in the lowest kilometre (FRICTION LAYER) of the atmosphere but may extend to much higher levels above hilly country. See also AIR POCKETS, CLEAR-AIR TURBULENCE.

buoyancy: The buoyancy of a balloon or airship is the total load, including the envelope and fittings, that can just be supported. This buoyancy arises from the difference between the density of the light gas inside the envelope and that of the heavier air outside. The vessel will just rest in equilibrium when the total weight is the same as that of the air displaced; the buoyancy of the vessel is thus the difference between the weight of the gas in the envelope and that of the volume of air displaced (principle of Archimedes).

Analogous buoyancy forces act on parcels or 'bubbles' of air which are at a different temperature from that of the surrounding air and are fundamental in the process of free CONVECTION. Where T' and T are the temperatures (kelvin) of the air parcel and environment, respectively, the buoyancy force acting per unit mass is given by $g[(T'/T) - 1]$. The force is reckoned positive upwards ($T' > T$) and negative downwards ($T' < T$).

buran: A strong north-easterly wind which occurs in Russia and central Asia. It is most frequent in winter, when it is very cold and often raises a drift of snow, but strong north-easterly winds in summer are also termed buran. The winter snow-bearing wind is also termed 'purga'.

Buys Ballot's law: A rule in synoptic meteorology, enunciated in 1857 by Buys Ballot, of Utrecht, which states that if, in the northern hemisphere, an observer stands with his back to the wind, pressure is lower on his left hand than on his right, whilst in the southern hemisphere the converse is true. This law implies that, in the northern hemisphere, the winds blow counter-clockwise round a depression, and clockwise round an anticyclone; the converse is true in the southern hemisphere. This is a statement of the direction of the GEOSTROPHIC WIND.

C

calendar: For meteorological purposes it is usual to adhere to the civil calendar, and to publish summaries of climatological data for ordinary civil months or weeks. For certain purposes, however, there are advantages in selecting an epoch other than 1 January as the commencement of the year. Thus the 'farmer's year' adopted for the *Weekly weather report* begins on the Sunday nearest to 1 March; a 'grower's year' beginning on 6 November is adopted for the Crop Weather Scheme; and a 'water year' beginning on 1 October has long been used for the presentation of data in the *Surface water year book of Great Britain** now prepared by the Water Resources Board.

The existing (Gregorian or 'New Style') civil calendar was introduced in Great Britain in 1752 (earlier in most other parts of Europe), replacing the Julian calendar which was based on a solar year of $365\frac{1}{4}$ days and which involved an error of one day in 128 years. In the existing calendar, centurial years are ordinary, as opposed to leap years, unless they are divisible exactly by 400, thus reducing the error to one day in 3323 years. The accumulated error of the Julian calendar was eliminated by calling the day, which followed 2 September 1752, 14 September.

See also SEASONS.

calibration: Originally the name given to the process of finding the calibre (area of cross-section) of a tube. When a tube is not of uniform cross-section, marks so spaced that the volume of the tube between consecutive marks is everywhere the same, are calibration marks.

The use of the word has now been extended to include the determination of absolute values appropriate to selected fixed points of an instrument, by comparison with primary or secondary STANDARD instruments.

calm: Absence of appreciable wind: On the BEAUFORT SCALE of wind force calm is accorded the figure 0 and has a wind speed equivalence of less than one knot.

calorie (or **gramme-calorie**): A unit of heat, being the heat required to raise the temperature of 1 g of water by 1 degC. This quantity of heat depends, however, on the initial temperature of the water. The 15°C calorie (cal_{15}), defined as the heat required to raise the temperature of 1 g of water from 14·5° to 15·5°C, and the International Steam Table calorie (IT cal) are the calories most commonly used. It was decided at the Ninth General Conference of Weights and Measures (1948) that the JOULE should replace the calorie as the unit of heat, or the joule equivalent be given. The dimensions are ML^2T^{-2}. See also KILOCALORIE.

$$1\ cal_{15} = 4·1855\ \text{joules}$$
$$1\ \text{IT cal} = 4·1868\ \text{joules}$$

calvus (cal): A CLOUD SPECIES (Latin, *calvus* bald or stripped).

'CUMULONIMBUS in which at least some protuberances of the upper part are

---

* London, Water Resources Board and Scottish Development Department. *Surface water year book of Great Britain*. London, HMSO.

beginning to lose their cumuliform outlines but in which no cirriform parts can be distinguished. Protuberances and sproutings tend to form a whitish mass, with more or less vertical striations.'* See also CLOUD CLASSIFICATION.

canalization: See FUNNELLING.

candela: The magnitude of the candela is such that the LUMINANCE of the black-body radiator, at the temperature of solidification of platinum, is 60 candela per square centimetre.

cap: A name frequently given to the transient patches of cloud which sometimes form on or just above the tops of growing cumulus clouds, and are soon absorbed into them. It is also used for clouds on hilltops. It is technically known as PILEUS.

capillary potential: A concept used in SOIL MOISTURE studies, being the force of attraction exerted by soils on contained water, or the equivalent force required to extract the water from the soil against the capillary forces (SURFACE TENSION) acting in the soil pores. It is generally expressed in the pressure unit of atmospheres or the equivalent height of a specified liquid column (mercury or water).

Since the capillary potential increases very rapidly with increasing dryness of soil, a logarithmic measure pF is often used, defined as the logarithm to base 10 of the capillary potential expressed in centimetres of water. Thus a capillary potential of 1 atmosphere corresponds to pF 3, since a pressure of 1 atmosphere supports a column of water about 1000 cm in length. See also FIELD CAPACITY, WILTING POINT.

capillatus (cap): A CLOUD SPECIES (Latin, *capillatus* having hair).

'CUMULONIMBUS characterized by the presence, mostly in its upper portion, of distinct cirriform parts of clearly fibrous or striated structure, frequently having the form of an anvil, a plume or a vast, more or less disorderly mass of hair. Cumulonimbus capillatus is usually accompanied by a shower or by a thunderstorm, often with squalls and sometimes with hail; it frequently produces very well-defined VIRGA.'* See also CLOUD CLASSIFICATION.

carbon-dating: A technique of estimating the age of carbon-containing fossil materials, based on the measurement of their RADIOACTIVITY per unit mass and comparison with that of materials of known age. The technique is relevant, for example, to problems of past climatic changes. It is based on the fact that assimilation of RADIOACTIVE CARBON (^{14}C) ceases at the time of death of living material, the radioactivity of the material then decreasing at a rate which is determined by the radioactive HALF-LIFE of ^{14}C. Thus, for example, a period of 5500 years would be assumed to have elapsed since the death of a fossil sample whose specific radioactivity is one-half that of a living sample. Basic assumptions, inherent in the method, are (i) that a steady state has long existed between the rate of production of ^{14}C and its rate of disappearance, and (ii) that the strong latitudinal variation of production of ^{14}C by cosmic rays is eliminated in its lifetime by world-wide mixing in the atmosphere and oceans.

A modification of the basic technique has been used in order to estimate a possible SECULAR change of CARBON DIOXIDE concentration in the atmosphere. The method adopted in this case is to measure the specific radioactivity of a recent or living sample and also that of wood, for example, of known greater age since death

---

* Geneva, World Meteorological Organization. International cloud atlas, Vol. 1. Geneva, WMO, 1956, p. 14.

(say, 100 years). From these measurements the relative proportions of ^{14}C and CO_2 in the atmosphere at the two epochs are inferred.

Available data suggest that the basic assumption of a steady state between ^{14}C production and disappearance has recently been invalidated, to an extent which is not negligible, by the amount of ^{14}C produced in the atmosphere by nuclear-weapon testing. For this reason, the use of plant or animal materials which have been alive during the period since about 1950 is best avoided in the application of the techniques.

carbon dioxide: A gas, of chemical formula CO_2, which comprises 0·03 and 0·05 part per 100 parts of dry air by volume and weight, respectively. It is created by animal life and by the oxidation of carbon compounds (as in fuel burning), and is used by plants; it is also destroyed by photochemical processes in the high atmosphere. Owing to a ready but temperature-dependent solubility of CO_2 in water, the oceans act as a great reservoir of the gas. The measured amount in the atmosphere is generally considered to have increased significantly during this century. Because of the important part played by CO_2 in the radiative equilibrium of the atmosphere, such SECULAR changes of CO_2 amount are advanced as a possible contributory factor in effecting climatic changes. The CO_2 absorption band of chief importance lies between 12·5 and 17·5 μm, with peak absorption at about 15 μm.

carbon monoxide: A poisonous and colourless gas, of chemical formula CO, which is formed by the incomplete combustion of carbon-containing material. Its presence in minute concentration in the atmosphere has been observed spectroscopically.

Cartesian co-ordinates: A system of co-ordinates in which the x, y, z axes are mutually at right angles (rectangular system). In meteorology, a 'right-handed' system is usually employed in which the xy plane is horizontal, positive x and y to east and north, respectively, and positive z vertically upwards.

cascade impactor: A device used for sampling the aerosol content of gases. The gas is forced through a series of nozzles of decreasing size, each one being directed on to a slide which captures part of the aerosol content of the gas. The distribution of size on each slide is related to the nozzle diameter and the free-stream velocity; the smaller nozzles lead to more efficient capture of smaller particles by virtue of the greater impaction speed.

castellanus (cas): A CLOUD SPECIES, previously termed 'castellatus' (Latin, *castellum* castle).

'Clouds which present, in at least some portion of their upper part, cumuliform protuberances in the form of turrets which generally give the clouds a crenelated appearance. The turrets, some of which are taller than they are wide, are connected by a common base and seem to be arranged in lines. The castellanus character is especially evident when the clouds are seen from the side.

This term applies to CIRRUS, CIRROCUMULUS, ALTOCUMULUS and STRATOCUMULUS.'* See also CLOUD CLASSIFICATION.

CAT: Abbreviation for CLEAR-AIR TURBULENCE.

catchment area: Defined for administrative purposes as the area within the jurisdiction of a Catchment Board under the Land Drainage Act of 1930. The term is also commonly used with the same meaning as DRAINAGE AREA.

---

* Geneva, World Meteorological Organization. International cloud atlas, Vol. 1. Geneva, WMO, 1956, p. 12.

ceiling: The 'ceiling' of a specified mass (e.g. balloon, aircraft, thermal) is the maximum height in the atmosphere which the mass can attain.

The term is also used in the United States of America, and fairly generally in aviation circles, to specify the lowest height above the ground at which all cloud layers at and below that level cover more than half of the sky.

celestial sphere: That imaginary sphere, concentric with the earth, on the inner surface of which the heavenly bodies appear to lie, the observer being situated at the centre of the sphere. See also EQUATOR, POLE.

Celsius scale: A scale of temperature based on one introduced in 1742 by Celsius, a Swedish astronomer and physicist, who divided the interval between the freezing- and boiling-points of water into 100 parts, the lower fixed point being marked 100. The present system, whereby the freezing-point is marked 0 and the boiling-point 100, was introduced by Christin, of Lyons, in 1743. This latter scale is now generally referred to as the Celsius scale; alternative names are the centigrade scale and, less commonly, the centesimal scale. See TEMPERATURE SCALES.

centigrade: An alternative, though now less favoured, name for the Celsius scale of temperature.

centre of action: A term, introduced by Teisserenc de Bort in 1881, which generally signifies an area covered by a large-scale low- or high-pressure system, which dominates the circulation, and so has a big influence on weather conditions, over a large area for a considerable period of time.

The term has, however, also been used with other meanings. G. T. Walker, for example, defined a centre of action as an area in which conditions of pressure, temperature, rainfall or ice were strongly correlated with similar conditions in other parts of the world; an 'active' centre was one in which the conditions were highly correlated with conditions which occurred at other, 'passive' centres at a later time.

centrifugal force: A body rotating in a circle round a central point is subject to a 'centrifugal force' acting outwards from the centre, of magnitude $\omega^2 R$ or V^2/R per unit mass (ω = angular velocity, V = linear velocity, R = radius of curvature of path).

This force arises in meteorology in two main ways:
  (i) The observed force of GRAVITY is the vector sum of the force of gravitation directed towards the earth's centre and the centrifugal force acting on the earth and atmosphere due to rotation round the earth's axis.
  (ii) Air moving on a curved path with respect to the earth's surface is subject to a centrifugal force, and corresponding CENTRIPETAL ACCELERATION, giving rise to the cyclostrophic component of the GRADIENT WIND.

centripetal acceleration: That ACCELERATION of a body moving on a curved path, equal and opposite to the CENTRIFUGAL FORCE per unit mass, which is directed to the instantaneous centre of curvature of path of the body.

ceraunometer: An instrument designed to count and to give warning of the occurrence of LIGHTNING flashes within a specified area.

c.g.s. system: A system of units based on the centimetre, the gramme and the second as FUNDAMENTAL UNITS. This is the system in which the various derived units have usually been expressed in the sciences, including meteorology; for example, the unit of force is in this system defined as that which, applied to a mass of 1 g, produces an acceleration of 1 cm/s² and is termed the dyne.

The c.g.s. system is being replaced by SI UNITS.

chaff: Very fine needles or ribbon (of copper, metallized nylon or glass fibre) used for wind measurement. If large numbers of these needles are ejected, for example from a METEOROLOGICAL ROCKET, they form a cloud which acts as a passive wind sensor and which can be tracked by radar. The finest chaff, only 0·025 mm in diameter, is suitable for wind measurement up to 85 km.

chain lightning: An alternative for PEARL-NECKLACE LIGHTNING.

chance expectation: The probability of a certain result, given that the processes producing the result have no known, recognizable or definable connection with any of the possible values of this result. See PROBABILITY.

Chapman layer: An atmospheric layer conforming in properties to a model proposed by S. Chapman, who investigated the distribution in height of the product (electrons, molecules, free radicals, etc.) of a process depending on absorption of solar RADIATION.

The formation of a particular product in the high atmosphere by incident solar radiation is subject to two counteracting influences in that, with increasing depth of penetration, the amount of gas capable of absorbing the radiation increases because of increasing gas density downwards, while the amount of radiation decreases because of absorption at higher levels. There is thus a level, for a particular absorbing gas, at which the rate of formation of the product is a maximum and at which is centred a layer of characteristic 'shape'.

characteristic: See BAROMETRIC CHARACTERISTIC.

Charles's law: At constant pressure, the volume of a given mass of gas (or the SPECIFIC VOLUME) is directly proportional to the absolute temperature.

chemosphere: Term sometimes applied to that region of the ATMOSPHERE, extending mainly over the height range 40 to 80 km, in which PHOTOCHEMISTRY is important.

chinook: A warm and dry west wind, of the FÖHN type, which occurs on the eastern side of the Rocky Mountains. Its arrival is usually sudden, with a consequent large temperature rise and rapid melting of snow.

chi-square distribution: If $x_1, \ldots, x_n$ are items chosen at random from a NORMAL FREQUENCY DISTRIBUTION of mean zero and standard deviation unity, then

$$\sum_{j=1}^{n} x_j^2$$

falls into the chi-square distribution with $n-1$ DEGREES OF FREEDOM. This distribution has been studied extensively, and tables of the important PERCENTILES have been published, e.g. by Fisher and Yates, to facilitate the application of the CHI-SQUARE TEST and other tests of SIGNIFICANCE.

chi-square statistic:

$$\chi^2 = \Sigma[(O-E)^2/E].$$

For description see CHI-SQUARE TEST.

chi-square test: A statistical test of the agreement between an observed distribution of frequencies and the distribution expected according to some hypothesis. If O is the observed and E the expected frequency in each of n cells, then the CHI-SQUARE STATISTIC, often denoted by χ^2, is

$$\Sigma[(O-E)^2/E].$$

On certain assumptions, of which the most important is that the lowest expected frequency should not be less than 5, the value of χ^2 can be shown to fall into the CHI-SQUARE DISTRIBUTION, and the published tables can be used to estimate the SIGNIFICANCE of the value found. The appropriate number of DEGREES OF FREEDOM is found by subtracting from the number of cells the number of constants found from the data that are used in estimating the expected frequencies.

The chi-square test has two special advantages, namely:
(i) Apart from some adjustments in a 2×2 table, it is valid whatever the statistical distribution of the elements in question, provided only that the lowest expected frequency exceeds 5.
(ii) The contributions of the cells to χ^2 tend to be very unequal in size; thus when a significant value of χ^2 is found, it is easy to identify the cells making the important contributions.

A significantly high value of χ^2 casts doubt on the hypothesis used for finding the expected frequencies, and examination of the contribution to χ^2 made by the various cells will often point the way to a more satisfactory hypothesis. A significantly low value of χ^2, i.e. one likely to be exceeded by chance in over 95 per cent of cases, is a much rarer occurrence but is just as effective in casting doubt on the original hypothesis; it usually signifies that the frequencies of certain cells are not independent, but are linked in some way.

If the application of the chi-square test is complicated by the occurrence of cells with expected frequencies of less than 5, the difficulty can often be overcome by pooling the observed and expected frequencies of such cells with others.

chromosphere: That part of the atmosphere of the SUN above the REVERSING LAYER. Some 10 000 km thick, it consists of faintly luminous gases. Though visible directly only during a solar eclipse it is studied at other times by means of a spectroheliograph and is found to be a region of various types of disturbance including the SOLAR FLARE.

circle of inertia: In horizontal motion of a body on the rotating earth, subject to no force except gravity, the path described by the body, except near the equator, is approximately a circle which is known as the 'circle of inertia'. Inward acceleration (V^2/r) is balanced by the deviating force (fV) due to the earth's rotation and the path is thus of radius r, given by $r = V/f$, where V is the air velocity and f the CORIOLIS PARAMETER. Circular flow ($r = $ constant) is not strictly adhered to because of the variation of f with latitude.

circular frequency distribution: A distribution in a series of vector quantities (e.g. winds) such that, when the individual vectors are drawn on a polar diagram, the lines of equal frequency of the vector end-points are circles centred on the end point of the vector mean wind of the series. If the frequency of wind components about the mean is distributed in accordance with the NORMAL FREQUENCY DISTRIBUTION, the distribution is termed 'normal circular'. Frequency distributions of wind data often depart drastically from the circular form, which should be used only after its appropriateness has been confirmed.

circulation: The circulation (C) round a closed curve is defined as the line integral round the curve of the velocity vector component along the curve, i.e.

$$C = \oint \mathbf{V} \cdot ds = \oint V_T ds$$

where V_T is the component of the velocity vector **V** tangential to the curve. Cyclonic circulation is reckoned positive, anticyclonic negative. The dimensions are $L^2 T^{-1}$.

circulation, general atmospheric: See GENERAL CIRCULATION.

circulation index: A numerical measure of the strength of the atmospheric circulation as, most commonly, in the ZONAL INDEX.

circumzenithal arc: A HALO phenomenon in the form of a short arc centred on the zenith and convex to the sun, at or near the highest point of the 46° halo (if present). The arc is formed by REFRACTION in suitably orientated ice crystals and may show vivid rainbow colouring.

cirrocumulus (Cc): One of the CLOUD GENERA.
'Thin, white patch, sheet or layer of cloud without shading, composed of very small elements in the form of grains, ripples, etc., merged or separate, and more or less regularly arranged; most of the elements have an apparent width of less than one degree.'* See Plate 6. See also CLOUD CLASSIFICATION.

cirrostratus (Cs): One of the CLOUD GENERA.
'Transparent, whitish cloud veil of fibrous (hair-like) or smooth appearance, totally or partly covering the sky, and generally producing halo phenomena.'* See Plate 7. See also CLOUD CLASSIFICATION.

cirrus (Ci): One of the CLOUD GENERA (Latin, *cirrus* lock or tuft of hair).
'Detached clouds in the form of white, delicate filaments or white or mostly white patches or narrow bands. These clouds have a fibrous (hair-like) appearance, or a silky sheen, or both.'* See Plate 5. See also CLOUD CLASSIFICATION.

Clausius–Clapeyron equation: The equation which expresses the change of pressure with temperature in a state of equilibrium between two phases of the same substance.
The form of the equation familiar in meteorology is

$$\frac{1}{e'} \frac{de'}{dT} = \frac{L}{R_v T^2}$$

where e' is the saturation water vapour pressure, T the temperature, L the latent heat and R_v the specific gas constant for water vapour. This equation is integrated to give the saturation vapour pressure as a function of temperature. See VAPOUR PRESSURE.

clear-air turbulence (CAT): Air TURBULENCE of a type other than that associated with airflow close to rough ground or that encountered in or near cumulonimbus clouds.
Clear-air turbulence has been observed mainly in the high troposphere and low stratosphere, especially in the vicinity of JET STREAMS. Its chief practical significance lies in the acceleration, varying in intensity up to several times g (acceleration of gravity), which may be imparted to high-speed aircraft. Investigation of the horizontal and vertical anomalies in the flow pattern that constitute the turbulence is made difficult by their small scale. Evidence suggests, however, that favourable conditions for the development of such anomalies include high static stability,

---

* Geneva, World Meteorological Organization. International cloud atlas, Vol. 1. Geneva, WMO, 1956, p. 10.

and large horizontal and vertical wind shear; orographic effects may also be an important contributory factor.

clear ice: An alternative for GLAZE. See also ICE FORMATION ON AIRCRAFT.

clear sky, day of: In a resolution adopted at the International Meteorological Meeting at Utrecht in 1874, a 'day of clear sky' was defined as one on which the average CLOUDINESS at the hours of observation is less than two-tenths of the sky.

On 1 January 1949 OKTAS were first adopted, and since then a 'day of clear sky' has been defined in Great Britain as one on which the average cloudiness at hours of observation is less than two oktas.

climagram: A climatic diagram comprising a plot of monthly values of two selected meteorological elements (ordinate and abscissa), the plotted points being joined by a line which represents the annual variation of the relationship of the elements.

climate: The climate of a locality is the synthesis of the day-to-day values of the meteorological elements that affect the locality. Synthesis here implies more than simple averaging. Various methods are used to represent climate, e.g. both average and extreme values, frequencies of values within stated ranges, frequencies of weather types with associated values of elements. The main climatic elements are precipitation, temperature, humidity, sunshine, wind velocity, and such phenomena as fog, frost, thunder, gale; cloudiness, grass minimum temperature, and soil temperature at various depths may also be included. Climatic data are usually expressed in terms of an individual calendar month or season and are determined over a period (usually about 30 years) long enough to ensure that representative values for the month or season are obtained.

The climate of a locality is mainly governed by the factors of (i) latitude, (ii) position relative to continents and oceans, (iii) position relative to large-scale atmospheric circulation patterns, (iv) altitude and (v) local geographical features. A broad classification is made into (*a*) 'continental climate', which is found mainly in the interior and eastern parts of continents and is characterized by low rainfall and humidity and large diurnal and seasonal ranges of temperature, and (*b*) 'maritime climate', typical of oceanic islands and the western parts of continents and characterized by high rainfall and humidity and relatively uniform temperature, diurnally and seasonally. Among well-known classifications of greater refinement (though inevitably imperfect and so subject to criticism) are those of W. Köppen (1923) and C. W. Thornthwaite (1931 and 1948).

climatic changes: Geological and botanical evidence of past climates is to the effect that, on the geological time-scale, periods when there was little or no ice anywhere in the world have alternated with much shorter periods when glaciation was widespread—see ICE AGE. The difference between mean world temperatures in these extreme conditions is estimated at about 8 degC, the range being greatest in high latitudes.

Since the last GLACIAL PHASE of the Quaternary Ice Age, varying with locality from about 8000 to 40 000 years ago, there have been large fluctuations in climate in what is, technically, yet an 'interglacial' ice-age period. Conspicuous among the fluctuations of European climate, paralleled to a variable extent elsewhere, are the CLIMATIC OPTIMUM of about 5000 to 2000 B.C. and the 'Little Ice Age' of about A.D. 1550 to 1850. Other probable fluctuations are a warm, dry period in the sixth to eighth centuries A.D. and a stormy period in the twelfth to fourteenth centuries A.D. The evidence in the historical period includes: in Europe, fluctuations of Alpine glaciers and in the traffic across Alpine passes; in Asia, variations in the level of the Caspian Sea and other salt lakes; in North America, variations in the rate of growth of the sequoias of California, some of which are over 3000 years old.

It is now generally agreed that in most parts of the world instrumental observations of the past 100 to 200 years contain SECULAR TREND(s) of statistical significance, implying real climatic change(s) during this period. The most conspicuous feature has been a warming tendency from about 1850 to around 1940 on a scale which is almost world-wide and averages about 1 degC. It is supported by evidence such as the recession of glaciers and of Arctic sea ice; in the British Isles the warming was particularly marked in the winter months but was reversed after 1940 by the occurrence of a number of very cold winters, outstanding among them that of 1962–63, the coldest in central England since 1739–40.

Among the suggested causes of climatic variation are changes of solar radiation, astronomical (earth orbit) changes, CONTINENTAL DRIFT, POLAR WANDERING, mountain building, volcanic eruptions, changes of CARBON DIOXIDE content of the atmosphere, and changes of heat storage by the oceans. It is yet quite uncertain which of these or other factors is most important in effecting climatic changes.

climatic optimum: That period, lasting from about 5000 to 2000 B.C., when average temperatures are considered to have reached their highest level, probably on a world-wide scale, since the last ICE AGE. During this period European temperatures are thought to have averaged up to about 2 or 3 degC higher than at present. Periods equally warm are, however, believed to have occurred in each of the several interglacial phases of the last million years.

climatic zones: The word CLIMATE is derived from a Greek word meaning 'to incline' and the original zones of climate were zones in which the inclination of the sun's rays at noon was the same, that is, zones of latitude. The accumulation of meteorological data has shown that winds and rainfall, as well as temperature, have a zonal arrangement, but that the true climatic zones do not run strictly parallel to lines of latitude. Eight principal zones are distinguished: near the equator a zone of tropical rain climate, then two subtropical zones of STEPPE and DESERT climate, then two zones of temperate rain climate, then, in the northern hemisphere only, an incomplete zone of BOREAL CLIMATE with a great annual range of temperature and finally, two polar caps of snow climate. The equatorial zone is divided into the equatorial rain-forest zone, which extends over the Atlantic and Pacific Oceans as the DOLDRUMS, with rain in all seasons, and a belt of SAVANNA climate on either side with a well-marked alternation of dry and rainy seasons, the latter occurring in the 'summer' months. The subtropical zones include most of the world's great deserts—the Sahara, Arabia, Arizona, Kalahari, and the deserts of South America and Australia; over the oceans they include the TRADE WIND belts and the HORSE LATITUDES. The temperate zones are divided into the Mediterranean climates with mild, rainy winters and hot, dry summers, and the temperate rain belts with rain in all seasons. On the eastern margins of the continents, especially in Asia, the subtropical desert zone and the Mediterranean climate are replaced by areas with a MONSOON climate.

climatological station: A STATION from which climatological data are obtained. Such stations are classified by the World Meteorological Organization as 'principal' (hourly readings taken or observations made at least three times daily in addition to hourly tabulations from autographic records), or 'ordinary' (observations made at least once daily, including daily readings of extreme temperature and of amount of precipitation). Stations may also be classified as 'precipitation stations' (in normal British terminology, 'rainfall stations') where this is the only element observed; as 'climatological stations for specific purposes' where only a specific element or elements are observed; or as 'agricultural meteorological stations' where meteorological and biological data, or data otherwise contributing to the establishment of the relationship between the weather and the life of plants and animals, are concerned.

climatology: The study of CLIMATE.

climatotherapy: The treatment of disease by suitable climatic environment, i.e. applied BIOCLIMATOLOGY.

clinometer: An instrument for measuring the angle of elevation of a surface or of an object seen from the observing point. The angle is read by reference to a spirit-level or a small plummet.

closed system: A closed (thermodynamic) system is one in which there is no exchange of matter between the system and its environment though there is, in general, exchange of energy. The atmosphere as a whole may, to a high degree of approximation, be considered a closed system.

cloud: An aggregate of very small water droplets, ice crystals, or a mixture of both, with its base above the earth's surface. The limiting liquid-particle diameter is about 200 μm, larger drops than this comprising DRIZZLE or RAIN.

With the exception of certain rare types (NACREOUS and NOCTILUCENT) and the occasional occurrence of CIRRUS in the lower stratosphere, clouds are confined to the troposphere. They are formed mainly as the result of vertical motion of air, as in convection, or in forced ascent over high ground, or in the large-scale vertical motion associated with depressions and fronts. Cloud may result, in suitable lapse-rate and moisture conditions, from low-level turbulence or from mixing or from other minor causes.

At temperatures below 0°C cloud particles frequently consist entirely of super cooled water droplets down to about $-10°C$ in the case of layer clouds and to about $-25°C$ in the case of convective clouds. At temperatures below these very approximate limits and above about $-40°C$ (temperature of HOMOGENEOUS NUCLEATION), many clouds are 'mixed' but with ice crystals predominating in the lower part of the temperature range.

cloud amount: Amount of sky estimated to be covered by a specified cloud type (partial cloud amount), or by all cloud types (total cloud amount). In either case the estimate is made to the nearest OKTA (eighth) and is reported on a scale which is essentially one of the 'nearest eighth', except that figures 0 and 8 on the scale signify a completely clear and completely cloudy sky, respectively, with consequent adjustment to other figures near either end of the scale. See also CLOUDINESS.

cloud base: That lowest zone in which the type of obscuration perceptibly changes from that corresponding to clear air or haze to that corresponding to water droplets or ice crystals. (Conference of Directors of Members of the International Meteorological Organization, 1947.)

Height of cloud base is reported as height above ground level. In forecasts it is often expressed in PRESSURE ALTITUDE.

cloud-base recorder: Automatic cloud-base recorders use either a triangulation method or one in which the time interval is measured between the transmission of a pulse of light (using a spark-gap or laser) from ground level and the reception of this pulse reflected from the cloud base.

Equipment in general use in the Meteorological Office is based on the former method and employs a modulated searchlight beam which is made to sweep in a vertical plane. A receiver sited about 100 m away from the searchlight looks vertically upwards and detects the presence of modulated light scattered from the cloud base as the transmitted beam passes overhead. Cloud information is presented on a strip-chart recorder which may be sited up to about 1 km from the searchlight. With additional equipment cloud information may be transmitted up to about 25 km over a pair of telephone wires.

cloud-burst: A popular term for a very sudden and very heavy shower, often accompanied by thunder and hail. It is associated with strong upward and downward currents.

cloud chamber: The 'Wilson cloud chamber' is an apparatus in which supersaturation and condensation are produced, for example in moist air, by sudden ADIABATIC cooling of the air contained in a chamber. In its normal use as a tool for studying the tracks of ionizing radiations, use is made of the fact that, in dust-free moist air, condensation occurs more readily on charged IONS than on uncharged molecules. Moist air in the chamber is freed of dust particles (condensation nuclei) by repeated expansion. The dust-free air is then suddenly expanded to a controlled degree, which produces immediate condensation only on ions. A photograph synchronized with the expansion reveals the tracks of ionizing rays passing through the chamber.

cloud classification: Various methods of CLOUD classification are made, as follows:
 (i) In the publication *International cloud atlas*,* division is made into CLOUD GENERA (ten basic characteristic forms) with sub-division as required, into (*a*) CLOUD SPECIES (cloud shape and structure); (*b*) CLOUD VARIETIES (cloud arrangement and transparency); (*c*) Supplementary features and accessory clouds—see INCUS, MAMMA, VIRGA, PRAECIPITATIO, ARCUS, TUBA, PILEUS, VELUM and PANNUS; (*d*) Growth of a new cloud genus from a 'mother-cloud', indicated by the addition of 'genitus' to the new cloud and mother-cloud genera, in that order, if a minor part of the mother-cloud is affected, and of 'mutatus' if much or all of the mother-cloud is affected, e.g. stratocumulus cumulogenitus, or stratus stratocumulomutatus.
 (ii) A classification is made in terms of the level (étage)—high, middle, or low—at which the various cloud genera are usually encountered. In temperate regions the approximate limits are: high, 5–13 km (16 500–45 000 ft); middle, 2–7 km (6500–23 000 ft); low, earth's surface–2 km (0–6500 ft). The high clouds are Ci, Cc and Cs (see CLOUD GENERA for significance of abbreviations); the middle clouds are Ac and As (the latter often extending higher), and Ns (usually extending both higher and lower); the low clouds are Sc, St, Cu, and Cb (the last two often also reaching middle and high levels).
 For synoptic purposes, a ninefold cloud classification is made in each of these three latter divisions of cloud genera, the corresponding codes being designated C_H, C_M and C_L respectively. The purpose is to report characteristic states of the sky rather than individual cloud types.
 (iii) Less formal classifications are made (*a*) in terms of the physical processes of cloud formation, notably into HEAP CLOUDS and LAYER CLOUDS (or 'sheet clouds') and (*b*) in terms of cloud composition, namely ICE-CRYSTAL CLOUDS, WATER-DROPLET CLOUDS and MIXED CLOUDS.

---

* Geneva, World Meteorological Organization. International cloud atlas. Vol. 1. Geneva, WMO, 1956, p. 10.

cloud discharge: A lightning flash confined within a thundercloud. See LIGHTNING.

cloud genera: The 10 characteristic cloud types, comprising CIRRUS (Ci), CIRROCUMULUS (Cc), CIRROSTRATUS (Cs), ALTOCUMULUS (Ac), ALTOSTRATUS (As), NIMBOSTRATUS (Ns), STRATOCUMULUS (Sc), STRATUS (St), CUMULUS (Cu), and CUMULONIMBUS (Cb). See CLOUD CLASSIFICATION.

cloudiness: Amount of sky covered by cloud, irrespective of type, i.e. total CLOUD AMOUNT.
   Charts showing the distribution of mean cloudiness over the earth in each month and the year are given in the '*Manual of meteorology*'.*

cloud physics: 'The study of the physical processes which govern the formation, nature, size, size distribution and number (per unit volume) of the individual particles which together constitute CLOUD, FOG or PRECIPITATION.' (A. C. Best, 1957.†)

cloud searchlight: Light projected vertically, by remote control, from a searchlight at distance L (about 300 metres) from an observer causes a spot of light to be thrown on the lowest cloud overhead. The observer measures the angle of elevation (E) of the light spot by means of an ALIDADE and deduces the cloud base (h) from the relationship $h = L \tan E$. Fixed and portable versions of the instrument are used in the Meteorological Office.

cloud seeding: The attempted modification of the physical processes occurring within natural clouds by injecting the clouds with a seeding agent such as (i) 'dry ice' (solid carbon dioxide) at very low temperature to cause local cooling commonly to below $-40°C$ and spontaneous freezing; (ii) silver iodide to act as ICE NUCLEI; (iii) hygroscopic salt nuclei or a spray of fine water drops to stimulate COALESCENCE.
   Injection of the seeding agent into the cloud has been carried out from aircraft, balloons, rockets and shells, and from the ground. In general, the purpose of the cloud seeding has been to stimulate precipitation. Attempts have also been made to clear stratus clouds and to inhibit hail and thunderstorms. Some success has been claimed in at least the first two aims, but the existence of a method of cloud seeding of commercial value has not yet been established.

clouds, particle distribution in: Measurements have shown water droplets in clouds to have a 'median volume diameter' (also termed 'mean effective diameter') of about 15 μm, with a range from about 1 μm to 100 μm; median volume diameter is defined as the drop diameter such that half the total water present is contained in drops of larger diameter. Rather larger median values have been found in convective clouds (15–20 μm) than in layer clouds (10–15 μm). Still larger systematic differences with locality have been reported (larger drops in air free from pollution), probably because of associated differences in concentration of effective condensation nuclei.
   Measured water-droplet concentrations are generally in the range 1×10^8 to $4 \times 10^8/m^3$, but with smaller values in altocumulus clouds.
   ICE CRYSTALS occur in clouds in various forms, determined by such conditions as temperature and degree of supersaturation with respect to ice. Clouds composed of ice crystals are very tenuous in comparison with water clouds; a typical range for the concentration of individual particles or clusters of particles is 1×10^5 to $5 \times 10^5/m^3$. See also RAINDROPS.

---

* SHAW, SIR N.; Manual of meteorology. Vol. IV, Comparative meteorology. Cambridge University Press, 1928, p. 146.
† BEST, A. C.; Physics in meteorology. Applied physics series. London, Sir Isaac Pitman and Sons Ltd, 1957.

cloud species: A sub-division of CLOUD GENERA in terms of cloud shape and structure. The fourteen species comprise FIBRATUS, UNCINUS, SPISSATUS, CASTELLANUS, FLOCCUS, STRATIFORMIS, NEBULOSUS, LENTICULARIS, FRACTUS, HUMILIS, MEDIOCRIS, CONGESTUS, CALVUS and CAPILLATUS.

None, one, or more than one of the cloud species may be allotted to any specific example of cloud genus.

See also CLOUD CLASSIFICATION.

cloud street: An extended line of cumulus cloud parallel to the wind direction, often in an otherwise lightly clouded sky. Various sources of thermals spaced across wind may give rise to parallel cloud streets. It appears that such streets may also be produced in an air mass in which the convection layer has a well-marked top and in which the wind direction in the layer is almost constant.

clouds, water content of: The amount of water, in the liquid or solid state, which is contained in unit volume of cloud. It is usually expressed in grammes/metre^3.

The water content of convective cloud may be computed on the basis of various assumptions. It is normally assumed that the water which is condensed on ADIABATIC expansion is retained within the rising air. The computed values increase with the temperature of the CONDENSATION LEVEL and decrease with the amount of ENTRAINMENT of ambient air which is assumed. In each theoretical model the computed maximum values occur towards the top of the cloud; such a value for mid-latitude summer conditions is, on the assumption that there is no mixing of cloud and environment, about 5 g/m^3. Measurements have supported the theoretical distribution of water content with height in a convective cloud. While measured values have approached the maximum theoretical values in some instances, most measured values have been much lower. Because of the large dependence of water content on cloud-base temperature and degree of vertical development of cloud, average water-content values of convective clouds are of little significance.

Median values of water content of low-level layer clouds in middle latitudes are about 0·2 g/m^3, and of medium-level clouds about 0·1 g/m^3. Variability is, however, rather great since values up to about five times the median values have been measured.

cloud varieties: A sub-division of CLOUD GENERA in terms of the arrangement of the cloud elements and the cloud transparency. The nine varieties comprise INTORTUS, VERTEBRATUS, UNDULATUS, RADIATUS, LACUNOSUS, DUPLICATUS, TRANSLUCIDUS, PERLUCIDUS and OPACUS.

None, one, or more than one of the cloud varieties may be allotted to any specific example of cloud genus.

See also CLOUD CLASSIFICATION.

cloudy day: Defined as a day on which the average CLOUDINESS at the hours of observation is more than six OKTAS. (Prior to 1 January 1949 the criterion was based on an average cloudiness greater than eight-tenths of the sky.) Such days have sometimes been described in Great Britain as 'overcast days'.

coalescence: In meteorology, usually used to denote the growth of water drops by collision. The term is also often used for the growth of an ice particle by collision with water drops ('accretion'). See PRECIPITATION.

coalescence efficiency: That fraction of the number of collisions between water drops which results in the union of the drops and the formation of larger drops. See also COLLISION EFFICIENCY, COLLECTION EFFICIENCY.

PLATE 5 Cirrus.

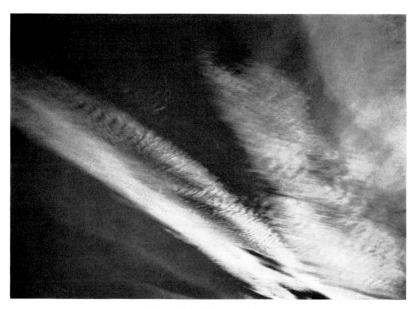

PLATE 6 Cirrocumulus.

PLATE 7   Cirrostratus with cumulus humilis below.

PLATE 8   Altocumulus.

PLATE 9 Altostratus.

PLATE 10 Nimbostratus.

PLATE 11 Stratocumulus.

PLATE 12 Stratus.

PLATE 13 Cumulus.

PLATE 14 Cumulonimbus with mamma.

PLATE 15 Cumulonimbus with anvil.

PLATE 16 Altocumulus lenticularis.

PLATE 17 Cumulus protruding through stratocumulus.

PLATE 18 Noctilucent cloud: Perthshire.

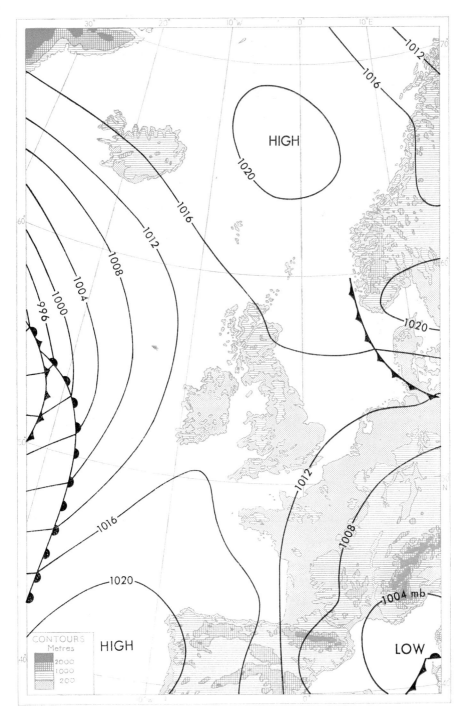

FIGURE 9—Col over the British Isles, 06 GMT, 29 April 1961.

coherence: A term used for the tendency, almost universal in meteorology, for measurements which are close together in space and time to be more alike than other similar but more widely separated measurements.

If the strictly periodic components of variation, derived from the diurnal or annual changes, are removed the meteorological data are yet found to exhibit coherence on every possible scale in space and time. This fact must be allowed for if statistical methods developed to deal with non-coherent data are to give reliable results. See also PERSISTENCE.

coherent system of units: A system of units such that the product or quotient of any two unit quantities in the system is the unit of the resultant quantity; for example, in any coherent system, unit velocity results when unit length is divided by unit time. See SI UNITS.

col: That atmospheric pressure distribution, saddle-backed in shape, which occurs between two anticyclones and two depressions, arranged alternately (see Figure 9). A col which is markedly elongated along the high-pressure axis is an 'anticyclonic col', along the low-pressure axis a 'cyclonic col'; where there is no marked elongation, the col is 'neutral'.

Light, variable winds are a feature of all types of col. The weather of the neutral col is dominated by the characteristics of the particular air mass but is often thundery in summer and dull or foggy in winter. The general characteristics of the anticyclonic and cyclonic cols are those of the anticyclone and depression, respectively.

Since small pressure changes suffice to cause appreciable movement of a col, it is generally not an abiding feature of the synoptic pressure distribution. The cyclonic col occupies a region through which an approaching depression may readily pass.

colatitude: The colatitude at any point is the complementary angle of the LATITUDE (φ), i.e. $90° - \varphi$; it is also termed 'polar distance'.

cold anticyclone: See ANTICYCLONE.

cold dome: A term applied, in upper air analysis, to a closed centre of low pressure on a FRONTAL CONTOUR CHART. Such a region on the chart indicates the isolation of cold air at that level from the main body of cold air which is seen on the chart, usually at higher latitudes.

The term is also applied to a mass of cold air resting on the earth's surface in the general shape of an inverted bowl.

cold front: A FRONT whose movement is such that the colder air mass is replacing the warmer.

The passage of a cold front is normally marked at the earth's surface by a rise of pressure, a fall of temperature and dew-point, and a veer of wind (in northern hemisphere). Rain occurs in association with most cold fronts and may extend some 100 to 200 km ahead of or behind the front. Some cold fronts give only a shower at the front, while still others give no precipitation. Thunder, and occasionally a LINE-SQUALL, may occur at a cold front.

The average slope of a cold frontal surface is about 1 in 50. A cold front moves, on average, at about the speed of the GEOSTROPHIC WIND component normal to the front and measured at it. See also ANAFRONT, KATAFRONT.

cold-front wave: A secondary wave DEPRESSION which forms on an extended COLD FRONT along which there is an appreciable thermal wind more or less parallel to the front. Such wave depressions are common and may be members of a depression 'family', each member of which may go through a typical life cycle of development and decay. The formation of a wave is aided by the distortion of the front, as for example by a range of hills or by movement towards a COL. Initial movement of the wave depression is parallel to the warm-sector isobars.

cold-occlusion depression: A SECONDARY DEPRESSION which forms at the point where a cold and a warm front unite to form a cold OCCLUSION. The formation of this type of secondary has been linked with a THICKNESS pattern in which the main features are a weak gradient near the centre of the primary depression and marked DIFFLUENCE at the point of occlusion. There is no well-defined pattern of behaviour of the secondary after formation.

cold pole: That region of the earth's surface, one in each hemisphere, where the lowest air temperature has been measured. For the northern hemisphere the location is usually taken as Verhojansk in north-east Siberia (67° 33′ N, 133° 23′ E, altitude 122 m) where a temperature of $-68°C$ ($-90°F$) was measured on 5 and 7 February 1892. The same temperature was measured at Oymjakon (63° 16′ N, 143° 15′ E, altitude 800 m) on 6 February 1933. During the period (from 1957) of regular measurement of air temperature in the interior of the Antarctic continent, several values between $-85°C$ and $-90°C$ ($-121°F$ and $-130°F$) have already been reported at and near the Soviet base Vostok (78° 27′ S, 106° 52′ E).

The cold pole is sometimes alternatively defined as the location of lowest mean monthly temperature, or of lowest mean annual temperature, or of coldest air in the troposphere. In the last case it is usually indicated by the area of lowest THICKNESS on the chart of 1000–500-mb thickness or other representative thickness chart.

cold pool: A closed centre of low THICKNESS on a thickness chart, so called because it represents a region of low mean temperature in the isobaric layer concerned.

cold sector: That part of a DEPRESSION occupied by cold air on the earth's surface; it usually comprises about half to three-quarters of a recently formed depression, and the whole of an old one.

cold trough: A pressure TROUGH (or trough on an isobaric contour chart) in which temperature is generally lower than in adjacent areas.

cold wave: A term which is used in a technical sense by the United States Weather Bureau to signify a fall of temperature at a given place by at least a specified amount in 24 hours, to at least a specified minimum. The specified amount and minimum vary with season and locality.

collection efficiency: That fraction of the total number of water drops (cloud or rain), moving on an initial collision path with other drops, which actually collide and unite with such drops to form larger drops. It is given by the product of the COLLISION EFFICIENCY and COALESCENCE EFFICIENCY.

collision efficiency: That fraction of the total number of water drops (cloud or rain), moving on an initial collision path with other drops, which actually collide. The efficiency is a function mainly of the relative motion and the particle sizes involved. In general, it is less than unity because of the deflexion of the streamlines of the particles on their near approach, in such a way as to avoid collision. In certain cases, for example that of particles of about equal size, the efficiency approaches zero. On the other hand, an efficiency greater than unity occurs in some conditions, due to the sucking of particles into the wakes of others.

The collision efficiency is an important factor in the COALESCENCE process of PRECIPITATION. See also COALESCENCE EFFICIENCY and COLLECTION EFFICIENCY.

collision frequency: The collision frequency (v) of a particle (e.g. molecule or electron) is the number of collisions it makes per second. Its inverse is the 'collision interval' (τ), which is the interval (in seconds) between collisions. Collision frequency is a function of gas pressure and density and is related to the MEAN FREE PATH (l) and the mean particle speed (v) by the relation $v = v/l$.

colour temperature: The colour temperature of a radiating body is that temperature obtained by insertion of the observed wavelength of peak radiation into Wien's formula. See RADIATION.

comfort zone: In relation to EFFECTIVE TEMPERATURE, the range of effective temperature within which most people can work with maximum efficiency. For acclimatized persons the comfort zone is considered to be 66–76°F (19–24°C) with an optimum of 69°F (21°C).

compass: The magnetic or mariner's compass consists in its simplest form of a graduated card at the centre of which a magnetized steel needle is pivoted so that it may turn freely in a horizontal plane. The card is divided into 32 equal parts of $11\frac{1}{4}°$ each, these being N, N by E, NNE, NE by N, NE and so on (compass 'points'). The compass needle points to magnetic north which does not, in general, coincide with geographical north. See DECLINATION, MAGNETIC and GEOMAGNETISM.

Magnetic compasses have now largely given way to the gyrostatic compass. This compass is one of several devices which utilize the property, possessed by a spinning fly-wheel, of maintaining its axis in space. The gyro is maintained electrically, and has been used in aeroplanes and on board ship; it has the advantage of being independent of both magnetic and electrical disturbances.

compensation of instruments: An instrument designed to measure changes in a particular physical quantity (e.g. pressure) may be affected also by some other influence (e.g. temperature). To eliminate or minimize the influence of the disturbing element, a device may be introduced for the purpose of rendering the instrument insensitive to changes in the latter, in which case it is said to be 'compensated'. Thus, chronometers and aneroid barometers are ordinarily compensated for temperature.

composite forecast chart: A forecast chart so constructed as to depict, for a specified aircraft route, time of departure and airspeed, the meteorological conditions which it is expected the aircraft will encounter at each point of the route.

computational instability: A phenomenon, which arises in numerical computations employing approximate methods, in which the error of the computation increases rapidly as the computation proceeds.

A particular effect of this kind, originally discovered by Courant, Friedrichs and Lewy, is of special importance in relation to NUMERICAL WEATHER FORECASTS. The effect is such that if the selected grid size is less than the distance travelled in the selected time interval by the fastest waves permitted by the equation, small errors of computation and of numerical approximation grow in successive time intervals to such an extent as to swamp the physical solution of the equation.

concrete minimum temperature: A reading obtained from a standard grass minimum thermometer which is exposed to the air with its bulb in contact with a concrete slab; the slab lies horizontally in an open situation and with its top almost flush with the ground.

Such readings, introduced at Meteorological Office stations on 1 December 1968, are less subject to very local influences than are GRASS MINIMUM TEMPERATURES. They are also more relevant to such operational problems as the incidence of ice on roads.

condensation: The process of formation of a liquid from its vapour: in meteorology, the formation of liquid WATER from WATER VAPOUR.

Since the capacity of air to hold water in the form of vapour decreases with temperature, cooling of air is the normal method by which first SATURATION, then condensation, is produced. Such cooling is effected by three main processes:

(i) adiabatic expansion of ascending air,
(ii) mixing with air at lower temperature, and
(iii) contact with earth's surface at lower temperature.

The water vapour condenses as cloud in (i), as fog or cloud in (ii), and as dew or hoar-frost in (iii).

Condensation in the atmosphere occurs at or near the temperature appropriate to the saturation VAPOUR PRESSURE, which is defined in terms of equilibrium between the vapour and a plane water surface, because of the presence in all parts of the troposphere of an adequate supply of 'condensation nuclei', which are hygroscopic. In the absence of such nuclei a high degree of supersaturation (several hundred per cent) would be required to produce condensation—so-called 'homogeneous condensation'.

The main factors governing the rate of growth of a cloud droplet formed by direct condensation of water vapour on a condensation NUCLEUS (radius > 0·1 μm, approximately) are the nuclei concentration and sizes, the degree of saturation or supersaturation, the excess saturation vapour pressure over a spherical droplet relative to a plane surface, the reduction of saturation vapour pressure associated with the solution which a hygroscopic salt forms with water, and the warming of the growing cloud droplets by the latent heat of condensation which is released. Calculation shows that the drop grows quickly when small and increasingly slowly with increase of drop size. The process of direct condensation explains well the formation of normal cloud particles (diameter about 15 μm) but proceeds much too slowly to account for the size of drop often associated with PRECIPITATION (diameter about 1000 μm).

condensation level: That level (geometric or pressure) at which CONDENSATION occurs in the atmosphere.

The term is normally used in relation to one or other of the processes of lifting, convection, or vertical mixing. The appropriate condensation level is found by means of an AEROLOGICAL DIAGRAM.

The 'lifting condensation level' is, for an air sample at any height, that (isobaric) level at which the dry ADIABATIC through the temperature intersects the HUMIDITY MIXING RATIO line drawn through the dew-point of the sample.

The 'convective condensation level', applied only to surface air, is the same as the lifting condensation level for surface air if the intersection of the dry adiabatic through the temperature and mixing ratio line through the dew-point lies to the right of the environment curve. If this is not the case, the convective condensation level is the level at which the mixing ratio line through the surface dew-point intersects the environment curve.

The 'mixing condensation level' is the lowest level at which condensation occurs (if at all) as the result of complete vertical mixing throughout a given layer. Such mixing produces constant POTENTIAL TEMPERATURE and mixing ratio throughout the layer; the mixing condensation level is the level of intersection of the dry adiabatic and mixing ratio lines appropriate to these constant values.

condensation nucleus: See NUCLEUS, CONDENSATION.

condensation trail: An initially thin trail of water droplets or ice crystals produced by an aircraft engine exhaust when the humidifying effect of the water vapour exceeds the opposed heating effect of the exhaust air ('exhaust trail'). Such trails generally broaden rapidly and evaporate as mixing with drier air proceeds; when the atmosphere is already moist, however, they may be very persistent.

Theory indicates that for a given engine there is for a given pressure a critical air temperature above which a trail cannot form. A 'mintra' line corresponding to these critical pressure–temperature conditions is shown on the Meteorological Office TEPHIGRAM form. It is found in practice that the appearance of a trail is delayed until the temperature is several degrees below the corresponding mintra value. The critical temperature is such that over Great Britain trails rarely form below about 8·5 km in summer and 6 km in winter. In tropical areas the heights are greater but in very cold conditions, such as winter in central Canada, trails can form at ground level.

Non-persistent condensation trails more rarely form, sometimes at low levels, in near-saturated air at aircraft wing tips and propeller tips, because of the aerodynamically produced pressure falls which occur at these points ('aerodynamic trail').

conditional instability: See STABILITY.

conduction (of heat): The process of heat transfer through matter by molecular impact from regions of high temperature to regions of low temperature without transfer of the matter itself. It is the process by which heat passes through solids; its effects in fluids are usually negligible in comparison with those of CONVECTION.

conductivity, thermal: That property of matter whereby thermal CONDUCTION occurs. Two quantities are defined: (i) thermal conductivity (k), by the equation $q = -k\, \partial T/\partial x$ (q, flux of heat per unit area, is proportional to $\partial T/\partial x$, the thermal gradient, and acts in the down-gradient direction, and k is the constant of proportionality); (ii) thermometric conductivity, or thermal diffusivity (a), by the relation $a = k/c\rho$, where c is the specific heat and ρ the density of the substance.

For most soils the value of a is of the order 10^{-2} to 10^{-3} cm²/s, though with appreciable dependence on density and water content. For still air a is about 0·2 cm²/s.

In a turbulent atmosphere (its normal state) the vertical heat transfer effected by molecular conduction is swamped by that effected by eddies. See EDDY CONDUCTIVITY.

confluence: The nearer approach to each other of adjacent STREAMLINES in the direction of flow. See DIVERGENCE.

confluent thermal ridge: A pattern of thickness lines which is concave towards high THICKNESS and in which the thickness lines crowd together in the direction of the THERMAL WIND. According to the theorem of DEVELOPMENT, cyclogenesis (C) may be expected to occur behind and to the right, anticyclogenesis (A) ahead and to the left of the pattern, as illustrated in Figure 10.

confluent thermal trough: A pattern of thickness lines which is concave towards low THICKNESS and in which the thickness lines crowd together in the direction of the

THERMAL WIND. According to the theorem of DEVELOPMENT, cyclogenesis (*C*) may be expected to occur ahead and to the right, anticyclogenesis (*A*) behind and to the left of the pattern, as illustrated in Figure 11.

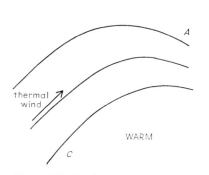 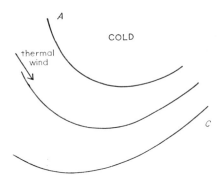

FIGURE 10—Confluent thermal ridge.   FIGURE 11—Confluent thermal trough.

congestus (con): A CLOUD SPECIES (Latin, *congestus* piled up).
'CUMULUS clouds which are markedly sprouting and are often of great vertical extent; their bulging upper part frequently resembles a cauliflower.'* See also CLOUD CLASSIFICATION.

Congress: The Congress of the WORLD METEOROLOGICAL ORGANIZATION is the general assembly of delegates representing Members and as such is the supreme body of the Organization. It meets at intervals not exceeding four years.

conjunction, astronomical: A planet or other heavenly body is said to be in conjunction when it is in line with the earth and sun, and on the same side of the earth as the sun.

conservation: In statistics, a term which is sometimes used with the same meaning as PERSISTENCE or (time) COHERENCE.

conservative property: A conservative air-mass property is one that remains unchanged, or almost unchanged (quasi-conservative), in a specified process or processes. Thus, for example, POTENTIAL TEMPERATURE is strictly conservative in a dry adiabatic process, is quasi-conservative in unsaturated vertical motion in the atmosphere, but is not conservative in a non-adiabatic process such as radiational cooling at constant pressure. See AIR-MASS ANALYSIS.

constancy of winds: Owing to varying wind direction, the magnitude of the VECTOR mean (V_R) of a series of wind observations is less than that of the corresponding SCALAR mean (V_S). The quantity q, defined by $q = 100\ V_R/V_S$, gives a measure of the constancy of wind direction. Values of q range between 0 (winds equally strong and frequent in all directions) and 100 (winds unvarying in direction but not necessarily in speed).

q is controlled in part by synoptic events and in part by the constraints placed

---

* Geneva, World Meteorological Organization. International cloud atlas, Vol. 1. Geneva, 1956, WMO, p. 14.

on wind direction by local topography. Typical values of q for surface winds over and near the British Isles are within the range 15 to 50. Interpretation of a value of q is, however, unambiguous only on the assumption of a normal and CIRCULAR FREQUENCY DISTRIBUTION of the winds about the vector mean wind.

constant absolute vorticity trajectory (CAVT): The TRAJECTORY of an air parcel, evaluated on the assumption that absolute VORTICITY is conserved during the motion in which there is no shear.

The basis of such trajectories is the POTENTIAL VORTICITY THEOREM which is, in polar co-ordinates,

$$\frac{\frac{V}{r} + \frac{\partial V}{\partial r} + 2 \Omega \sin \varphi}{\triangle p} = \text{constant}.$$

Under the restrictive conditions of a broad current of air moving horizontally at constant speed ($V =$ constant) with little lateral WIND SHEAR ($\partial V/\partial r \approx 0$) and in a thin layer at a level of non-DIVERGENCE ($\triangle p =$ constant), the above equation states that the air will move in such a way that latitudinal (φ) changes of vorticity are compensated by curvature (V/r) changes. A wave-like motion between two parallels of latitude results: the air passes through its mean latitude without curvature of path and acquires maximum anticyclonic curvature in its highest latitude and maximum cyclonic curvature in its lowest latitude. The wavelength and amplitude of the oscillation depend on the wind speed, the mean latitude and the angle at which the trajectory intersects this mean latitude.

Tables for computing CAVT are contained in '*Weather analysis and forecasting*'.*

constant-pressure chart: A SYNOPTIC CHART, relating to a surface of specified constant pressure (isobaric surface), on which contours of GEOPOTENTIAL height of the surface are drawn. Other elements, e.g. temperatures and winds observed at the given pressure, are usually entered on the same chart. See also GEOSTROPHIC WIND.

constituent body: The constituent bodies of the WORLD METEOROLOGICAL ORGANIZATION are Congress, Executive Committee, Regional Associations and Technical Commissions.

contessa del vento: A type of eddy cloud frequently found to the lee side of isolated mountains. It is often observed over Mount Etna with a westerly wind. In its most characteristic form it consists of a rounded base or disc, sometimes with several vertically stacked discs, surmounted by a protuberance directed upwind.

continental climate: A type of CLIMATE, characteristic of the interior of large land masses of middle latitudes; the main distinguishing features are large annual and diurnal ranges of air temperature, with low rainfall a further characteristic feature. The most extreme form of continental climate is found in eastern Asia where there is an annual range of temperature (July mean minus January mean) of about 60 degC. See CONTINENTALITY.

continental drift: A hypothesis, generally linked with the names of F. B. Taylor and A. Wegener, of displacements, on a geological time-scale and by distances of the order of thousands of kilometres, of various parts of the earth's surface relative to other parts. According to this hypothesis continents now separated were once joined

---

* PETTERSSEN, S.; Weather analysis and forecasting. Vol. 1, Motion and motion systems, New York, McGraw-Hill Book Co., 1956, p. 414.

together as, for example, South America with Africa, and North America with Europe; in general, changes in both latitude and longitude are implied.

Evidence advanced in support of the hypothesis includes similarities of shape and topographical features and of plant and animal life of land masses supposed once to have been joined. The hypothesis has been used to explain, with limited success, climatic changes on a world-wide scale. Recent evidence strongly advanced in support concerns the history of the earth's magnetic field as inferred from rock magnetism (see GEOMAGNETISM).

The hypothesis of continental drift has been highly controversial. Various geophysicists have held that there is no conceivable mechanism which can give rise to such an effect; in addition, various items of palaeontological and palaeoclimatological evidence have been held to be in conflict with, rather than in support of, the theory. The hypothesis has nevertheless now gained wide acceptance. See also POLAR WANDERING.

continentality: In meteorology, a measure of the extent to which the climate at any place is subject to land, as opposed to maritime, influences. Because of the low specific heat of land relative to water and, more especially, the fact that heat exchanges by day and night, summer and winter, are confined to a shallow layer of a land surface, the diurnal and seasonal temperature changes of a land mass and its overlying air are great. Various measures of continentality have been suggested, most of them based on the observed annual range of temperature but depending also on latitude (the INSOLATION factor).

At a given latitude continentality increases, in general, with distance from the sea, but maximum continentality occurs at an appreciable distance downwind (relative to the PREVAILING WIND) from the centre of a land mass. A large range of mountains across the prevailing wind produces a marked increase of continentality to leeward of the mountains, mainly due to the clearer skies and consequently greater radiational heating in summer and cooling in winter that prevail there.

See also CONTINENTAL CLIMATE.

continental polar air: See POLAR AIR.

continental tropical air: See TROPICAL AIR.

continuity, equation of: An equation which expresses the law of conservation of mass and states that the mass of air entering an elementary volume is equal to the increase of mass within it.

The equation has various forms, for example:

$$\text{div}\,\mathbf{V} = -\frac{1}{\rho}\frac{d\rho}{dt},$$

$$-\frac{\partial w}{\partial z} = \text{div}_H\mathbf{V} + \frac{1}{\rho}\frac{d\rho}{dt}$$

where $\mathbf{V}$ is the wind vector, ρ the air density, w the vertical component of velocity, and z the height.

The latter equation states that if an air parcel contracts vertically it may expand horizontally and increase in density. Since local density change is small in the atmosphere the equation has the approximate form:

$$-\frac{\partial w}{\partial z} \approx \text{div}_H\mathbf{V}.$$

Assuming that the vertical variation of pressure is governed by the HYDROSTATIC EQUATION, the equation of continuity in PRESSURE CO-ORDINATES has the form

$$\text{div}_p \mathbf{V} = -\frac{\partial}{\partial p}\left(\frac{dp}{dt}\right).$$

See also DIVERGENCE, EULERIAN CHANGE, LAGRANGIAN CHANGE.

contour (contour line): In synoptic meteorology, a line of constant height of an ISOBARIC SURFACE. The terms 'absolute contour' and 'isohypse' are sometimes used. Contours represent STREAMLINES in strict GEOSTROPHIC flow.

contrail: A common abbreviation for CONDENSATION TRAIL.

contrast threshold of the eye: A ratio ε, important in measurements of VISIBILITY, defined by the relation

$$\varepsilon = \frac{B - H}{H}$$

where B is the apparent brightness of an object which is just visible against a background of apparent brightness H.

The value of ε varies considerably with the observer and with viewing conditions. The value of 0·02 was originally assumed for normal observers in ordinary daylight illumination but experimental determinations have ranged from about 0·01 to 0·06. The WORLD METEOROLOGICAL ORGANIZATION has recommended the use of the value 0·05 and this figure should be used for the purposes of conversion of attenuation measurements to corresponding visibility. See VISIBILITY METER, TRANSMISSOMETER.

convection: A mode of heat transfer within a fluid, involving the movement of substantial volumes of the substance concerned. The convection process frequently operates in the atmosphere and is of fundamental importance in effecting vertical exchange of heat and other air-mass properties (water vapour, momentum, etc.) throughout the troposphere.

Two distinct, though not mutally exclusive, types of convection occur in the atmosphere. In 'forced' convection the vertical air motion is produced by mechanical forces, as in the passage of air over rough or high ground, and the vertical transport of properties is effected by 'eddies' (see EDDY). In 'free' or 'natural' convection BUOYANCY forces operate, in the absence of static STABILITY of the air, to effect vertical mixing through the agency of convection cells or 'bubbles'.

convective condensation level: See CONDENSATION LEVEL.

convective rain: Rainfall which is caused by the vertical motion of an ascending mass of air which is warmer than its environment; the horizontal dimension of such an air mass is generally of the order of 15 kilometres or less and forms a typical CUMULONIMBUS cloud. Convective rain is typically of greater intensity than either of the other two main classes of rainfall (cyclonic and orographic) and is sometimes accompanied by thunder.

The term is more particularly used in those cases in which the precipitation covers a large area as the result of the agglomeration of cumulonimbus masses.

convective region: An alternative for ADIABATIC REGION.

convergence: See DIVERGENCE.

Coriolis acceleration: An apparent acceleration which air possesses by virtue of the earth's rotation, with respect to axes fixed in the earth. It is sometimes termed 'geostrophic acceleration'.

The Coriolis acceleration is a three-dimensional vector which is given in vector notation by the expression $-2\Omega \wedge \mathbf{V}$, where Ω is the earth's angular velocity and $\mathbf{V}$ the wind velocity (see ACCELERATION). It is therefore everywhere at right angles to the earth's axis (in the plane of the equator) *and* to the air velocity; in the northern (southern) hemisphere it acts to the right (left) when viewed along the motion. The effect is often expressed in terms of the equal inertia force per unit mass termed the 'Coriolis force', or 'deviating force', or 'geostrophic force'.

In a Cartesian co-ordinate system in which x and y are in the horizontal plane to east and north respectively, and z is vertically upwards, the Coriolis acceleration has components

$$\text{C.A.}_x = 2\,\Omega\,(v \sin \varphi - w \cos \varphi)$$
$$\text{C.A.}_y = -2\,\Omega\, u \sin \varphi$$
$$\text{C.A.}_z = 2\,\Omega\, u \cos \varphi$$

where u, v, w are the component wind velocities in the x, y, z directions, respectively, and φ the latitude.

Only the component of the Coriolis acceleration which acts in the horizontal (xy) plane is significant in meteorological dynamics; the vertical component is of comparable magnitude, but is negligible compared with the large forces (gravity and pressure gradient) which act in this direction. For quasi-horizontal air motion ($w = 0$), of velocity V, the horizontal Coriolis acceleration is $2\,\Omega\,V \sin \varphi$.

The component of the earth's rotation in the horizontal plane is such that the earth simulates a flat disc rotating anticlockwise in the northern hemisphere and clockwise in the southern hemisphere (see ANGULAR VELOCITY OF THE EARTH). Air moving horizontally outwards from the centre of the disc appears to an observer stationed there to be deflected to the right in the northern hemisphere and to the left in the southern hemisphere, i.e. in a frame of reference fixed in the rotating earth the air has horizontal acceleration in these respective directions. The effect is negligible if the scale of the motion considered (radius of disc) is of the order of only a few kilometres.

Coriolis force: See CORIOLIS ACCELERATION.

Coriolis parameter: A quantity, denoted f (sometimes l), defined by the equation $f = 2\,\Omega \sin \varphi$, where Ω is the magnitude of the earth's angular velocity and φ the latitude.

corona: A series of coloured rings surrounding the sun or moon. The space next to the luminary is bluish white, while this region is bounded on the outside by a brownish-red ring, the two forming the 'aureole'. In most cases the aureole alone appears but a complete corona has a set of coloured rings surrounding the aureole —violet inside, followed by blue, green, yellow to red on the outside. The series may be repeated more than once, but the colours are usually merely greenish and pinkish tints.

The corona is produced by DIFFRACTION of the light by water drops. Pure colours indicate uniformity of drop size. The radius of the corona is inversely proportional to drop size; thus growth of a corona indicates decrease of drop size.

A corona is distinguished from a HALO (caused by REFRACTION) by its reversal of colour sequence, the red of the halo being inside, that of the corona outside; the dull red, which is the first notable colour in the aureole, ranks as outside the bluish tint near the luminary. An alternative criterion is that the colours of a halo are at the inner edge of a luminous area, while those of a corona are at the outer edge. For auroral corona see AURORA.

corona, solar: The outer atmosphere of the SUN, extending to great distances and comprising feebly luminous and highly ionized gases at a temperature of about 10^6K. Directly visible only during a solar eclipse it is studied at other times by means of the coronograph. Roughly circular in shape at SUNSPOT maximum, it is much extended along the sun's equator at sunspot minimum.

corposant: See ST ELMO'S FIRE.

correlation: The correlation coefficient or more formally the product moment (r_{xy}) is the best known measure of association between pairs of variables. If (x_1,y_1), ..., (x_n,y_n) are n pairs then

$$r_{xy} = \frac{\sum_{j=1}^{n}(x_j-\bar{x})(y_j-\bar{y})}{\sqrt{\left\{\sum_{j=1}^{n}(x_j-\bar{x})^2 \sum_{j=1}^{n}(y_j-\bar{y})^2\right\}}} = \frac{\text{cov}(x,y)}{\sigma_x \sigma_y}$$

where $\bar{x}, \bar{y}$ are the MEANS and σ_x, σ_y the STANDARD DEVIATIONS of the variables and cov(x,y) their COVARIANCE.

The value of the correlation coefficient can range from 1, (indicating complete functional dependence) through zero (independence) to -1 (implying complete dependence in opposing directions). Its significance can be judged from published tables provided that (i) the variables concerned are drawn from the NORMAL DISTRIBUTION and (ii) each pair of variables is independent of the rest. See INDEPENDENCE.

When, as usual with meteorological data, these conditions are not known to be satisfied, the correlation coefficient must be used with caution. It also suffers from the objection that it gives an exaggerated impression of the level of association between the variables; thus a value of $r = 0.5$ which appears to be half way between independence and functional dependence corresponds to quite a low degree of association. In judging the value of a forecasting technique, for example, the percentage reduction of VARIANCE is a much more realistic measure of success.

A significant value of r is to be regarded as implying a measure of causal relationship only if this view is supported by strong physical reasoning.

correlogram: A diagram in which r_L, the AUTOCORRELATION at lag L, is plotted as ordinate, with $L = 1, 2, 3$, etc., for abscissa. The interpretation of a correlogram requires care, but it forms the basis of many of the advanced techniques of analysis of TIME SERIES.

cosmic radiation (or **rays**): Very high energy radiation which originates outside the earth and probably, in large part, outside the solar system.

The cosmic 'primary' radiation consists mainly of positively charged nuclei of hydrogen (protons) and heavier elements, which are deflected by the earth's magnetic field on near approach to the earth and enter the earth's atmosphere in numbers that increase with increasing geomagnetic latitude. Predominance of a positive charge in the primary radiation is inferred from the preponderance of entry of radiation from a westerly direction at all latitudes ('east-west asymmetry').

The primary radiation reacts to an appreciable extent with the nuclei of atmospheric atoms on penetrating to a level some 15–20 km above ground, and so gives rise to a variety of secondary, tertiary, etc., products (especially various types of meson), some very penetrating ('hard'), others less penetrating ('soft'). Continuous recording of the intensity of the radiation which reaches the ground is made in many parts of the world. The intensity and average energy of cosmic radiation increase rapidly with height above mean sea level.

Before the era of nuclear explosions, measurement suggested that about one-fourth of the conductivity of the air near the ground was due, on average, to the ionizing power of cosmic radiation; at low levels over the sea, and in all regions at a level greater than about 2 km, nearly all the conductivity of the air was due to cosmic radiation.

Among the time variations of cosmic radiation are: a solar-cycle variation more or less in phase with SUNSPOT NUMBER; a 27-day recurrence tendency; decreases associated with magnetic storms and increases associated with intense solar flares.

Cosmic radiation intensity measured at the ground is affected by atmospheric conditions in a way which, though not completely understood, proceeds, to a first approximation, as follows. Since atmospheric pressure at any point is a measure of the mass of air vertically above the point, the pressure at which 'frequent' inter-action between the primary radiation and the atmosphere occurs, i.e. pressure at the level of production of secondary radiation, will be approximately constant. Since, also, the probability of absorption or radioactive decay of an individual secondary particle increases with depth of layer between height of production and mean sea level, measured intensity of radiation decreases both with high mean temperature in the layer between mean sea level and level of production, and with high mean-sea-level pressure (high thickness between the upper constant-pressure level and mean sea level being implied in either case). While approximate corrections for these meteorological influences are required for the purpose of most analyses, the uncorrected observations may provide valuable information concerning meteorological phenomena. A notable example was the interpretation of an apparent lunar time variation of cosmic radiation intensity as a vertical motion of the mean pressure level of generation of secondary radiation, i.e. as a measure of the amplitude of the lunar tidal motion in the high atmosphere.

counterglow: The narrow coloured band or bands which are seen at TWILIGHT above the blue SHADOW OF THE EARTH thrown by the sun just above the horizon (eastern horizon at sunset, western at sunrise). The light from the bands is scattered back to the observer by dust particles and is predominantly soft red in colour, corresponding to the colour acquired by the incident solar beam as it passes through the atmosphere and suffers preferential scattering of the blue wavelengths by the air molecules. In some cases there are well-marked higher bands of orange, yellow and green corresponding to a decrease in redness acquired by the incident light on passage through less dense regions of the atmosphere. The phenomenon is also termed 'countertwilight', or 'ANTITWILIGHT', or 'antitwilight arch'.

The term 'counterglow' is also otherwise used as an alternative for GEGENSCHEIN.

counter sun: An alternative for ANTHELION.

countertwilight: An alternative for COUNTERGLOW.

Courant–Friedrichs–Lewy condition: See COMPUTATIONAL INSTABILITY.

covariance: If $(x_1y_1)(x_2y_2), \ldots, (x_ny_n)$ are n pairs of variables, then

$$\text{cov}(x,y) = \frac{1}{n} \sum_{j=1}^{n} (x_j - \bar{x})(y_j - \bar{y}).$$

The covariance depends on the association between the variables, and has an expected value of zero if they are independent. When divided by the product of the STANDARD DEVIATIONS, $\sigma_x \sigma_y$, it gives the CORRELATION coefficient.

PLATE 19 Crepuscular rays.

crachin: Conditions of drizzle combined with low stratus, mist, or fog that occur at times between January and April in south China and in the coastal area between about Cape Cambodia and Shanghai. The conditions are caused by an interaction between the maritime tropical air and the maritime polar air circulating round the eastern side of the Asiatic anticyclone. Orographic and coastal lifting are also significant factors.

CRDF: Abbreviation for cathode-ray direction finding. A system, used in the Meteorological Office for the locating of thunderstorms, in which visual representations of lightning flashes are displayed on a calibrated cathode-ray tube so that their bearings are indicated. See also SFERICS FIX.

crepuscular rays: These include three similar classes of phenomenon:
(i) Sunbeams penetrating through gaps in a layer of low cloud and rendered luminous by water or dust particles in the air (phenomenon termed 'sun drawing water' or 'Jacob's ladder').
(ii) Pale blue or whitish beams diverging upwards from the sun hidden behind cumulus or cumulonimbus clouds. The well-defined beams are separated by darker streaks which are the shadows of parts of the irregular cloud. See Plate 19.
(iii) Red or rose-coloured beams, diverging upwards at twilight from the sun below the horizon. The light is scattered to the observer by atmospheric dust; the beams are separated by greenish-coloured regions which are the shadows of clouds or hills below the horizon.

In cases (ii) and (iii) the beams and shadows may persist across the sky before converging at the ANTISOLAR POINT (anticrepuscular rays). The apparent divergence and convergence of the rays is an optical illusion produced by perspective.

critical frequency: In radio sounding of the IONOSPHERE, that radio frequency at which the wave just penetrates a specified layer. The square of the critical frequency is directly proportional to the maximum electron concentration in the layer.

critical temperature: That temperature above which a specified gas cannot, and below which it can, be liquefied by pressure alone.

cross-section: As commonly used in meteorology, this term means the representation in schematic form of conditions prevailing, or expected to prevail, in the atmosphere at a specified time in a vertical plane from the surface up to any desired height along a line from one point to another. The information plotted on the cross-section depends upon the purpose for which it is plotted—whether, for example, as an aid in analysis or as a representation of expected conditions along an air route.

cumulonimbus (Cb): One of the CLOUD GENERA (Latin, *cumulus* heap, *nimbus* rainy cloud).

'Heavy and dense cloud, with a considerable vertical extent, in the form of a mountain or huge towers. At least part of its upper portion is usually smooth, or fibrous or striated, and nearly always flattened; this part often spreads out in the shape of an anvil or vast plume.

Under the base of this cloud which is often very dark, there are frequently low ragged clouds either merged with it or not, and precipitation sometimes in the form of VIRGA.'* See Plates 14 and 15. See also CLOUD CLASSIFICATION.

---

* Geneva, World Meteorological Organization. International cloud atlas, Vol. 1. Geneva, WMO, 1956, p. 11.

cumulus (Cu): One of the CLOUD GENERA (Latin, *cumulus* heap).

'Detached clouds, generally dense and with sharp outlines, developing vertically in the form of rising mounds, domes or towers, of which the bulging upper part often resembles a cauliflower. The sunlit parts of these clouds are mostly brilliant white; their base is relatively dark and nearly horizontal.

Sometimes cumulus is ragged.'* See Plates 13 and 17. See also CLOUD CLASSIFICATION.

cup-contact anemometer: See ANEMOMETER, ANEMOGRAPH.

cup-counter anemometer: See ANEMOMETER.

cup-generator anemometer: See ANEMOMETER, ANEMOGRAPH.

curie: A measure of the activity of a radioactive substance, defined as that quantity of a radioactive substance which decays at the rate of $3 \cdot 7 \times 10^{10}$ disintegrations per second.

curl: See VORTICITY.

current, ocean: A general movement, of a permanent or semi-permanent nature, of the surface water of the ocean. The term must not be used of tidal streams, which change direction and speed hour by hour. A 'drift current' is a drift of the surface water which is dragged along by a wind blowing over it. A drift current which is deflected by an obstruction, such as a shoal or land, forms a 'stream current'. A stream current may be formed as a counter current to a primary current, replacing the water displaced by the primary current. The 'set' of a current or tidal stream is the direction in which it is going.

Owing to the high specific heat of water, the main ocean currents are of great climatic importance. Ocean currents result from several causes, the most important being wind and differences of density resulting from differences of temperature and salinity.

Among the best-known currents are those of the North Atlantic, notably the GULF STREAM, North Atlantic Drift, and the Labrador Current.

curve-fitting: The representation of experimental data or frequencies by a curve, generally an algebraic form determined by certain coefficients or parameters. The parameters are chosen to provide the 'best' or at least an acceptable fit to the data, using the LEAST SQUARES or other appropriate criterion. Curve-fitting is simplified if the data can be transformed so that curves of the appropriate form are represented by straight lines. For example, if the relationship is of the nature $y = Ax^n$, a plot on logarithmic paper yields the values of A and n; if of the nature $y = Ae^{bx}$, a plot on semi-logarithmic paper yields the values of A and b.

A simple curve form involving few parameters nearly always produces a more stable fit than a more complicated form. It is also possible for a fit to be too good, as judged by the CHI-SQUARE TEST, and hence unlikely.

cut-off high: A warm ANTICYCLONE, separated from the subtropical anticyclones, which is usually associated with a marked interruption of the zonal westerlies (BLOCKING). This feature forms initially as a ridge in the upper westerlies, intensifies into a closed circulation, and extends downwards towards the surface.

cut-off low: A DEPRESSION which lies equatorwards of the main mid-latitude belt of westerly winds. The feature starts as a trough in the upper westerlies, deepens into a closed circulation and extends downwards towards the surface.

cyanometer: An instrument for measuring the blueness of the sky.

cycle: A term for a recurrent phenomenon which is best reserved for changes of strictly periodic origin; the diurnal cycle of cloud or the annual temperature cycle are acceptable usages. The use of a neutral term such as OSCILLATION is preferable for changes which are not directly dependent on astronomical processes.

cyclogenesis: The initiation of cyclonic circulation, or its strengthening around an existing CYCLONE or DEPRESSION.

cyclolysis: The disappearance or weakening of cyclonic circulation around an existing CYCLONE or DEPRESSION.

cyclone: That atmospheric pressure distribution in which there is a low central pressure relative to the surroundings. It is characterized on a synoptic chart by a system of closed isobars, generally approximately circular or oval in form, enclosing a central low pressure (see Figure 12 under DEPRESSION). 'Cyclonic circulation' is counter-clockwise round the centre in the northern hemisphere, clockwise in the southern hemisphere; in either case the sense of rotation about the local vertical is the same as that of the earth's rotation.

A cyclone of middle and high latitudes is called a 'depression'. A TROPICAL CYCLONE of moderate intensity is a 'tropical storm'; if of great intensity, a tropical cyclone in the Indian Ocean, Arabian Sea or Bay of Bengal is termed a 'cyclone', in the western Pacific a 'typhoon', in Western Australia a 'willy-willy', in most other tropical latitudes a 'hurricane'.

See also DEPRESSION.

cyclonic rain: Rainfall which is caused by the large-scale vertical motion associated with synoptic features such as DEPRESSIONS and FRONTS. It is one of a broad three-fold classification of rainfall, the other classes being 'orographic' and 'convective'.

cyclostrophic wind: A class of winds, of markedly curved flow, in which the CORIOLIS ACCELERATION is negligible in comparison with the CENTRIPETAL ACCELERATION. In such winds the cyclostrophic term is predominant in the expression for the GRADIENT WIND. A tropical cyclone provides an example in which the wind is mainly of this type.

D

daily variation: An alternative for DIURNAL VARIATION.

Dalton's law: See PARTIAL PRESSURE.

dawn: (OE, *dagian* to become day) the time when light appears (*daws*) in the sky in the morning or the interval between the first appearance of light and the rising of the sun. See TWILIGHT.

dawn chorus: A type of radio disturbance which consists of a chorus of overlapping, rising tones, mainly in the middle audio-frequency range, and is most intense at local dawn. The disturbance is rarely heard at geomagnetic latitudes less than about 50°. Since it is correlated with geomagnetic disturbance it is considered to be initiated by extraterrestrial charged particles. It is also, however, sometimes related to the occurrence of SFERICS and WHISTLERS.

day: A 'solar day' is the SYNODIC interval of time between successive occasions on which the sun is in the MERIDIAN of any fixed place (sun 'transits'). A 'sidereal day' is the corresponding interval between successive transits of a distant fixed star. Since the earth moves in an orbit round the sun with the same sense of rotation as that about its own axis, the length of the solar day is slightly greater than that of the sidereal day.

Owing to the eccentricity of the earth's orbit and to the inclination of the equator to the ecliptic, solar days are of slightly unequal lengths at different times of the year. The average length of the solar day is 86 400 seconds. This is called the 'mean solar day' and is taken as the length of the 'civil day', or the 'day' of ordinary parlance. The sidereal day contains 86 164 seconds nearly.

Other small measured variations of length of day, on various time-scales, are of geophysical interest. (i) A secular lengthening (about 10^{-3} s/100 years) is attributed mainly to tidal friction. (ii) Fluctuations of the order of decades are related to slow changes in the earth's magnetic field and are attributed to transfers of angular momentum between the core and mantle of the earth. (iii) An annual variation (length of day 2×10^{-3} s greater in February than in August) is mainly attributed to an annual variation of wind velocity on a global scale, involving angular momentum changes of the air opposite to those of the surface of the earth and so maintaining constant the total angular momentum of the earth–atmosphere system.

As used in geophysics, a 'lunar day' is the interval between successive transits of the moon; it varies in length from about 24 h 40 min to 25 h.

dayglow: The day-time AIRGLOW emission, thought to be rather more intense than the NIGHTGLOW but not observable at low levels against the intense background of scattered solar radiation.

daylight: The intensity of daylight illumination on a horizontal surface is recorded at some stations in the British Isles. The method consists of the measurement of the current emitted by a photocell on which radiation from the sun and sky falls after passing through a protective hemispherical dome of clear glass, a horizontal diffusing surface of translucent material, and a filter. The instrument, commonly called an illuminometer, is so designed that its spectral sensitivity is similar to that of the human eye.

The 'duration of daylight', or 'length of day', is the interval between SUNRISE and the following SUNSET. Tabulated values are contained in the *Smithsonian meteorological tables.**

débâcle: The breaking up in the spring of the ice in the rivers. The term is chiefly applied to the great rivers of Russia and Siberia and of the North American continent. Débâcle lasts from two to six weeks; during the period the rivers often overflow their banks, inundating the surrounding country. In southern Russia débâcle begins about the middle of March, in latitudes 55°–60°N it begins early in April, but in the north it does not begin until May and in the extreme north of Siberia not until June. In Canada the débâcle in Ontario takes place in March and the water is free by April; in the St Lawrence it is a little later, the river being free of ice in May.

debriefing, meteorological: Post-flight interrogation of aircrew in order to ascertain the meteorological conditions experienced during a flight.

decad: A period of ten consecutive days. See also PENTAD.

decibel: Unit of relative measure of two flux densities, increasingly employed in various contexts such as electric power density, sound or light intensity—see, for example, NEBULE.

The decibel is one-tenth of a bel. Two flux densities (I_1, I_2) are said to differ by n bels when

$$\frac{I_1}{I_2} = 10^n, \text{ i.e. when } n = \log_{10} \frac{I_1}{I_2}.$$

Thus, flux densities differ by N decibels (unit normally employed in preference to bel) when

$$N = 10 \log_{10} \frac{I_1}{I_2}.$$

declination: Angular distance of a body north or south of the celestial EQUATOR. In meteorology, seasons are controlled by the sun's declination which varies from about 23° 27′ N at the June solstice, through 0° at the March and September equinoxes, to about 23° 27′ S at the December solstice.

declination, magnetic: The departure (degrees east or west) of a compass needle from true (geographical) north. This angle, termed by mariners 'magnetic variation', varies in space and time. Mean values of westerly declination for 1967, and rates of annual decrease of westerly declination at this epoch were: Lerwick (Shetland Islands) 9° 14·2′ and 3·4′ per year; Eskdalemuir (Dumfriesshire) 9° 52·1′ and 4·3′ per year; Hartland (Devon) 9° 20·3′ and 4·8′ per year.

deepening: In synoptic meteorology, 'deepening' of a depression signifies a decrease of pressure at the centre of the system with time. The converse term is 'filling'.

---

* LIST, R. J.; Smithsonian meteorological tables. *Smithson Misc Collns, Washington*, **114**, 1968, pp. 506–512.

deformation: In hydrodynamics, the change of shape of a small element of fluid produced by space variations of the fluid velocity.

In meteorology, the term is used mainly in respect of the kinematical development of FRONTOGENESIS and FRONTOLYSIS. Thus, flow associated with a COL represents a deformation field of motion. If persistent, such a flow produces a relative crowding of isotherms along the axis of outflow from the neutral point of the col, and a displacement of isotherms away from the axis of inflow. Mathematically, the deformation involves two terms

$$A = \frac{\partial u}{\partial x} - \frac{\partial v}{\partial y}$$

$$B = \frac{\partial v}{\partial x} + \frac{\partial u}{\partial y}.$$

Individually, these terms are not invariant; the quantity

$$F^2 = A^2 + B^2$$

is invariant and is usually taken as a measure of deformation in meteorology.

degree-day: See ACCUMULATED TEMPERATURE.

degrees of freedom: When a result is being tested for statistical SIGNIFICANCE, the number of degrees of freedom determines the standard of comparison, and hence the proper place to enter the statistical table of significance. The number of degrees of freedom is generally the number of items (or the number of cells in the CHI-SQUARE TEST) reduced by the number of constants calculated from the data in setting up the hypothesis to be tested.

dendrochronology: The interpretation of the varying width of the annual growth rings of certain trees ('tree rings') in terms of the corresponding year-to-year climatic fluctuations. This method of inferring past climatic fluctuations appears to have rather strict limitations owing to the various possible combinations of weather element values which may be associated with a given rate of growth. The most important of the elements involved are probably temperature and rainfall.

dendroclimatology: An alternative for DENDROCHRONOLOGY.

density: The mass of unit volume of a substance, at a specified temperature and pressure. The dimensions are ML^{-3}.

The density of the air is not measured directly but may be calculated in terms of the normally observed meteorological elements of pressure (p), temperature (T), and humidity, from the GAS EQUATION:

$$\rho = \frac{p}{RT} = \frac{Mp}{R^*T}$$

where M = molecular weight of air sample
R^* = universal gas constant
R = specific gas constant for air sample.

The value of M decreases with increasing percentage weight of water vapour in the air (regarded as a mixture of dry air and water vapour); relatively moist air is therefore less dense than relatively dry air, at the same pressure and temperature.

The value of M appropriate to a given sample is less readily obtained than the VAPOUR PRESSURE (e'). Air density is therefore more readily obtained from the formula

G

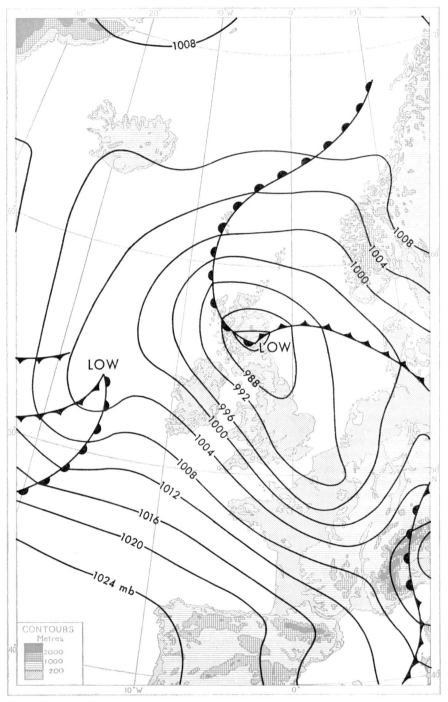

FIGURE 12—Depression over the British Isles, 06 GMT, 13 July 1961.

$$\rho = \rho_0 \frac{p - \tfrac{3}{8} e'}{p_0} \cdot \frac{T_0}{T}$$

where ρ is the density of the sample of air to be computed
ρ_0 is the density of dry air at pressure p_0 and absolute temperature T_0
p is the pressure of the sample
T is the absolute temperature of the sample
e' is the vapour pressure of the sample.

The density of dry air at a pressure of 1000 mb and temperature of 290 K is 1·201 kg/m³. Thus,

$$\rho \text{ (kg/m}^3) = 1\cdot201 \times \frac{p - \tfrac{3}{8} e'}{1000} \cdot \frac{290}{T},$$

p and e' being measured in millibars.

Air density in the British Isles is about 1·2 kg/m³ at the surface; mean values at 5 and 10 km, for example, are about 0·7 and 0·4 kg/m³, respectively. At heights beyond those attainable by balloons (about 40 km) air density has been inferred, for example, from FALLING SPHERE experiments, ROCKETSONDES, GRENADES SOUNDINGS and SATELLITE orbits. Systematic diurnal and annual variations of density have been confirmed, the amplitude of variation generally increasing with height. At satellite heights (above about 150 km) there are also large variations of air density which are associated with solar and geomagnetic activity.

departure: The amount, positive or negative, by which the value of a meteorological element departs from the normal, average or expected value for the place and time. It is also termed 'anomaly'.

deposit gauge: A gauge designed for the measurement of the deposited products of ATMOSPHERIC POLLUTION. The solid and liquid products which enter the collecting bowl of such a gauge within a specified period (generally a calendar month) are subjected to volumetric, gravimetric and chemical analyses in order to determine their amount and constitution.

depression: The term commonly applied to CYCLONES, of various intensities, in extratropical latitudes (see Figure 12). It is also used of a weak TROPICAL CYCLONE.

Central pressure of an extratropical depression varies from about 950 to 1020 millibars. It is described as 'shallow' or 'deep' if encircled by few or many isobars, respectively. It is said to 'deepen' or 'fill up' if the central pressure decreases or increases, respectively, with time. Its diameter varies from about 200 to over 4000 km. The associated weather is unsettled, often with much precipitation and strong winds or gales. It is, in general, a highly mobile feature of the synoptic chart, with speeds ranging up to about 50 knots; large depressions may, however, remain almost stationary for several days. The general direction of movement is eastwards, though any direction of movement may occur. The extratropical depressions largely control the rainfall distribution of these latitudes and are responsible for much of the interchange of air between high and low latitudes, which is an essential feature of the GENERAL CIRCULATION of the atmosphere.

The typical extratropical depression is associated with FRONTS in the manner illustrated in Figure 13, in which three distinct stages in the 'life cycle'— the 'wave depression', 'warm-sector depression', 'occluded depression'—are shown. Such a cycle usually lasts some three to four days, sometimes appreciably longer. Deepening and rapid movement of the depression are common up to the warm-sector stage, with subsequent filling up and slowing down. The formation of a fresh wave depression on the trailing cold front of a depression which is in a later stage of development is a very common occurrence. Typically there are four or five such members of a 'depression family'; the series normally ends with an incursion of cold air to unusually low latitudes.

In the early stages of a depression the troposphere is relatively warm near the centre (except in a shallow layer under the frontal surface) and the cyclonic circulation is shallow; in late stages the air is relatively cold throughout the troposphere and the cyclonic circulation is deep. Horizontal convergence and associated upward motion in the lower troposphere, and horizontal divergence in the upper troposphere, prevail near and in advance of the depression centre. The structure of horizontal divergence and vertical motion is reversed in the rear of the depression.

Important but less common types of extratropical depression which are not associated with fronts are the POLAR-AIR DEPRESSION, THERMAL DEPRESSION, LEE DEPRESSION.

Shallow TROPICAL CYCLONES (depressions) are a relatively common feature of low-latitude non-desert regions and accentuate cloudiness and showeriness. They are often associated with horizontal convergence, especially at the INTERTROPICAL CONVERGENCE ZONE. Infrequently, such a depression greatly intensifies and is termed a HURRICANE, TYPHOON, or CYCLONE, depending on locality.

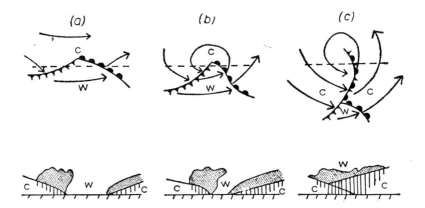

FIGURE 13—Plan views (above) of three stages of development of a frontal depression with (below) corresponding cross-sections along the horizontal dashed line.
  Letters 'C' and 'W' signify cold and warm air masses, respectively.
  In the lower diagrams the vertical lines indicate precipitation and the dotted areas show frontal cloud.

depression, angle of: The angular depression of an object measured by an observer, with reference to the horizontal plane through the observer.

descendent: See GRADIENT.

desert: A region in which rainfall is insufficient, in relation to rate of evaporation, to support vegetation. The main desert regions are in latitudes lower than 50° and are marked by relatively large diurnal and seasonal ranges of temperature. In terms of synoptic meteorology, the main causes of desert conditions are either the presence of a persistent anticyclone, as in northern Africa where the Sahara coincides with the average position of the subtropical belt of high pressure, or a configuration of ground which shuts out the moisture-bearing winds as, for example, Gobi in central Asia.

For a formula used by W. Köppen and R. Geiger for the limit of rainfall which constitutes a desert climate, see ARID CLIMATE.

desiccation: The permanent disappearance of water from an area due to a change of climate and especially a decrease of rainfall. Large areas in central Asia, Africa and western America have been desiccated since the last GLACIAL PHASE, but there does not appear to have been much progressive desiccation during the past 2000 years. See also EXSICCATION.

detachment: In meteorological literature, usually refers to the process of ejection of ELECTRONS from negative IONS which occurs in the high atmosphere by various types of particle collision and by the action of light photons ('photo-detachment').

development: In synoptic meteorology, the intensification of circulation, cyclonic or anticyclonic.

A strong measure of 'compensation' exists in the atmosphere such that, for example, the surface pressure fall associated with strong surface cyclonic development is a small residual between horizontal convergence at low levels and horizontal DIVERGENCE at high levels (with upward motion through much of the troposphere). The converse situation obtains in strong anticyclonic development. In R. C. Sutcliffe's 'development theorem', which is based on these conceptions, the relative divergence between selected low and high levels, e.g. surface and tropopause, is expressed, on various simplifying assumptions, in terms of the THERMAL WIND and VORTICITY, both of which may be measured from charts, i.e.

$$\text{div}_p \mathbf{V}_1 - \text{div}_p \mathbf{V}_0 = -\frac{1}{f}\left(\mathbf{V}'\frac{\partial f}{\partial s} + \mathbf{V}'\frac{\partial \zeta'}{\partial s} + 2\mathbf{V}'\frac{\partial \zeta_0}{\partial s}\right)$$

where div_p is the divergence in an isobaric surface, $\mathbf{V}_1$ and $\mathbf{V}_0$ the wind vectors at the upper and lower pressure levels, f the CORIOLIS PARAMETER, $\mathbf{V}'$ the thermal wind vector between the selected levels and ζ' its vorticity, ζ_0 the vorticity at the lower level, and $\partial/\partial s$ denotes differentiation in the direction of the thermal wind. A positive (negative) result implies cyclonic (anticyclonic) development.

Of the three terms on the right-hand side of the above development equation, the first involves the variation of f (i.e. of latitude) in the direction of the thermal wind and is generally small. The second is the so-called 'development term' and is proportional to the strength of the thermal wind and to its variation of vorticity along its own direction. Contributions to the latter vorticity are made by the curvature and shear of the thermal or THICKNESS pattern, as illustrated, for example, in the CONFLUENT THERMAL RIDGE. The third term is the THERMAL STEERING term which is proportional to the thermal wind and to the variation of surface vorticity in the direction of the thermal wind ($\partial \zeta_0/\partial s$). This term predominates when the pattern of surface vorticity is well marked and the thermal wind almost zonal and without horizontal shear, the first two terms then being small.

deviation: A general term for the difference between an observed value and what is in some sense the true, expected or average value for the place and time, for example STANDARD DEVIATION, deviation of a light ray from a straight path, etc. The angle between the surface wind and the isobars is sometimes termed the 'deviation of the wind'.

dew: Condensation of water vapour on a surface whose temperature is reduced by radiational cooling to below the DEW-POINT of the air in contact with it. Of two recognized processes of dew formation the more common occurs in conditions of calm (wind at 10 metres less than one knot) when water vapour diffuses from the soil upwards to the exposed cooling surface in contact with it (e.g. grass) and there condenses. The second of the processes is one of 'dewfall' when, in conditions of light wind, downward turbulent transfer of water vapour from the atmosphere to the cooled surface occurs. A night of clear skies, moist air and sufficient but not excessive wind is such as gives maximum dewfall rates (about 3 mg/cm² h in the British Isles). Such nights are relatively uncommon and are estimated by J. L. Monteith[*] to produce in the British Isles a total of only about 2·5 to 5·0 mm (0·1 to 0·2 in) dewfall per year. See also DROSOMETER, GUTTATION.

dewbow: A RAINBOW formed on the ground, usually in the shape of a hyperbola, by the REFRACTION and REFLECTION of the sun's rays in dew-drops.

dew-point: The dew-point temperature (T_d) of a moist air sample, commonly termed simply the dew-point, is that temperature to which the air must be cooled in order that it shall be saturated with respect to water at its existing pressure and HUMIDITY MIXING RATIO.

T_d is that temperature for which the saturation VAPOUR PRESSURE with respect to water (e'_w) is identical with the existing vapour pressure (e') of the air, i.e.

$$e' = e'_w \text{ at } T_d.$$

Dew-point may be measured indirectly from wet-bulb and dry-bulb temperature readings with the aid of humidity tables (see PSYCHROMETER), or directly with a 'dew-point HYGROMETER'.

dew-pond: A pond on high ground on chalk downs, artificially constructed with watertight bottom. It is found that, despite the watering of cattle, such ponds retain water during all but the most prolonged droughts, after ponds at lower levels are dried up.

Observation does not support the theory, which gave rise to the name, that such ponds are replenished to a significant extent by night dewfall. The explanation for their persistence probably lies in the relatively large amounts of precipitation that fall at high levels and in the occurrence of FOG PRECIPITATION.

diabatic: A diabatic thermodynamic process is one in which heat enters or leaves the system. Meteorological examples are evaporation, condensation, turbulent mixing, heat conduction, and emission and absorption of radiation. The established equivalent term 'non-adiabatic' is generally preferred because it better emphasizes the nature of the process involved. See also ADIABATIC.

diathermancy: The ability of a substance to transmit radiant energy. Oxygen and nitrogen are diathermanous, while water vapour, ozone and carbon dioxide absorb heat radiation of certain wavelengths; the atmosphere is therefore only partially diathermanous.

dielectric constant: The ratio (ε) of the capacity of a condenser with a given substance as dielectric to that of the same condenser with a vacuum as dielectric is the dielectric constant of the substance.

The dielectric constant is a property of a medium which determines the curvature of the path of an electromagnetic wave through it and so is closely related to the REFRACTIVE INDEX (n) of the medium. For a given substance, ε is a function of temperature and frequency. At frequencies and for a substance in which there is no absorption of the wave energy the relationship between them is $\varepsilon = n^2$. For air, the relatively small departures of ε from unity are determined mainly by the amount of water vapour present.

---

[*] MONTEITH, J. L.; Dew. *Q J R Met Soc, London*, **83**, 1957, pp. 322–341.

differential thermal advection: Differential thermal advection signifies a change in the vertical temperature structure, and so in the static STABILITY, of the air at a particular place, brought about by horizontal ADVECTION of air. Thus, the advection of relatively warm air at lower levels, or of cold air at higher levels, or both, is differential thermal advection of the type that leads to a decrease of static stability. In general, in northern hemisphere temperate latitudes veering (backing) of wind with increase of height signifies warm (cold) air advection in the atmospheric layer concerned.

diffluence: The separation of adjacent STREAMLINES in the direction of flow. See DIVERGENCE.

diffluent thermal ridge: A pattern of THICKNESS lines which is concave towards high thickness and in which the thickness lines separate in the direction of the THERMAL WIND. According to the theorem of DEVELOPMENT, cyclogenesis (*C*) may be expected to occur behind and to the left, anticyclogenesis (*A*) ahead and to the right, of the pattern, as illustrated in Figure 14.

diffluent thermal trough: A pattern of THICKNESS lines which is concave towards low thickness and in which the thickness lines separate in the direction of the THERMAL WIND. According to the theorem of DEVELOPMENT, cyclogenesis (*C*) may be expected to occur ahead and to the left, anticyclogenesis (*A*) behind and to the right, of the pattern, as illustrated in Figure 15.

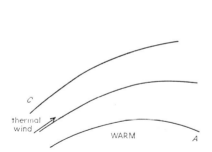

FIGURE 14—Diffluent thermal ridge.

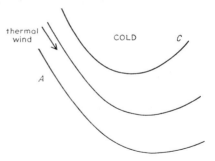

FIGURE 15—Diffluent thermal trough.

diffraction: The bending of light rays produced by an obstacle in the path of the radiation, such as to produce an 'interference' pattern within the geometrical shadow region of the obstacle. The amount of the bending varies with the wavelength; resolution into spectral components therefore occurs in the case of visible light. Diffraction is explained by the wave theory of light, but not on the simple assumption that light travels in straight lines.

The phenomenon is responsible for a number of effects in atmospheric optics—see, for example, BISHOP'S RING, CORONA, GLORY. See also BABINET'S PRINCIPLE.

diffuse front: A FRONT at which the changes in air-mass properties (temperature, humidity, wind velocity) from one side of the front to the other are not concentrated in a narrow belt but are spread through a wide 'frontal zone', which may be up to several hundred kilometres.

diffuse radiation: RADIATION which is received simultaneously from very many directions. In meteorology, it often signifies the solar radiation received (in particular, at the earth's surface) after SCATTERING by atmospheric molecules, cloud, dust, etc., as opposed to the parallel radiation received by direct incidence. It is also termed 'sky radiation'.

diffusion: Molecular diffusion is the process by which contiguous fluids mix slowly, despite differences in their density. The process follows laws similar to those of thermal CONDUCTION (heat diffusion). It is, however, so slow a process in the atmosphere as to be negligible in comparison with the mixing effected by turbulent eddies—'EDDY DIFFUSION'. See EDDY.

Attention is given in the atmospheric problem to the diffusion of such properties as momentum, heat, water vapour, and gaseous or particulate matter, the aim being to relate observed diffusion to the state of the atmosphere, as expressed by the temperature, humidity and wind structure. See also DIFFUSIVITY.

diffusive equilibrium: That state of equilibrium, also termed 'gravitational equilibrium', in which the concentrations of different gases in a mixture of gases change with height according to the molecular weights of the gases involved. Efficient mixing prevents this state from being reached in the atmosphere below a level of about 80 km. See also GRAVITATIONAL SEPARATION.

diffusive separation: See GRAVITATIONAL SEPARATION.

diffusivity: A coefficient (K), of dimensions L^2T^{-1}, which measures the rate of DIFFUSION of a property (E). In the case of simple (Fickian) diffusion, K is defined by the equation

$$\frac{\partial E}{\partial t} = K \nabla^2 E$$

where ∇^2 is the LAPLACIAN.

In meteorology, the term refers to a measure of the rate of diffusion effected by atmospheric eddies, that effected by molecular processes being negligible in comparison. The coefficients involved are termed EXCHANGE COEFFICIENTS. That pertaining to the diffusion of matter is the 'eddy diffusivity', while the exchange coefficients for momentum and heat are the 'eddy viscosity' (see VISCOSITY) and 'EDDY CONDUCTIVITY', respectively. A definition of the coefficients in terms of the Fickian diffusion equation is of limited applicability in meteorology on account of wide space and time variability of the exchange coefficients; the appropriate equation which defines K in these circumstances involves the space co-ordinate l and is

$$\frac{\partial E}{\partial t} = \frac{\partial}{\partial l} \left\{ K(l) \frac{\partial E}{\partial l} \right\}.$$

dimensions (of units): The powers to which the FUNDAMENTAL UNITS of mass (M), length (L), time (T) have to be raised to express fully the units of a physical quantity; e.g. pressure = force/unit area = mass × acceleration/unit area, has the dimensions $MLT^{-2}L^{-2}$, i.e. $ML^{-1}T^{-2}$. A pure number is assigned the dimension unity; temperature is allotted the special dimension θ.

Dines compensation: That property of the atmosphere whereby the sign of the DIVERGENCE is reversed at least once between troposphere and stratosphere. This implies that the integrated divergence from the earth's surface to the top of the atmosphere, and the associated surface pressure tendency, are small residuals of much larger contributions.

This fundamental atmospheric property was inferred by W. H. Dines from his studies of the correlations between atmospheric variables at low and high atmospheric levels. See DEVELOPMENT, PRESSURE TENDENCY EQUATION.

dip, magnetic: Angle of inclination of a magnet with respect to the horizontal. This angle, termed also 'inclination', varies in space and time. Mean values for 1967 were: Lerwick (Shetland Islands) 72° 48·3'; Eskdalemuir (Dumfriesshire) 69° 33·8'; Hartland (Devon) 66° 31·6'. At this epoch, magnetic dip in Great Britain was decreasing at the approximate rates of 0·5' per year in the north and 1·2' in the south.

In so far as the earth's magnetic field resembles that of a dipole, the angle of dip (I) is related to geomagnetic latitude φ (see GEOMAGNETISM) by the relation

$$\tan I = 2 \tan \varphi.$$

direct circulation: A circulation in which the POTENTIAL ENERGY represented by the juxtaposition of relatively dense and light air masses is converted into KINETIC ENERGY as the lighter air rises and the denser air sinks. Land- and sea-breezes are examples of such a circulation. In an 'indirect circulation' the converse holds and the potential energy increases as kinetic energy decreases.

discontinuity: As a rule, the fundamental atmospheric variables of wind velocity, temperature and humidity are continuous functions of space and time, while pressure is invariably so. Occasionally, the space rate of change at a fixed time (or time rate of change at a fixed place) of wind velocity, temperature, humidity and also of pressure tendency is so abnormally great that the distribution of these variables may be regarded as discontinuous. Examples occur at well-marked FRONTS and surfaces of SUBSIDENCE, more particularly, however, in association with SHOCK WAVES.

dishpan experiment: One of a series of experiments, notably by D. Fultz, in which the motions generated in a rotating liquid have been studied as an aid in the understanding of atmospheric motions. The liquid is contained in a cylindrical vessel ('dishpan') which is rotated about a vertical axis; heat is supplied at the walls of the vessel and removed at the centre, simulating the equatorial heat source and polar heat sink, respectively.

dispersion: In physics, the separation of RADIATION into its component wavelengths, due to wavelength dependence of such processes as REFRACTION, DIFFRACTION, and SCATTERING.

In statistics, the 'scatter' of a group of values. The most common measure of dispersion is the STANDARD DEVIATION; other measures are the 'mean deviation' (mean departure from arithmetic mean, independent of sign) and the RANGE of the values.

dissipation trail: An effect, opposite to the CONDENSATION TRAIL, in which the passage of an aircraft through cloud is marked by the appearance of a clear lane. The phenomenon, which is relatively rare, is found under conditions when the effect of the heat of combustion of the fuel released by the aircraft exhaust is sufficient to outweigh that of the released water vapour to the extent of causing tenuous cloud in the wake of the aircraft to evaporate. Occasionally, the effect may be produced by an aircraft flying in relatively dry air, just above a thin cloud layer, by the dragging down and mixing of the dry air with the cloud in the wake of the aircraft. A similar (spurious) effect is sometimes produced when the shadow of a condensation trail is cast on a thin cloud layer.

dissociation: In geophysics, the breakdown of atmospheric molecules into component atoms by the action of ultra-violet radiation in the high atmosphere. The minimum energy required to effect dissociation of a molecule is its 'dissociation energy'.

Dissociation in the atmosphere becomes increasingly important with increase in height above 80 km, mainly in respect of oxygen (molecular to atomic oxygen).

distrail: A common abbreviation for DISSIPATION TRAIL.

disturbance: Sometimes used instead of DEPRESSION or TROUGH of low pressure, especially in broadcast GENERAL INFERENCES.

diurnal variation: The changes of value, for example of a meteorological element, within the course of a (solar) day, more especially the systematic changes that occur within the average day.

A systematic diurnal variation (or 'daily variation') may be revealed by determining the average values, in a large number of days, of the selected element at 0h, 1h, 2h,..., 24h, and removing the non-cyclic change (if any) which is given by the difference between the average values obtained for 0h and 24h. Irregular fluctuations present on individual days are eliminated by averaging over a sufficient number of days.

The systematic diurnal variation of atmospheric pressure is dominated in most latitudes by a 12-hourly oscillation which proceeds according to LOCAL TIME (maxima at 10h and 22h, minima at 4h and 16h local time) and whose amplitude decreases polewards from the equator; at 0° and 50°, for example, approximate values of the swing either side of the mean are 2·0 and 0·7 mb, respectively. In low latitudes the variation is prominent enough to be easily visible on a barogram, even with the near approach of a tropical storm (see Figure 32 under TROPICAL CYCLONE). In middle latitudes the variation is normally obscured by non-periodic pressure changes, but is easily revealed by averaging. In high latitudes the 12-hourly local-time oscillation is very small and the main feature of the diurnal variation is a small 24-hourly UNIVERSAL TIME oscillation.

Temperature and relative humidity have systematic local-time diurnal variations, nearly opposite in phase, and of greater amplitude in summer than in winter. Minimum temperature, for example, occurs at about sunrise, and maximum temperature some two hours in winter, three hours in summer, after noon.

Systematic diurnal variations within the lunar day—'lunar diurnal variations'—are obtained by averaging the magnitudes of elements arranged according to lunar time. Pressure variations of this type have been determined for many parts of the world, and an associated temperature variation in low latitudes has been discovered.

divergence: The divergence of the flux of a quantity (e.g. radiation or momentum) expresses the time rate of depletion of the quantity per unit volume. Negative divergence is termed 'convergence' and relates to the rate of accumulation.

In meteorology, divergence (or convergence) is mostly used in relation to the velocity vector and so refers to the flux of air particles themselves. The 'divergence of velocity' is a three-dimensional property which expresses the time rate of expansion of the air per unit volume. The following relation holds:

$$\text{div}\,\mathbf{V} \text{ (or } \nabla \cdot \mathbf{V}) = \frac{\partial u}{\partial x} + \frac{\partial v}{\partial y} + \frac{\partial w}{\partial z}$$

where u, v, w are the components of the velocity vector **V** in the Cartesian x, y, z directions, respectively.

In the atmosphere, div **V** is small and is of little direct interest. The main concern is with the 'horizontal divergence of velocity' (often referred to as the 'horizontal divergence' or even, misleadingly, the 'divergence') which, denoted $\text{div}_H \mathbf{V}$ or $\text{div}_2 \mathbf{V}$, etc., is identified with the sum $(\partial u/\partial x + \partial v/\partial y)$, and expresses the time rate of horizontal expansion of the air per unit area.

Since div **V** ≈ 0,

$$\text{div}_H \mathbf{V} \equiv \frac{\partial u}{\partial x} + \frac{\partial v}{\partial y} = -\frac{\partial w}{\partial z},$$

i.e. horizontal divergence of air is closely associated with vertical contraction (usually termed 'vertical convergence') of an air column centred at the level concerned. Similarly, horizontal convergence is linked with 'vertical divergence' $(\partial w/\partial z)$.

Contributions to $\text{div}_H \mathbf{V}$ may be made by 'diffluence' (separation of streamlines in the direction of flow) and by increase of wind downstream. Since, however, diffluence is generally associated with decrease of wind downstream (and 'confluence' with increase of wind downstream), the respective contributions to $\text{div}_H \mathbf{V}$ are generally of opposite sign. Therefore, $\text{div}_H \mathbf{V}$ is a small remainder which is difficult to determine, even as to sign, in the free atmosphere from the substitution of wind observations in the expression $(\partial u/\partial x + \partial v/\partial y)$. It is more accurately determined from the VORTICITY EQUATION and from the distribution of vertical motion.

The dimensions of div **V** are T^{-1}. Values of $\text{div}_H \mathbf{V}$ in large-scale motions in the free atmosphere generally range from zero up to about 10^{-5} s^{-1}. (It can be considerably greater locally in frontal zones.) On average, divergence (or convergence) is least at a level of about 600 mb—the so-called 'level of non-divergence'.

D-layer: That part of the IONOSPHERE, situated between about 65 and 80 km, which is mainly responsible for the absorption of radio energy reflected from higher levels, but from which radio waves of very low frequency may also be reflected. It is sometimes referred to only as the 'D-region' because it has apparently no peak value of electron concentration. The extra ionization caused at this level by SOLAR FLARE radiation gives rise to various types of 'sudden ionospheric disturbance' (see SID).

Dobson spectrophotometer: An instrument used in the routine measurement of atmospheric OZONE. The method consists of isolating, in the solar radiation spectrum, two wavelengths in the region of partial ozone absorption in the near ultraviolet (0·30 to 0·33 μm), allowing them to fall in rapid succession on a photomultiplier connected through an a.c. amplifier and rectifier to a galvanometer, and reducing, by means of a calibrated optical wedge, the intensity of the longer, less-absorbed wave until the photomultiplier outputs are equal and the galvanometer records no current. The amount of atmospheric ozone in a vertical column is obtained from the position of the wedge and the solar zenith angle at the time of observation.

Incorporated in the calibration of the instrument is the ratio of the intensities of the selected wavelengths outside the atmosphere. This is obtained, once for all, from a series of instrumental observations at various solar zenith angles and extrapolation to zero path length through the atmosphere. The extrapolation is based on the assumption, supported by photochemical theory, that there is little or no systematic diurnal variation of total ozone.

The standard measurements are those made against direct sunlight. Measurements are also made against the zenith sky, clear or cloudy, and against moonlight, using empirical relations with the direct sunlight observations.

The instrument may be used to obtain the vertical distribution of ozone, using the UMKEHR METHOD.

doctor: A popular name given to the HARMATTAN of West Africa. The term is also used for the KATABATIC WIND that blows most evenings from the Blue Mountains over the Liquanea Plain in Jamaica.

doldrums: The equatorial oceanic regions of light variable (mainly westerly) winds, accompanied by heavy rains, thunderstorms and squalls. These belts, variable in position and extent, have a systematic north and south movement some 5° either side of a mean position, following the sun with a lag of one or two months.

Doppler effect: The observed frequency of a source moving with velocity v towards an observer, and emitting waves (e.g. radio, sound or light) of wavelength λ, is greater than the emitted frequency f by the amount $\delta f = v/\lambda$—the 'Doppler effect'. For a receding source the observed frequency is less than f by the same amount. If, on average, emitting sources approach and recede from an observer in equal numbers, 'Doppler broadening', without shift, is observed. Frequency and spectral measurements may thus be employed to measure the velocity of fast-moving sources.

The Doppler-shift technique may be used to obtain a measure of the vertical velocities of precipitation elements. The frequency shift suffered by a radio wave on reflection from moving objects ($f = v/\lambda$) gives a measure of the mean 'line of sight' velocity of the elements, while variations of the intensity of the radar echo may be used to infer motions possessed by the various elements relative to one another.

Doppler radar: If a RADAR target is moving towards or away from the radio receiver, the frequency of a reflected signal is, because of the DOPPLER EFFECT, slightly different from the transmitted frequency. Doppler radar is designed to interpret this effect in terms of the radial velocities of targets. It is employed in meteorology to deduce various types of information: for example, horizontal air motion at various heights in a precipitation region, the speeds of fall of precipitation particles and, on certain assumptions, vertical air motion within precipitation regions.

drag coefficient: A non-dimensional coefficient (C_D), also termed the 'skin-friction coefficient', which is defined by the equation $\tau = C_D \rho \bar{U}_s^2$, where τ is the REYNOLDS STRESS (surface shearing stress), ρ the air density, and $\bar{U}_s$ the wind speed observed near the surface.

C_D is a conventional, as opposed to a unique, parameter of surface roughness. Its value depends on the height at which wind speed is observed and also on the stability of the air. Wind speed observed at a height of 1 or 2 m is generally employed.

drainage area: The area whose surface directs water towards a stream above a given point on that stream. See also CATCHMENT AREA and GATHERING GROUND.

drainage gauge: See PERCOLATION.

drainage wind: An alternative for KATABATIC WIND.

drift: In oceanography, the velocity of an ocean current.

drifting dust or sand: 'Dust or sand, raised by the wind to small heights above the ground. The visibility is not sensibly diminished at eye level.'*

drizzle: Liquid precipitation in the form of water drops of very small size (by convention, with diameter between about 200 and 500 μm, i.e. 0·2 and 0·5 mm).
Drizzle forms by COALESCENCE of droplets of stratus cloud. It falls from stratus cloud of low base, in which the widespread upcurrents which form the cloud have a velocity smaller than the terminal velocity of the drops concerned. High relative humidity below the cloud base is also required to prevent the drops from evaporating before reaching the earth's surface.
For synoptic purposes, drizzle is classified as 'slight', 'moderate', or 'heavy': slight drizzle corresponds to negligible run-off from roofs, heavy drizzle to a rate of accumulation greater than 1 mm/h.

drop, droplet: Terms generally applied in meteorology to the spherical particles of water which constitute liquid PRECIPITATION and CLOUD elements, respectively. The differentiation, while not precise, is essentially one of size, the limiting diameter being about 200 μm (0·2 mm).

drosometer: An instrument for measuring the amount of DEW deposit. In the Duvdevani dew-gauge, the size, form and distribution of dew-drops deposited overnight on a rectangular block, carrying a special paint and exposed at a standard height, are compared with standard photographs relating to a dew deposit of known amount. In other gauges, some of them recording, dew deposited (e.g. on a hygroscopic surface) is obtained by weighing. A basic uncertainty in all cases is the relationship between the amount of dew deposited on the gauge and that deposited on natural surfaces.

drought: Dryness due to lack of RAINFALL. Drought and partial drought were defined in *British Rainfall* 1887† in order to obtain comparable statistical information for inclusion in each annual volume of this publication. An 'absolute drought' was a period of at least 15 consecutive days to none of which was credited 0·2 mm, or 0·01 in, or more of rainfall. A 'partial drought' was a period of at least 29 consecutive days the mean daily rainfall of which did not exceed 0·2 mm, or 0·01 in. A 'dry spell' was first defined and used in *British Rainfall* 1919, a dry spell being a period of at least 15 consecutive days to none of which was credited 1·0 mm, 0·04 in, or more of rainfall. The occurrences of partial droughts at selected stations were tabulated in *British Rainfall* until 1957, and absolute droughts and dry spells until 1960. A new method of presenting rainfall deficiencies and excesses was introduced with the volume for 1961.

dry adiabatic (or dry adiabat): Line on an AEROLOGICAL DIAGRAM representing the dry adiabatic lapse rate. See ADIABATIC.

dry air: In physical meteorology this term generally signifies air which is completely dry, i.e. which contains no water vapour. In synoptic meteorology and climatology the term usually refers to air of low RELATIVE HUMIDITY. In the BEAUFORT NOTATION the letter 'y' signifies 'dry air', precisely defined as having a relative humidity of less than 60 per cent.

dry- and wet-bulb hygrometer: An alternative for PSYCHROMETER.

---

* Geneva, World Meteorological Organization. International cloud atlas, Vol. 1. Geneva, WMO, 1956, p. 71.
† SYMONS, G. I.; *British Rainfall*, 1887, p. 15.

dry ice: A common term for solid carbon dioxide, frequently used in CLOUD SEEDING experiments. The substance is generally discharged from an aircraft into supercooled cloud in particles of about 10 mm diameter at the rate of a few kilogrammes per kilometre. The particles vaporize at a temperature of $-78 \cdot 5°C$. The local cooling of the ambient air effected by the sublimation process produces myriads of minute ice crystals, which may then act as ice nucleating agents.

dry season: A period of a month or more, recurring every year, which is marked in a given region (generally tropical or subtropical) by the complete or almost complete absence of precipitation. Winter is the dry season in most tropical regions; summer is, however, the dry season in the MEDITERRANEAN-TYPE CLIMATE.

dry spell: See DROUGHT.

duplicatus (du): One of the CLOUD VARIETIES (*Latin, duplicatus* doubled).
'Superposed cloud patches, sheets or layers, at slightly different levels, sometimes partly merged. This term applies mainly to CIRRUS, CIRROSTRATUS, ALTOCUMULUS, ALTOSTRATUS and STRATOCUMULUS'.* See also CLOUD CLASSIFICATION.

dust: The atmosphere carries in suspension, often for long distances, solid particles of varying character and size. The chief sources of these particles are volcanic eruptions, meteors, dust and sand raised by winds, and the smoke produced in industrial and domestic combustion processes and in forest fires.

An important meteorological effect of atmospheric dust is its depletion of solar radiation by scattering and, to a smaller degree, by absorption. When present in appreciable quantity it gives atmospheric HAZE. Most dust particles are of a size sufficiently small to cause differential SCATTERING of sunlight and so produce, for example, highly coloured sunsets—see SUNRISE AND SUNSET COLOURS.

The concentration and size distribution of solid particles suspended in the atmosphere, and their effectiveness as condensation nuclei, are discussed under NUCLEUS.

dust counter: An instrument for counting the dust particles in a known volume of air. In Aitken's dust counter, condensation is made to occur on the nuclei present by ADIABATIC expansion of air, and the number of drops is ascertained. In Owen's dust counter, a jet of damp air is forced through a narrow slit in front of a microscope cover glass. The fall of pressure due to expansion of the air after passing the slit causes the formation of a film of moisture on the glass, to which dust adheres, forming a record which can be studied under a microscope.

dust, or sand, haze: A suspension in the air of dust or small sand particles, raised from the ground prior to the time of observation by a duststorm or SANDSTORM.

duststorm: See SANDSTORM, DUSTSTORM.

dust whirl (or devil): A WHIRLWIND, in which dust and sand are carried aloft from the ground by very strong convection from a hot, sandy region. The rotation may be in either direction round the centre, which is itself often free from dust. Heights of about a kilometre have been reported but are generally less than 30 m. Speeds at which dust whirls move vary from less than 5 knots to over 25 knots. The phenomenon is sometimes termed 'sand pillar'.

---

* Geneva, World Meteorological Organization, International cloud atlas, Vol. 1. Geneva, WMO, 1956, p. 16.

D-value: The difference, D, between the actual height, Z above mean sea level, of a particular pressure surface, and the PRESSURE ALTITUDE, Z_p (the height of the same pressure surface in the ICAO STANDARD ATMOSPHERE), i.e. $D = Z - Z_p$.

A system of analysis based on the cross-section presentation of D-values is a method which well illustrates the vertical structure of pressure systems.

dynamical forecast: See NUMERICAL WEATHER FORECAST.

dynamical meteorology: The study of the causes and nature of motion of the atmosphere.

dynamic heating: The frictional heating experienced by a body due to its rapid motion through the air.

dynamic metre: See GEOPOTENTIAL.

dynamic pressure: In meteorology, the force exerted by the wind on unit area of the windward face of a surface. At points where the air is brought to rest relative to the surface it is given by the quantity $\frac{1}{2}\rho V^2$, where ρ is air density and V the relative wind velocity.

dynamic stability: A term sometimes used, in a restricted sense, as a synonym of INERTIA STABILITY. More generally this term, or the alternative 'hydrodynamic stability', signifies an atmospheric state characterized by there being no tendency for small wave-like perturbations of the flow to grow. Various influences operate in the atmosphere to cause the contrary state of (hydro)dynamic instability, e.g. INERTIA, GRAVITY, WIND SHEAR, or a strongly BAROCLINIC atmosphere. See also STABILITY.

dynamic temperature change: A change associated with the compression (warming) or expansion (cooling) of a gas. See ADIABATIC.

dynamo theory: See GEOMAGNETISM.

dyne: The unit of force in the C.G.S. SYSTEM of units. It is the force that produces an acceleration of 1 cm/s² when applied to a mass of 1 g. The dimensions are MLT^{-2}.

$$1 \text{ dyne} = 10^{-5} \text{ NEWTONS}.$$

E

earth: The earth is a spheroid which is somewhat flattened at the poles relative to the equator because of rotation about its axis—the so-called 'equatorial bulge'. It is of mean radius 6371·229 km, of angular velocity $7·292 \times 10^{-5}$ rad/s, of mean distance from the sun $1·4968 \times 10^8$ km, of mass $5·975 \times 10^{24}$ kg and of mean density relative to water 5·5.

The earth's interior comprises a thin outer 'crust' of rocks which is part of a solid 'mantle' extending to a depth of about 3000 km; below this is probably a metallic fluid 'outer core' of thickness 2200 km and, near the earth's centre, an 'inner core' of radius 1300 km which is probably solid. The estimated temperature at the earth's centre is in the range 2000 to 10 000K and the estimated pressure of the order 4×10^6 atmospheres.

The respective percentages of water and land on the earth's surface are: in the northern hemisphere 60·7 and 39·3; in the southern hemisphere 80·9 and 19·1; for the earth as a whole 70·8 and 29·2.

earth currents: The term 'earth currents' (or 'telluric currents') is generally taken to comprise the world-wide systems of electric currents (with large local differences) which flow in the earth's crust and display both systematic and irregular variations which are closely related to variations of the earth's magnetic field (see GEOMAGNETISM) and to allied phenomena. On some occasions the irregular variations are large enough to interrupt land-line communications. Such small and local currents as are associated with chemical effects in the earth's surface and the large but short-lived current surges associated with lightning flashes are excluded from consideration.

Systematic recording of earth currents is made in terms of the potential gradient between pairs of electrodes which are generally separated by a distance of at least 2 km. A typical 'quiet' value is about 0·2 volt/km, and a 'disturbed' value upwards of 1 volt/km. Precise measurements are difficult because of the difficulties of electrode contact potential, and insulation, etc. Determination of earth resistivity (found to vary over a wide range with type of soil or rock and to vary greatly also with moisture content) allows, in conjunction with the potential-gradient measurements, the earth-current density to be calculated.

The earth potential gradient undergoes regular solar and lunar diurnal variations, the lunar variations having about one-fifth the range of the solar. The latter show considerable differences in type and amplitude at different stations; two maxima and minima occur, however, at most places. The range of the solar variation is least in winter and greatest near the equinoxes.

Earth-current and magnetic storms accord closely in time and agree well in their general form. The systematic slower diurnal variations of earth currents are also related to magnetic variations, but in a more complex way. The cause of both phenomena lies in electrical effects in the high atmosphere. Complications are introduced, however, by local differences in the conductivity of the earth's crust, by differences in speed of the high atmospheric variations on different occasions, and by reactions of the earth currents themselves on the recorded magnetic variations.

earthlight: See MOON.

earthquake: A natural movement of the ground, originating below the surface, due to the build-up and sudden release of strain energy in a small underground region, termed the 'focus' or 'hypocentre' of the earthquake. The point of the earth's surface vertically above the focus is termed the 'epicentre'. The focus of most earthquakes is at a depth of less than 20 km. Earthquakes of much deeper focus (up to 700 km depth) occur in some areas, more especially in the Pacific.

About 100 destructive earthquakes occur each year; some damage is done by ten times as many, while many more are felt by human beings and still more are recorded by seismographs.

Major earthquakes are generally followed, over a period, by 'aftershocks' whose foci are in the region of the original focus. No practical method of prediction of earthquakes has yet been devised. See also SEISMOLOGY.

earth temperature: If the temperature of the earth's surface were constant, heat would be conducted continually upwards from the earth's very hot interior to the earth's surface and there dissipated, mainly by radiation; a steady state would be reached, with temperature increasing with depth at an almost steady rate. This simple régime is much changed in the top layer of the earth by the fact that the earth's surface is raised, by absorption of solar radiation, to a higher temperature than that appropriate to the above conditions. The diurnal and annual temperature waves observed at the surface proceed downwards by heat conduction, but with rapid reduction of amplitude and progressive increase of time-lag relative to the surface waves. The diurnal wave extends to a depth of about 0·5 m and the annual wave extends to a depth of about 10 m; these values are approximately as the square root of the ratio of the periods (1 and 365 days, respectively), in accordance with the theory of heat conduction in solids.

earth thermometer: A thermometer for measuring the temperature of the ground at a known depth. The commonest form (Symons's) consists of a mercury thermometer with its bulb embedded in micro-crystalline wax, suspended in a steel tube at a standard depth, usually 30, 50, 100, 150 and 300 cm, although in the United Kingdom measurements are still made at 122 cm (4 feet). For depths of a few centimetres only, a mercury thermometer, with its stem bent at right angles for convenience in reading, is used; this type is termed a bent-stem soil thermometer.

Permanently buried mercury-in-steel thermometers have proved reliable and convenient as earth thermometers. Such thermometers have 5 m or more of steel capillary tubing and display their readings either on dials which may be placed at eye level or as recordings on a chart. Both bulb and capillary are enclosed in stainless steel and may thus be left undisturbed for many years.

easterly wave: A shallow trough disturbance in the easterly current of the tropics, more in evidence in the upper-level winds than in surface pressure, whose passage westwards is followed by a marked intensification of cloudy, showery weather.

ecliptic: That GREAT CIRCLE on the CELESTIAL SPHERE which the sun appears to describe, with the earth as centre, in the course of a year; the plane in which the apparent orbit of the sun lies is the 'plane of the ecliptic'.

In terms of actual motion, the ecliptic is the path described by the earth round the sun in the course of a year.

The 'obliquity of the ecliptic' is the inclination of the plane of the ecliptic to that of the earth's equator. This value controls the limits of variation of the sun's DECLINATION in the course of the year and gives rise to the meteorological seasons. The obliquity of the ecliptic was close to 23° 26′ 40″ in 1960 and was then decreasing at the rate of about 0·47″ per year. The limits of variation of the value of the obliquity and the period or periods involved are yet uncertain but are very probably significant factors in long-period CLIMATIC CHANGE.

ecoclimatology: That branch of ecology which is concerned with the study of living matter (animal and plant) in relationship to its climatic environment.

eddy: A term in fluid motion for which, like that of the closely associated term 'turbulence' a brief comprehensive definition is impossible. Its essential characteristics are, however, covered by its definition as a mass of fluid which retains its identity for a limited time while moving within the surrounding fluid.

Atmospheric eddies range in size from less than one centimetre (microscale turbulence) to some hundreds of kilometres (large depression or anticyclone) or more. The smaller-scale eddies play a vital part in effecting vertical mixing of such atmospheric properties as momentum, heat and water vapour ('eddy flux'), while the large-scale eddies are responsible for much of the meridional transport implicit in the GENERAL CIRCULATION of the atmosphere. See, for example, TURBULENCE, VISCOSITY, DIFFUSIVITY, EDDY SPECTRUM.

eddy conductivity: EXCHANGE COEFFICIENT (K), of dimensions L^2T^{-1}, relating to the transfer of heat effected by eddies. For transfer in the vertical, for example, the coefficient may be regarded as being defined by the equation

$$- \overline{w'T'} = K_z \frac{\partial \bar{T}}{\partial z}$$

where w' and T' are corresponding fluctuations of vertical velocity and temperature from mean values. K varies greatly in space and time.

See, for example, EDDY, DIFFUSIVITY, VISCOSITY, K-THEORY.

eddy diffusion: The mixing of atmospheric matter and properties which is effected by eddies. It is also termed 'turbulent diffusion'. See EDDY, DIFFUSION, DIFFUSIVITY.

eddy flux: The rate of transport of atmospheric properties and matter (heat, momentum, water vapour, etc.) which is effected by atmospheric eddies. Frequently the vertical transport is implied. It is also termed 'turbulent flux'. See EDDY.

eddy shearing stress: An alternative for REYNOLDS STRESS.

eddy spectrum: Specification of the character of TURBULENCE in terms of the partition of kinetic energy between eddies of various sizes. It is normally obtained by correlogram or Fourier analysis of the time fluctuations of motion observed at a particular point. It is also termed 'turbulence spectrum', 'energy spectrum' (of turbulence), or 'power spectrum' (of turbelance).

effective height of an anemometer: The height over open level terrain in the vicinity of an anemometer which, it is estimated, would have the same mean wind speeds as those actually recorded by the anemometer.

No definite rules are laid down for obtaining the effective height of an anemometer. The nature, extent, height and distance of local obstructions and the height of the anemometer itself must be taken into account. The effective height may be different for different wind directions. See also EXPOSURE.

effective radiation: An alternative for NOCTURNAL RADIATION.

effective temperature: As defined by the American Society of Heating and Ventilating Engineers, the temperature of saturated motionless air which would produce the same sensation of warmth or coolness as that produced by the temperature, humidity and air motion under observation.

The definition implies that effective temperature is dependent on the amount of clothing worn. Nomograms which relate effective temperature to dry-bulb and wet-bulb temperatures and wind speed, for two categories of clothing, have been published by the Air Ministry.* See also COMFORT ZONE.

Ekman spiral: An equiangular spiral which is the locus of the end points of the (idealized) wind vectors, starting from a common origin, within the FRICTION LAYER. The centre of the spiral corresponds to the GEOSTROPHIC WIND, while the surface wind is backed 45° from this direction (see Figure 16). Such a spiral, first derived by V. W. Ekman for the ocean currents produced by surface stress, is, however, obtained under assumed conditions that are rarely realized in the atmosphere—namely, no variation of eddy viscosity with height and no horizontal temperature gradient, within the friction layer.

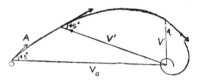

FIGURE 16—Equiangular spiral such that the ageostrophic wind component (V') makes an angle of 45° with the tangent at all points of the curve.
    V' decreases exponentially to zero with increasing height within the friction layer. V_G represents the geostrophic wind and OA the direction of the surface wind.

E-layer: The lowest regularly observed layer of the IONOSPHERE occurring at a height of about 100 to 120 km, and observable in normal ionospheric sounding mainly during the day. The maximum electron density is of the order 10^5 per cm³. The structure of the E-region is at times complicated by the appearance of various 'ledges' of ionization and of intense 'sporadic E' (E_s) ionization which exists in extensive patches and gives rise to anomalous reception of high-frequency radio waves.

electrical units: The relationship between the practical (m.k.s.), electromagnetic, and electrostatic systems of electrical units is shown in Table III.

TABLE III—*Relationship between electrical units.*

	Practical m.k.s. and SI units	Electro-magnetic c.g.s.	Electro-static c.g.s.
Quantity of electricity	coulomb	10^{-1}	3×10^9
Electric current	ampere		
Potential	volt	10^8	1/300
Resistance	ohm	10^9	$1/(9 \times 10^{11})$
Induction	henry		
Energy	joule	10^7	10^7
Power	watt		
Capacity	farad	10^{-9}	9×10^{11}

* London, Air Ministry. Handbook of preventive medicine, 2nd edn. London, HMSO, 1959.

electricity: See ATMOSPHERIC ELECTRICITY.

electricity of precipitation: The main features of surface measurements of the electric charge carried by solid and liquid PRECIPITATION are as follows:
  (i) Charges of both signs are carried by all kinds of precipitation.
  (ii) Rapid fluctuations of sign occur within short intervals of time.
  (iii) A net positive charge is, on average, carried to ground by all types of precipitation.
  (iv) The charges carried per cubic centimetre of precipitation and by individual drops are of the general order 3×10^{-10} and 3×10^{-7} coulombs, respectively.
  (v) Thunderstorm-type rain and snow carry, on average, larger charges than steady rain.

These observations, together with the results of similar measurements made from aircraft, show that the processes involved are highly complex. As with the generation and separation of charges associated with a THUNDERSTORM, various theories have been advanced to explain the charges carried by precipitation. They include: effects associated with a mixture of water in the solid and liquid phases; the separation of charge on the break-up of large raindrops; and the selective capture by precipitation particles of positive ions which stream upwards by BRUSH DISCHARGE from the ground in the disturbed surface POTENTIAL GRADIENT conditions which obtain in the vicinity of rain clouds. The complexity of the observed effects makes it probable that several processes are important.

electric storm: Little-used alternative for a THUNDERSTORM.

electrojet: The term applied to the narrow belt of intense 'dynamo' electric currents which flow in the low ionosphere near the equator and which produce at the earth's surface in low latitudes a large augmentation of the solar and lunar diurnal variations of the magnetic elements, compared with other latitudes. See GEOMAGNETISM.

electromagnetic radiation: The propagation of energy in which, according to classical 'electromagnetic theory', a transverse wave motion exists in the form of periodic fluctuations of the strengths of electric and magnetic fields which act at right angles to the direction of propagation of the energy. The energy may be propagated through a medium or vacuum; the speed of propagation in the latter, and to a close approximation in air, is about 3×10^8 m/s.

Electromagnetic radiation is divided into various classes which differ only in wavelength. In order of increasing wavelength (decreasing frequency) they comprise GAMMA RADIATION, X-RAYS, ULTRA-VIOLET RADIATION, VISIBLE RADIATION, INFRA-RED RADIATION, and RADIO WAVES. See also RADIATION.

electrometeor: A little-used generic term for a visible or audible manifestation of atmospheric electricity, e.g. THUNDER, LIGHTNING, ST ELMO'S FIRE, AURORA.

electrometer: An instrument with a high input impedance (of the order of 10^9 to 10^{11} ohms) for measuring electric potential difference. The quadrant electrometer has largely been superseded by versatile modern electronic instruments which can measure voltages of one millivolt while drawing less than 10^{-15} A of current.

electron: The elementary particle that carries the indivisible (unit) negative electric charge of $4 \cdot 803 \times 10^{-10}$ electrostatic units ($1 \cdot 6 \times 10^{-19}$ coulombs).

electronvolt: The electronvolt (eV) is the unit of energy employed in discussing processes of ION formation; it is the energy acquired by an electron on passing through a potential rise of 1 volt. $1 \text{ eV} = 1 \cdot 6 \times 10^{-19}$ J.

electroscope: An instrument for indicating the presence of electrification, usually by some simple action, such as the mutual repulsion of two strips of foil.

elevation: In meteorology and in aviation, term used to denote the height of the ground above mean sea level.

elevation, angle of: The angular height of an object measured by an observer, with reference to the horizontal plane through the observer.

Elsasser's diagram: See RADIATION CHART.

emagram: A thermodynamic diagram used to depict conditions in the upper atmosphere, the name being a contraction for 'energy-per-unit-mass diagram'. The co-ordinate axes (rectangular or oblique) are T and $\log p$. See AEROLOGICAL DIAGRAM.

energy: A quantity, of dimensions ML^2T^{-2}, defined as the capacity for doing work. The unit of energy in the C.G.S. SYSTEM is the ERG, in the M.K.S. SYSTEM and SI UNITS the JOULE.
The various forms of energy include, for example, POTENTIAL ENERGY, KINETIC ENERGY, and heat, radiant, electric, magnetic and chemical energy.

energy balance: In meteorology, a concept generally applied in the form of an 'energy balance equation' which relates the net radiation flux at a portion of the earth's surface to the heat lost or gained by conduction to or from below, the heat lost or gained to or from the atmosphere by molecular and eddy processes, and the heat lost or gained at the surface by evaporation or condensation, respectively. See also RADIATION BALANCE.

energy spectrum: See EDDY SPECTRUM.

enthalpy: A thermodynamic quantity (H) which represents the 'total heat' content per unit mass of a substance; the units normally employed are joules per kilogramme.
A change in enthalpy (dH) of a mass of gas is the heat gained or lost by the gas in an isobaric process. It is given by $dH = c_p dT$, where c_p is the specific heat of the gas at constant pressure, and dT is the change in temperature of the gas. The isobaric transport of enthalpy is an important item in the over-all heat balance of the atmosphere.
In recent meteorological literature enthalpy is frequently referred to as 'sensible heat' as opposed to latent ('hidden') heat. See also HEAT.

entrainment: A term applied to the mixing of environment air into the updraught of a cumuliform cloud. The effects of such mixing (and also of the mixing into the cloud of environment air trapped by the rising cloud turrets) on the temperature, moisture and momentum properties of the cloud are progressively spread through the cloud. This general mixing process is a significant factor which is not taken into account in the PARCEL and SLICE METHODS of deducing the kinetic energy available in an atmosphere in which there is static instability.

entropy: If, in a reversible thermodynamic process, a substance absorbs a quantity of heat dQ at absolute temperature T, the ratio dQ/T represents the increase of entropy of the substance. Entropy per unit mass is normally measured in joules per kilogramme per kelvin and has the dimensions $L^2T^{-2}\theta^{-1}$.
Entropy is a function of the pressure, volume, and temperature of the substance but requires for its evaluation the arbitrary choice of a state of zero entropy.

Interest in the atmosphere is mainly confined to the changes of entropy to which air is subject in the course of a specified process. Of chief interest is the ADIABATIC process in which no heat is supplied to, or withdrawn from, the air ($dQ = 0$) and in which, therefore, the entropy (S) remains constant; such a process is called 'isentropic' and is characterized by a constant POTENTIAL TEMPERATURE (θ) since S and θ are related by

$$S = c_p \log \theta + \text{constant}$$

where c_p is the specific heat of air at constant pressure.

epicentre: That point of the earth's surface which is vertically above the focal region of an EARTHQUAKE from which the seismic waves originate.

equation of state: See GAS EQUATION.

equation of time: See TIME.

equations of motion: See MOTION, EQUATIONS OF.

equator: The earth's equator is that GREAT CIRCLE whose plane is perpendicular to the earth's axis of rotation (polar axis).
The celestial equator is that great circle whose plane is perpendicular to the line joining the celestial POLES.

equatorial air: The AIR MASS participating in the equatorial low-latitude circulation. Originating in the subtropical anticyclones, the air generally becomes moist and unstable when it stagnates in the DOLDRUMS region. See also TRADE WINDS.

equatorial bulge: The slight bulging of the EARTH in low relative to high latitudes. The earth's equatorial diameter is about 43 km greater than its polar diameter. As a result of the bulging the force of GRAVITY varies with latitude.

equatorial front: An alternative for INTERTROPICAL CONVERGENCE ZONE.

equatorial trough: The shallow TROUGH of low pressure, generally situated on or near the equator, marking the convergence zone of air which moves equatorwards from the subtropical anticyclones of either hemisphere. The trough over the oceans lies in the belt of the DOLDRUMS and has a north and south movement which follows the sun with a time-lag of one or two months. The trough is not, however, a permanent feature of the synoptic chart in all longitudes at all times of the year; in particular, it is generally absent over land areas in the northern hemisphere summer.

equatorial westerlies: The north-east or south-east TRADE WINDS are deflected on crossing the equator and acquire a westerly component to become the 'equatorial westerlies' when the INTERTROPICAL CONVERGENCE ZONE is farther than some 5° from the equator.
Over the eastern Indian Ocean and the Malaysian region westerly winds are found near the equator throughout the year. These are called equatorial westerlies although the turning of the wind usually begins before crossing, or without crossing, the equator.

equinoctial gales: The implication contained in this term, which is in fairly wide popular use, that GALES are either more frequent or more severe near the EQUINOXES than at other times is not supported by observations. In all parts of the British Isles, for example, the peak frequency of moderate or severe gales is near the winter solstice and the minimum frequency near the summer solstice (see Table VI, p. 118).

equinox: One of two periods of the year, which occur on about 21 March (spring or vernal equinox) and about 22 September (autumnal equinox), when the astronomical day and night are equal, each lasting 12 hours. (The sun is visible by refraction for a little longer than the duration of the astronomical day.)

At an equinox, when the sun is said to be 'on the equator', time of sunrise (or of sunset) is the same all along a MERIDIAN. Apart from the small refraction effect, the sun then rises exactly in the east and sets exactly in the west all over the world.

equiscalar surface: In meteorology, a surface along which a specified scalar quantity, e.g. pressure or temperature, is constant.

equivalent constant wind: An alternative for BALLISTIC WIND.

equivalent head wind: That uniform wind which, directed at all points along the track of an aircraft, results in the same average ground speed as that actually attained. A positive head wind is such as to oppose the flight of an aircraft. A negative head wind comprises a 'tail wind', i.e. groundspeed in excess of airspeed. The term 'equivalent tail wind' is also used.

equivalent potential temperature: See EQUIVALENT TEMPERATURE.

equivalent temperature: The (isobaric) equivalent temperature (T_e) of a moist air sample is the temperature that would be attained on the assumption of condensation at constant pressure of all the water vapour in the sample, all the latent heat released in the condensation being used to raise the temperature of the sample. T_e is given by the equation

$$T_e = T + Lr/c_p$$

where T is the dry-bulb temperature and r the mixing ratio of the sample, L the latent heat of condensation, and c_p the specific heat at constant pressure. Where r is expressed in grammes per kilogramme,

$$T_e \approx T + 2 \cdot 5r.$$

The 'equivalent potential temperature' (θ_e) is found on an AEROLOGICAL DIAGRAM by progressing along the dry adiabatic line from T_e to the 1000-mb level. See also PSEUDO-EQUIVALENT TEMPERATURE.

erg: The unit of work or energy in the C.G.S. SYSTEM of units. It is the work done by a force of 1 DYNE in moving its point of application 1 cm in the direction of the force. The dimensions are ML^2T^{-2}.

$$1 \text{ erg} = 10^{-7} \text{ JOULE} = 10^{-7} \text{ WATT second}.$$

error: An error of observation is the departure of a measured quantity from its true value. Such an error is in general partly 'systematic' and partly casual or 'random' in nature. The systematic component which may arise, for example, through errors of the scale or zero setting of the instrument, or from the PERSONAL EQUATION of the observer, can often be recognized (and corrected) by comparing averaged measurements made with the instrument in question with similar measurements made with an instrument of known accuracy, while the random component of error can be reduced, in proportion to the square root of the number of observations, by averaging several independent measurements of the same quantity.

error function: See NORMAL (FREQUENCY) DISTRIBUTION.

escape velocity: The minimum velocity which a molecular or atomic gas particle must attain in order that it may escape from the gravitational field of a specified planet and so from the planetary atmosphere. Such escape can take place only from the outer fringe (EXOSPHERE) of the atmosphere, where the particle MEAN FREE PATH is large and the probability of its collision with another particle correspondingly small.

The escape velocity, v m/s, is given by

$$v = \sqrt{\left(\frac{2GM}{a}\right)}$$

where G = constant of gravitation = 6.67×10^{-11} SI units
M = mass of planet (kg)
a = distance of level of escape from centre of planet (m).
For the earth, $v \approx 11.2$ km/s; for the moon, $v \approx 2.4$ km/s.

etesian winds: A Greek term for the winds which blow at times in summer (May to September) from a direction between north-east and north-west in the eastern Mediterranean, more especially in the Aegean Sea. The winds are termed 'meltemi' in Turkey.

Eulerian change: The time rate of change of an element at a fixed point in an Eulerian system of co-ordinates, i.e. the 'local change', designated $\partial/\partial t$. It is to be contrasted with the LAGRANGIAN CHANGE to which it is related by the velocity components and the gradients of the elements in the various component directions.

Eulerian wind: That class of winds, in the classification of H. Jeffreys, in which the earth's rotational term (involving the CORIOLIS FORCE) and the frictional term are unimportant relative to the acceleration term. The EQUATION OF MOTION reduces to

$$\frac{d\mathbf{V}}{dt} = -\frac{1}{\rho}\nabla_H p$$

where $\mathbf{V}$ is the wind velocity, ρ the air density, and $\nabla_H p$ the horizontal gradient of pressure.

The CYCLOSTROPHIC WIND of the tornado or tropical cyclone is an example of an Eulerian wind.

eustasy, glacial: The release or absorption of water in ice-caps, with consequent rise or fall of mean SEA LEVEL.

evaporation: In meteorology, the change of liquid water or ice to water vapour. In certain usages the term signifies only the liquid to vapour phase change, as distinct from SUBLIMATION which signifies the solid to vapour change. The rate of evaporation is controlled by the water and energy (mainly solar radiation) supplies and by the ability of the air to take up more water.

Since large amounts of energy are required to effect the above change of state, the evaporation that proceeds continuously from the earth's free water surfaces, soil, snow and ice fields, and vegetation, and from ice and liquid water within the atmosphere, is a fundamental item in the energy balance of the earth–atmosphere system. Evaporation also plays a basic role in the earth HYDROLOGICAL CYCLE and HYDROLOGICAL BALANCE.

The interpretation of direct measurements of evaporation presents certain problems. It is difficult to reconcile the results obtained from the different forms of EVAPORIMETER, or to relate them precisely to evaporation which occurs from a free natural water surface. Measurements of the changes in weight of a sample of soil are also used to a limited extent as a measure of evaporation, and conform better to natural conditions, but this type of apparatus presents many experimental difficulties.

Indirect measurements often use the relationship, evaporation equals measured RAINFALL minus measured RUN-OFF. In such cases the water storage by the soil is either experimentally held constant, or is assumed, by the use of long-period mean values, to be constant.

Indirect assessment of evaporation can also be made using a theoretical formula based on incoming and outgoing radiation, wind, and humidity conditions. Work on these lines has been done by H. L. Penman, C. W. Thornthwaite and others. Calculations made in this way are said to give the potential evaporation.

The extension of this work into the realms of agriculture and horticulture has led to the concept of potential TRANSPIRATION and EVAPOTRANSPIRATION.

evaporation fog: FOG which is formed by evaporation of relatively warm water into cool air. Examples are ARCTIC SEA SMOKE and FRONTAL FOG.

evaporimeter: An instrument for determining the rate of EVAPORATION of water into the atmosphere. Evaporimeters fall into two classes, (i) those employing a free water surface, and (ii) those in which evaporation takes place from a continuously wetted porous surface of blotting-paper, fabric or ceramic material. Class (i) is exemplified by the Symons evaporation tank commonly used in Great Britain. It consists of a galvanized iron tank usually six feet square and two feet deep (1 ft ≈ 30 cm), nearly filled with water and sunk into the ground with its rim projecting about three inches. The evaporation, expressed as depth in inches, is deduced from daily readings of level made with a micrometer gauge, allowance being made for rainfall. The level of the water is kept at two to three inches below the top of the tank by adding or removing water as required. Evaporimeters of class (ii) have been designed by Piche, Livingston, Owens and others, but they have not come into general use in this country as meteorological instruments.

evapotranspiration: The combined processes of EVAPORATION from the earth's surface and TRANSPIRATION from vegetation. 'Potential evapotranspiration' is the addition of water vapour to the atmosphere which would take place by these processes from a surface covered by green vegetation if there were no lack of available water.

The meteorological variables used in the estimation of potential evapotranspiration are air temperature, solar radiation or duration of sunshine, vapour pressure, wind speed and length of day. Mean annual values of estimated evapotranspiration range from about 37–43 cm in northern Scotland to 50–55 cm in southern England with, however, substantial regional and local variations.

Evapotranspiration over a particular period may also be estimated from a formula which contains the above variables. It is then possible, by allowing for the rainfall of the period, to estimate the soil moisture deficit and assess the amount of artificial irrigation which may be required to keep a particular crop growing in the most productive manner. Such methods are now widely used by farmers and horticulturists.

exchange coefficient: A general term for the coefficients of VISCOSITY (momentum), CONDUCTIVITY (heat), and DIFFUSIVITY (matter), which are defined, on analogy with molecular exchange processes, in respect of the vertical transfer effected by atmospheric eddies. See also K-THEORY.

Executive Committee: The Executive Committee of the WORLD METEOROLOGICAL ORGANIZATION is the executive body of the Organization. Its primary functions include implementing decisions taken by Members, administering the finances, considering and taking action on resolutions and recommendations of Regional Associations and Technical Commissions, and studying and making recommendations on any matter affecting international meteorology and the operation of Meteorological Services. All members act as representatives of the Organization and not as representatives of particular member states or territories. The Executive Committee normally meets at least once a year.

exosphere: The outermost 'fringe region' of a planetary atmosphere—in particular, that of the earth—in which the gas density is very small, MEAN FREE PATH large, and COLLISION FREQUENCY so small that particles moving upwards with sufficient velocity may escape from the atmosphere. For the earth, the exosphere is considered to extend upwards from about 700 km. See ESCAPE VELOCITY and ATMOSPHERE.

expansion: An increase in size of a sample of material; such an increase may be due to heat, or to the release of mechanical strain, or the absorption of moisture, or some other physical or chemical change.

The 'size' may be a length, an area, or a volume. Liability of materials to expansion on absorption of heat is expressed by a 'coefficient of (linear, areal, or volume) expansion', being the fractional increase in size per degree rise in temperature at a selected standard temperature. In barometer corrections important coefficients are those of linear expansion of brass (0·0000184 per degC or 0·0000102 per degF) and of volume expansion of mercury (0·000182 per degC or 0·000101 per degF), involving changes in the length of the brass scale and in the mercury density, respectively.

Gas expansion may be caused by a reduction of pressure or an increase of temperature. At constant pressure the coefficient of expansion is 0·00366 per degK, referred to 273 K as standard.

explosive warming: An alternative for SUDDEN WARMING.

exponential atmosphere: A hypothetical atmosphere in which pressure decreases exponentially with increasing height. The chief assumption made is that the atmosphere is isothermal, i.e. temperature (T) is constant with increasing height z; minor assumptions are that the gravitational acceleration (g) and the specific gas constant (R) remain constant with height and that the atmosphere is at rest. In such an atmosphere, the pressure (p) at height z above mean sea level is related to mean-sea-level pressure (p_0) by the equation

$$p = p_0 e^{-gz/RT},$$
$$\text{or } z = \frac{RT}{g}(\log_e p_0 - \log_e p).$$

exposure: In meteorology, the method of presentation of an instrument to that element which it is designed to measure or record, or the situation of the station with regard to the phenomena to be observed. If meteorological observations are to be of full value, attention must be paid to the exposure of the instruments. Details are to be found in the '*Observer's handbook*.'[*] Uniformity of exposure is of the greatest importance, and for that reason the pattern of the thermometer screen has been standardized in most countries, while in the British Isles, a standard height of one foot (30 cm) above ground for the rim of the rain-gauge has been fixed.

---

[*] London, Meteorological Office. Observer's handbook, 3rd edn. London, HMSO, 1969.

It is important, too, that the sites of the thermometer screen and rain-gauge should not be unduly shut in; on the other hand, a very open exposure, as on a bare moor, is undesirable for a rain-gauge, as is also a position on a slope, a roof, or near a steep bank. In these cases the catch is reduced by the effect of wind eddies due to the obstruction of the gauge itself. An ideal exposure for a SUNSHINE RECORDER requires that there should be no horizon obstruction in the direct line from sun to recorder at any time of the year.

The question of the exposure of ANEMOMETERS is one of great difficulty. The effect of the ground on a uniform current of air blowing above it is to reduce the velocity by an amount which increases as the ground is approached, and, at the same time, to introduce into the motion unsteadiness which is manifested by the creation of eddies in the air. The motion is then said to be turbulent (see TURBULENCE). An ANEMOGRAPH erected in a turbulent wind shows a large number of gusts and lulls corresponding with various parts of the eddy motion. The degree of GUSTINESS recorded by an anemograph is thus an indication of the exposure, the breadth of trace increasing with roughness of terrain. The ideal exposure is at the top of a pole 10 m (33 ft) high erected on a flat plain with no obstructions.

exsiccation: Drying by the draining away or driving away of moisture. The term implies some change, frequently the result of human agency, which decreases the quantity of moisture available without any appreciable change in the average rainfall. It is used in contrast with DESICCATION, which implies an actual drying up due to a change of climate. Examples of exsiccation are the washing away of the soil due to the cutting down of forests, with the consequent conversion of a fertile region into the semblance of a desert, the advance of sand dunes across cultivated ground, and the draining of swampy ground.

extinction coefficient: A term synonymous with 'attenuation coefficient' (see ATTENUATION) but often reserved as a measure of the combined effects of ABSORPTION and SCATTERING of wavelengths within the VISIBLE SPECTRUM. The extinction coefficient has the dimensions L^{-1}; its value in the atmosphere varies from about 10 per km in thick fog to 0·01 per km in air of very good visibility.

In relation to VISIBILITY, scattering is much the more important of the two extinction processes.

extremes: The highest and lowest values of meteorological elements in a specified period. The following definitions illustrate, with reference to maximum temperature, the nomenclature employed:

(i) Daily maximum temperature: highest temperature reached between two fixed times 24 hours apart.
(ii) Mean daily maximum temperature: mean of the 'daily maximum temperatures' observed during a given calendar month, either in a specified year or over a specified period of years.
(iii) Monthly maximum temperature: highest of the 'daily maximum temperatures' observed in the course of a given calendar month in a specified year.
(iv) Mean monthly maximum temperature: mean of the 'monthly maximum temperatures' observed during a given calendar month over a specified period of years.
(v) Absolute monthly maximum temperature: highest of the 'monthly maximum temperatures' observed during a given calendar month over a specified period of years.
(vi) Annual maximum temperature: highest of the 'daily maximum temperatures' observed during a given calendar year.

(vii) Mean annual maximum temperature: mean of the 'annual maximum temperatures' observed over a specified period of years.

(viii) Absolute maximum temperature: highest temperature observed during the whole period of observation ('absolute extreme').

The absolute extremes of temperature are: United Kingdom: highest 38·1°C (100·5°F) at Tonbridge, Kent, on 22 July 1868; lowest −27·2°C (−17°F) at Braemar, Aberdeenshire, on 11 February 1895. Surface of the globe: highest 57·8°C (136°F) at Azizia, Tripoli, Libya, on 13 September 1922; lowest −68°C (−90°F) at Verhojansk (67° 33′ N, 133° 23′ E, altitude 122 m) on 5 and 7 February 1892 and at Oymjakon (63° 16′ N, 143° 15′ E, altitude 800 m) on 6 February 1933. However, temperatures as low as −85°C to −90°C have been reported on several occasions at or near the Soviet Antarctic base Vostok (78° 27′ S, 106° 52′ E).

The extreme mean-sea-level pressures are: United Kingdom: highest 1055 mb at Aberdeen on 31 January 1902; lowest 925 mb at Ochtertyre, Perthshire, on 26 January 1884. World: highest 1079 mb at Barnaul, Siberia, on 23 January 1900; lowest 877 mb in the Pacific Ocean, about 600 miles north-west of Guam, on 24 September 1958.

The extreme annual total of rainfall for the United Kingdom is 6528 mm (257 in) at Sprinkling Tarn, Cumberland, in 1954. The heaviest known rainfalls in 24 consecutive hours are: United Kingdom, 279·4 mm (11·0 in) at Martinstown, Dorset, on 18 July 1955. World: 1870 mm (73·62 in) at Cilaos, Ile de Réunion, on 16 March 1952.

The extreme gust recorded in the United Kingdom is 101 kt (116 mile/h) at Tiree, Argyllshire, on 26 February 1961. The highest mean hourly wind speed is 69 kt (80 mile/h) at St Ann's Head, Pembrokeshire, on 23 November 1938. These extremes have been exceeded, however, at several hilltop and mountain sites. For example, a mean hourly speed of 74 kt (85 mile/h) and a gust of 106 kt (122 mile/h) were recorded at Lowther Hill, Lanarkshire (2412 ft (735 m) above mean sea level), on 12 February 1962. (1 kt ≈ 0·5 m/s.) See also RAINFALL.

eye of storm: The central region of an intense TROPICAL CYCLONE, usually some 10–15 miles (16–24 km) in diameter and fairly symmetrical, though subject to time fluctuations. The main features are absence of rain, small horizontal pressure gradient, light winds, high and turbulent sea, and layered clouds, often well broken.

eye of the wind: A nautical expression indicating the direction from which the wind blows.

F

facsimile transmission: A system of telegraphy providing reproduction, in the form of fixed images (photographic or otherwise), of the form and possibly of the depth of tone or of the colours, of an original document, whether written, printed or pictorial.

Fahrenheit scale: A scale of temperature introduced about 1709 by the German physicist Fahrenheit, who was the first to use mercury as the thermometric substance. Primary fixed points were the temperature of a mixture of common salt and ice and the temperature of the human body; with reference to these, the freezing-point of water was marked 32° and the boiling-point of water was marked 212°. See TEMPERATURE SCALES.

fair-weather cumulus: A common alternative for CUMULUS HUMILIS.

falling sphere: A device for measuring air density and winds in the stratosphere and mesophere. The acceleration of a rigid sphere is measured during its descent after ejection from a METEOROLOGICAL ROCKET. The density of the air is a function of the acceleration and drag coefficient of the sphere.

Two distinct forms exist. The first form consists of an inextensible balloon, 1 or 2 m in diameter. This is inflated to a pressure considerably in excess of the ambient pressure. The acceleration of the sphere is measured by tracking it very accurately with a high-precision radar, or optically with a high-precision theodolite. The second form is a rigid instrumented sphere, about 15 cm in diameter, which contains an accelerometer. The acceleration of the sphere is measured directly and is transmitted by radio to a ground station.

Air density measurements can be made by this method from 100 km downwards. If the sphere is tracked by radar its horizontal drift can be used to give winds below 65 km.

fallout: A term applied to both the process of deposition of solid material on the earth's surface and to the deposited material itself. It may be used in such a sense as to signify only 'dry deposition' (mainly the result of gravitational settling): in such a sense it is used in contrast to the term WASHOUT. The term fallout is, however, used mainly in respect of the radioactive debris which is associated with a nuclear explosion. The process of washout is then often important in the manner of deposition of the material, especially when the fallout is other than 'close-in'. See RADIOACTIVE FALLOUT.

fallstreak: A term often used for the supplementary cloud feature VIRGA.

false cirrus: A popular expression for CIRRUS SPISSATUS or anvil cirrus—see ANVIL CLOUD.

fata morgana: A rare and complex form of MIRAGE in which horizontal and vertical distortion, inversion and elevation of objects occur in changing patterns. The phenomenon occurs over a water surface and is produced by the superposition of several layers of air of different REFRACTIVE INDEX.

fetch: The fetch of an airstream is the length of its traverse across a sea or ocean area.

fibratus (fib): A CLOUD SPECIES (Latin, *fibratus* fibrous).

'Detached clouds or a thin cloud veil, consisting of nearly straight or more or less irregularly curved filaments which do not terminate in hooks or tufts.

This term applies mainly to CIRRUS and CIRROSTRATUS.'* See also CLOUD CLASSIFICATION.

Fickian diffusion: See DIFFUSIVITY.

FIDO: The term, abbreviated from 'Fog, Investigation Dispersal Operations', signifying the system, developed in Great Britain during the Second World War, of effecting temporary and local clearance of fog by the burning of petrol at intervals alongside an airfield runway.

fiducial point: A reference point of an instrument or scale, as, for example, the ivory tip of the Fortin barometer to which the mercury in the cistern must be brought before a reading is taken.

fiducial temperature: An alternative for STANDARD TEMPERATURE.

field capacity: The mass of water (per cent of dry soil) retained by a previously saturated soil when free drainage has ceased is known as the soil's field capacity or water-holding capacity. It varies from about 7 per cent in light sand to about 60 per cent in heavy clay and corresponds to a CAPILLARY POTENTIAL of about 500 cm of water ($pF = 2.7$). See SOIL MOISTURE.

field mill: A type of ELECTROMETER employed, in various forms, more especially in field measurement or recording of atmospheric POTENTIAL GRADIENT. The charge generated in a conductor which is alternately exposed to and sheltered from the atmospheric electric field is conveyed to a meter; d.c. or a.c. amplification is often used.

filling: See DEEPENING.

filtering: The processing of values of a variable to emphasize certain patterns of variation while suppressing others. Filtering of the basic hydrodynamic equations of motion may, for example, involve the assumption of QUASI-GEOSTROPHIC MOTION, while filtering of TIME SERIES, or of measurements at equal intervals in space, is carried out by means of weighted MOVING AVERAGES. The object is generally to isolate oscillations lying within a certain waveband, or range of frequencies, while reducing NOISE and oscillations in other wavebands.

The filtering of a time series is carried out as follows: if $x_1 x_2, \ldots x_N$ is a time series, and $(W_0 W_1 W_2, \ldots W_K)$ are the constants of a symmetrical filter of order K, then corresponding to any term x_n we find the value of

$$y_n = \sum_{j=0}^{K} \tfrac{1}{2} W_j (x_{n-j} + x_{n+j})$$

For $K \leqslant n \leqslant N-K$ the terms y_n, y_{n+1}, y_{n+2}, etc., can be found from the x series.

---

* Geneva, World Meteorological Organization. International cloud atlas, Vol. 1. Geneva, WMO, 1956, p. 12.

If
$$x_n = A\cos(2\pi fn + B)$$
then
$$y_n = A\cos(2\pi fn + B) . \sum_{j=0}^{K} W_j \cos 2\pi fj,$$

so the effect of filtering a harmonic wave by a symmetric filter is to leave the frequency and phase unchanged, and to multiply the amplitude by the magnification factor

$$M(f) = \sum_{j=0}^{K} W_j \cos 2\pi fj.$$

The problem of designing a filter to pass a particular waveband thus reduces to that of choosing the weights $W_0 W_1 ... W_K$ to make $M(f)$ take values near unity in a certain range of frequencies and values near zero for other frequencies.

finite-difference method: A method of approximation to space or time derivatives of a mathematical function, widely used in meteorology, in which the derivative is represented as the difference between values of the function at two points in space or time. The appropriate separation between the points depends on the scale of the phenomenon being studied.

The calculus of finite differences is a well-developed branch of mathematics dealing with quantities specified for discrete values of the variables and is much used in statistics.

fireball: This term is used to signify BALL LIGHTNING. It is also used of the larger type of METEOR, with the brightness of a first magnitude star or greater.

The rapidly expanding, white-hot ball of gas which is produced on explosion of a nuclear weapon is also termed a 'fireball'; the term is applied up to the time at which the volume of gas becomes, through adiabatic and radiational cooling, no longer incandescent.

firn: A German word meaning old snow which is in process of being transformed into GLACIER ice. The word is also used to denote an accumulation area of old snow above a glacier and is synonymous with the French word 'névé'. There is no corresponding word in English.

firn wind: An alternative for GLACIER WIND.

fitness figures (numbers): A scale of figures, developed in Great Britain during the Second World War, used as a measure of the meteorological fitness of a particular airfield for the landing of aircraft. The scale is governed mainly by visibility and observed cloud base in relation to the maximum height of obstructions near the airfield.

F-layer: A two-part layer of the IONOSPHERE. The lower (F_1) part is best observed in day-time in summer and has a maximum electron concentration, at about 160 km, of the order 3×10^5 per cm^3 with overhead sun. The higher (F_2) part, also called the 'Appleton layer', has a maximum electron concentration at a height which is very variable in space and time within the approximate range 220 to 400 km. The F_2-layer is, in terms of CHAPMAN LAYER theory, highly anomalous and is also the seat of 'ionospheric storm' phenomena.

float barograph: A (seldom used) type of recording SIPHON BAROMETER. See BAROGRAPH.

floccus (flo): A CLOUD SPECIES (Latin, *floccus* tuft).
'A species in which each cloud unit is a small tuft with a cumuliform appearance, the lower part of which is more or less ragged and often accompanied by VIRGA.
This term applies to CIRRUS, CIRROCUMULUS and ALTOCUMULUS.'* See also CLOUD CLASSIFICATION.

flux density: The rate of transport (flux) of a specified quantity (e.g. RADIATION) across unit area of a surface.

fluxplate: An instrument of thermo-electric design for recording heat flux electrically.

foehn: An alternative for FÖHN.

fog: Obscurity in the surface layers of the atmosphere, which is caused by a suspension of water droplets, with or without smoke particles, and which is defined, by international agreement, as being associated with VISIBILITY less than 1 km. In British practice, use of the term in forecasts for the general public relates to visibility less than 200 yards (180 m). 'Ice fog' is an obscurity produced by a suspension of numerous minute ice crystals.

While fogs which are entirely composed of smoke or dust particles do occur, the more persistent and thick fogs of industrial areas contain also water droplets. The term SMOG is sometimes used of such fogs. The frequent occurrence of fog in industrial areas is due in large measure to the plentiful supply of hygroscopic particles which are able to act as condensation NUCLEI when the relative humidity is less than 100 per cent.

Fogs which are composed entirely or mainly of water droplets are generally classified according to the physical process which produces saturation or near-saturation of the air; examples are RADIATION FOG, ADVECTION FOG, UPSLOPE FOG, and EVAPORATION FOG, this last including FRONTAL FOG and ARCTIC SEA SMOKE which is also known as 'steam fog' and has various other synonyms. Natural fogs are frequently the result of the combined action of two or more such physical processes.

Fog data relating to selected stations in the British Isles are shown in Table VI.

fogbow: A white RAINBOW of about 40° radius seen opposite the sun in fog. Its outer margin has a reddish, and its inner a bluish, tinge but the middle of the band is quite white. A supernumerary bow is sometimes seen inside the first and with the colours reversed. The bows are produced in the same way as the ordinary rainbow but owing to the smallness of the drops, the diameter of which is about 50 μm, the colours overlap and the bow appears white.

fog-point: The air (screen) temperature at which FOG forms.

fog precipitation: Fog precipitation, sometimes also termed 'fog drip', signifies the precipitation of liquid water from non rain-bearing clouds due to the interception of the cloud particles by trees and other vegetation.

The relatively few measurements of this phenomenon suggest that it is common in places which have a high frequency of orographic cloud and suitable vegetation. In measurements over the plateau of Table Mountain, Cape Town, for example, one day in three yielded fog precipitation in the absence of rainfall measured in the

---

* Geneva, World Meteorological Organization. International cloud atlas, Vol. 1. Geneva, WMO, 1956, p. 12.

TABLE IV—Average number of days of fog and thick fog at selected times (GMT) and stations in Great Britain, 1951–60

Station	Latitude	Longitude	Height above MSL	Visibility less than 1000 m							Visibility less than 180 m						
				April–September			October–March			April–September			October–March			
				3h	9h	15h	21h	3h	9h	15h	21h	3h	9h	15h	21h	3h	9h	15h	21h
			metres	number of days															
Lerwick	60° 08′ N	01° 11′ W	82	21·1	14·2	9·8	14·8	4·2	4·9	4·2	3·8	12·0	9·0	4·4	9·7	1·6	2·5	1·3	2·7
Stornoway	58° 13′ N	06° 20′ W	3	2·5	1·1	0·1	0·7	0·6	0·6	0·7	0·7	0·3	0·0	0·0	0·2	0·2	0·2	0·0	0·1
Aberdeen Airport	57° 12′ N	02° 12′ W	52	13·3	2·3	0·8	4·5	4·6	4·5	2·4	4·8	3·9	0·3	0·0	1·7	1·8	1·0	0·4	0·4
Renfrew	55° 52′ N	04° 24′ W	8	1·5	1·0	0·3	0·6	14·2	28·5	18·0	21·9	0·3	0·0	0·0	0·0	4·1	6·2	3·8	3·0
Acklington	55° 18′ N	01° 38′ W	42	11·5	3·4	0·6	6·0	9·5	8·1	4·6	6·9	3·3	0·8	0·1	1·8	2·8	1·9	0·3	1·6
Gorleston	52° 35′ N	01° 43′ E	2	4·5	3·5	2·2	2·5	7·5	13·9	6·5	8·1	1·2	0·8	0·9	0·8	3·8	5·3	2·2	3·1
Mildenhall	52° 22′ N	00° 28′ E	5	13·0	2·1	0·3	0·6	23·5	20·2	7·2	13·2	3·6	0·8	0·0	0·2	7·3	7·3	2·6	5·2
Birmingham Airport	52° 27′ N	01° 45′ W	99	6·2	1·8	0·1	1·1	24·3	28·5	11·7	22·7	1·7	0·5	0·0	0·1	10·0	9·5	2·1	5·6
London/Heathrow Airport	51° 29′ N	00° 27′ W	25	3·1	2·0	0·0	0·6	21·1	23·6	8·9	23·7	1·6	0·5	0·0	0·0	10·9	9·8	2·4	5·6
Manchester Airport	53° 21′ N	02° 16′ W	76	4·4	0·9	0·2	0·8	15·4	18·5	8·7	18·4	1·0	0·1	0·0	0·0	6·2	6·0	2·2	4·0
Holyhead (Valley)	53° 15′ N	04° 32′ W	8	4·6	2·5	2·0	3·0	3·5	4·3	1·8	4·0	0·6	0·0	0·0	0·6	0·8	0·9	0·0	0·6
Exeter Airport	50° 44′ N	03° 25′ W	32	13·1	0·8	0·0	1·6	17·7	11·8	1·7	9·4	4·1	0·4	0·0	0·4	8·2	4·9	0·7	4·0
Isles of Scilly (St Mary's)	49° 26′ N	06° 18′ W	48	9·2	7·3	5·2	8·0	6·0	6·3	4·8	5·9	4·7	3·2	1·2	3·7	2·7	2·8	1·8	2·3

normal way, and the total amount of fog precipitation measured in the year was nearly twice that of the measured rainfall.

föhn: A warm, dry wind which occurs to leeward of a ridge of mountains. The name originated in the Alps but is now used as a generic name for this type of wind.

When relatively moist air ascends a ridge of mountains, the ADIABATIC cooling of the air proceeds mainly at the saturated adiabatic lapse rate (about 0·5 degC per 100 metres). Provided water is removed by precipitation on the windward side, much of the adiabatic warming of the air on descent to leeward is at the dry adiabatic lapse rate (1 degC per 100 metres) and the air reaches the ground there as a warm, dry wind. It is often accompanied by lenticular clouds. A contributory factor in high leeward temperatures by day is often that cloud amounts are smaller to leeward than they are to windward.

There is good evidence that conspicuous föhn winds occur in conditions suitable for LEE WAVES. In such conditions the air reaching the ground may have descended from levels well above the mountain tops and higher temperatures to leeward may be observed in the absence of precipitation or cloud to windward.

föhn wall: Associated with the FÖHN effect, a mass of precipitating clouds often forms over the windward slopes of the hills. Cloud continues some way down the lee slope but evaporates in the descending current, terminating along a line parallel to the main ridge of hills.

foot-candle: An obsolescent unit of illumination, equal to 1 LUMEN per square foot. See LUX.

force: That which alters or tends to alter the state of rest or motion of a body. Force is a vector quantity with both magnitude and direction and has dimensions MLT^{-2}. In meteorology, 'specific force' (force per unit mass), with dimensions LT^{-2}, is used in place of total force. See also ACCELERATION.

forecast: The term, first applied in meteorology by Admiral FitzRoy, which signifies a statement of anticipated (meteorological) conditions for a specified place (or area, route, etc.) and period of time.

A threefold classification of forecasts, in terms of the period covered, is recognized:
  (i) 'Short-period' forecast for part or whole of a 24-hour period, often with a 'further outlook' for the following 24 hours.
  (ii) 'Medium-range' forecast for some two to five days, and
  (iii) 'Long-range' forecast for a period longer than about five days ahead, for example a month or season.

The short-period forecast generally contains information concerning wind velocity, weather (state of sky, precipitation, fog, frost, thunder, etc.) and temperature (relative to seasonal normal); for special purposes, additional information is given—for example, upper winds for aviation. In medium-range and more especially long-range forecasts the information given is in more general terms and is often confined to precipitation and temperature.

In short-period forecasting, the methods of SYNOPTIC METEOROLOGY are normally used to anticipate changes in surface pressure distribution and positions of fronts, on which, together with an appeal to physical reasoning and to precedent, the forecast is based. Alternatively, the changes in pressure distribution are derived by numerical methods (see NUMERICAL WEATHER FORECAST). Purely statistical methods of short-period forecasting are applied to a small extent. In medium-range forecasting, the synoptic method predominates, with concentration of attention on the dominant circulation features. In long-range forecasting, significant but limited success has been achieved with synoptic methods in which time-averaging of the pressure and circulation patterns is employed, and with a synoptic ANALOGUE method.

forked lightning: LIGHTNING in which many luminous branches from the main discharge channel are visible. Such branching occurs in response to local variations of SPACE CHARGE close to the main channel.

Fortin barometer: A form of mercury BAROMETER, the zero of whose scale is fixed by a pointer inside the cistern, which is made partly of leather. By adjustments of a screw, the level of mercury in the cistern is brought to the scale zero ('fiducial point') before each reading is taken.

Foucault pendulum: A pendulum designed by J. Foucault in Paris in 1851 to give experimental proof of the earth's rotation. The direction of swing of an oscillating large iron ball, suspended just above a tray of fine sand by a wire over 60 m in length, was observed to change clockwise (northern hemisphere) at a rate corresponding to $15°$ per sidereal hour $\times \sin \varphi$ (latitude). The observation is consistent with an apparent deviating force arising from the rotation of the earth about a polar axis.

Fourier analysis: An alternative for HARMONIC ANALYSIS.

Fourier series: A representation of any function of an independent variable in terms of sines and cosines of multiples of that variable. Such a series was first developed by J. Fourier in 1822. In symbols:

$$f(x) = A_0 + A_1 \sin x + A_2 \sin 2x + \ldots \\ + B_1 \cos x + B_2 \cos 2x + \ldots .$$

See HARMONIC ANALYSIS.

foyer: The place of origin of a group of ATMOSPHERICS.

f.p.s. system: A system of units, seldom used in meteorology, which is based on the foot, the pound, and the second as FUNDAMENTAL UNITS. In this system the unit of force is the 'poundal', which is the force required to give a mass of 1 pound an acceleration of 1 ft/s^2. The unit of work, the 'foot-pound', is the work done in raising a mass of 1 pound a distance of 1 foot against the force of gravity.

fractional volume abundance: The relative concentration by volume of a specified gaseous constituent in a mixture of gases, expressed in a suitable unit. For minor constituents it is generally expressed in parts per million (p.p.m.), signifying the number of volume units of the gas (e.g. cubic centimetres) which may be extracted from one million volume units of the mixture, both volumes being measured at the same temperature and pressure. It is identical with MOLE FRACTION. See AIR for tabulated values of fractional volume abundance, expressed as percentages. See also AVOGADRO'S LAW.

fractus (fra): A CLOUD SPECIES (Latin, *fractus* broken).
'Clouds in the form of irregular shreds, which have a clearly ragged appearance. This term applies only to STRATUS and CUMULUS.'* See also CLOUD CLASSIFICATION.

---

* Geneva, World Meteorological Organization. International cloud atlas, Vol. 1. Geneva, WMO, 1956, p. 13.

Fraunhofer lines: See SUN.

frazil ice: ICE which forms in spicules or small plates in rapidly flowing rivers, and at times in the sea, the movement of the water preventing the ice crystals from forming a solid sheet of ice. The formation has been best observed in the rivers of Canada; the word is from the French-Canadian *frasil*, meaning cinder, the frazil crystals being supposed to resemble forge-cinders.

free atmosphere: The atmosphere above the 'friction layer', i.e. above about 600 metres above ground, where the influence of surface friction on air motion is assumed negligible.

free lift: The free lift of a balloon is given by the excess load that would be required to make the balloon float at a constant level; it is the excess of the BUOYANCY force over the gross weight.

free period: That period of vibration of a system, determined by its physical characteristics, which the system adopts when set in motion by the application of an external force which is then removed; the corresponding frequency of vibration is termed its 'natural frequency'. According to the RESONANCE theory of ATMOSPHERIC TIDES the atmosphere has a free period very close to 12 hours, the period being governed by the variation of temperature with height. See ATMOSPHERIC TIDES, RESONANCE.

freeze, freezing: With reference to the weather, the term 'freezing' is used when the temperature of the air is below the freezing-point of water. Freezing conditions for an appreciable time over a widespread area are in America termed a 'freeze' (in Britain, a 'frost').

freezing drizzle, fog, rain: Supercooled water drops of drizzle (or fog or rain) which freeze on impact with the ground or other objects (in aviation, with an aircraft) to form GLAZED FROST or, in the case of the smaller droplets comprising fog, RIME. See SUPERCOOLING.

freezing level: That lowest height above mean sea level at which, for a given place and time, the air temperature is 0°C; it is now generally termed the height above mean sea level, or if appropriate the PRESSURE ALTITUDE, of the 0°C isotherm.

Over the British Isles the freezing level has average values of about 600 metres in winter and 3000 metres in summer with, however, large day-to-day variations from these values. The term 'melting level' is to be preferred from the physical point of view since water in the atmosphere is not necessarily frozen at temperatures below 0°C—see SUPERCOOLING.

freezing nucleus: See NUCLEUS.

freezing-point: That constant temperature at which the solid and liquid forms of a given pure substance are in equilibrium at STANDARD ATMOSPHERIC PRESSURE. For pure-water substance the temperature is 0°C and is termed the 'ice-point'. In meteorology, the term 'freezing-point' is often used to signify 'ice-point'.

In practice, a cooling liquid may not freeze at the freezing-point because of a pressure variation from standard atmospheric pressure, or the presence of impurities, or the phenomenon of SUPERCOOLING.

frequency: The statistical term for the number of occasions a variable takes a certain value or lies in a certain range of values.

In a wave vibration the frequency is the number of complete vibrations (cycles) per unit time. It is numerically equal to the velocity divided by the wavelength and is usually expressed in cycles per second (c/s), kilocycles (i.e. 10^3 cycles) per second (kc/s), or megacycles (i.e. 10^6 cycles) per second (Mc/s). For ELECTROMAGNETIC RADIATION the relationship is frequency (c/s) = 3×10^8/wavelength (m). The unit of frequency (1 c/s) is termed the 'hertz' (Hz).

frequency distribution: A term used rather loosely to cover both a graph of the actual frequencies of occurrence of values of a statistical variable in a sample, and what is better called a PROBABILITY DISTRIBUTION. When the variable can take only discrete values, the frequency distribution must refer to these. When, however, the variable is continuous, and the data are to be analysed mathematically, or represented in a HISTOGRAM, it is nearly always desirable to group them into ranges of equal width, e.g. 0–1, 1–2, 2–3, etc. Ranges of unequal width are inconvenient mathematically, although an open range containing, say, all cases above a certain value may be used in tables. Data grouped in equal ranges can readily be processed to find their MEAN and STANDARD DEVIATION.

friction: The mechanical force of resistance which acts when there is relative motion of two bodies in contact, or of a body in contact with a medium, or of adjacent layers of a medium, or of adjacent media. Within a fluid, the friction that arises from molecular collisions is termed VISCOSITY.

In meteorology, the effects of friction are important in the flow of air over the earth's surface ('surface friction') and also when there is WIND SHEAR. Surface friction affects the wind velocity within the FRICTION LAYER and is important in all scales of motion up to and including that implied in the general circulation of the atmosphere, in which it plays a vital part in the over-all momentum balance that is achieved. The effects of surface friction visible on the synoptic scale are a decrease of SURFACE WIND speed relative to that appropriate to the pressure gradient (GEOSTROPHIC WIND or GRADIENT WIND), and a 'frictional outflow' of surface air from higher to lower pressure. The magnitudes of these effects increase with the surface roughness and decrease with increasing height within the friction layer. Typical values for airflow over the ocean at surface wind level are a decrease of speed by about one-third and a 'cross-isobar' wind direction of about 20°.

Direct measurements of the surface friction effect at individual points have been made, employing 'drag plates'. Estimates of the larger-scale effects have also been made, using vertical wind profiles. See also EKMAN SPIRAL.

friction layer: The atmospheric layer, extending from the earth's surface to about 600 metres (2000 feet) above ground, in which the influence of surface FRICTION on air motion is appreciable.

friction velocity: That reference velocity (u_*), employed in the study of fluid flow over a rough surface, which is defined by the equation

$$u_* = \sqrt{(\tau_0/\rho)}$$

where τ_0 is the surface drag per unit area and ρ is the fluid density.

Generally in meteorology, τ_0 is little different from τ, the REYNOLDS STRESS in the fluid, within a shallow layer near the surface.

u_* increases with roughness of surface and with mean wind speed ($\bar{u}$). In meteorology, u_* is of the general order ($\bar{u}/10$).

fringe region: See EXOSPHERE.

front: A term introduced into synoptic meteorology by the Norwegian meteorologists in 1918. A 'frontal surface' is a sloping transition zone separating two air masses of different density and so of different temperature; a surface 'front' is the zone (usually represented on charts as a line) along which a frontal surface intersects the earth's surface.

The equilibrium slope of a surface of discontinuity ($\tan \theta$) was derived by M. R. Margules as

$$\tan \theta = \frac{f}{g} \frac{(\rho_1 v_1 - \rho_2 v_2)}{\rho_1 - \rho_2} \approx \frac{fT_m}{g} \frac{(v_1 - v_2)}{(T_2 - T_1)}$$

where ρ_1 and ρ_2 are the densities, v_1 and v_2 the geostrophic wind components parallel to the front, T_1 and T_2 the temperatures, in the cold and warm air masses, respectively, T_m the mean of T_1 and T_2, f the Coriolis parameter, and g the gravitational acceleration.

Frontal surfaces have very gentle slopes, of about 1 in 100, which, however, generally differ appreciably from the equilibrium values obtained from Margules's formula because the winds near the front are usually far from geostrophic. Horizontal convergence and associated vertical motion are essential features of a well-marked front; the upward motion results, especially within the warmer air mass, in the condensation and precipitation which are associated with a typical active front. Measurements in the free atmosphere show that the horizontal gradients of temperature within the separate air masses are not negligible (though, by definition, smaller than in the frontal zone), and that large local variations of temperature and humidity often occur near the edges of frontal cloud and precipitation.

A front necessarily lies in a trough of low pressure and is marked by discontinuities of wind velocity and, in general, of pressure tendency. The more important and extensive fronts, such as the 'polar front', may be traced from the earth's surface to the tropopause.

frontal contour chart: A synoptic chart of the contours (usually expressed in pressure units) of a selected frontal surface, i.e. a plan view of the intersection of the frontal surface with selected isobaric surfaces.

frontal fog: Fog which forms at and near a FRONT. Such fog forms when raindrops, falling from relatively warm air above a frontal surface, evaporate into cooler air close to the earth's surface and cause it to become saturated.

frontogenesis: The development or marked intensification of a FRONT. This process —the intensification of the horizontal temperature gradient in a restricted zone—is effected mainly by horizontal CONFLUENCE and/or CONVERGENCE in conditions of suitably orientated isotherms.

frontolysis: The disappearance or marked weakening of a FRONT (converse of FRONTOGENESIS). This process is effected mainly by horizontal DIVERGENCE of air from the frontal zone, usually accompanied by SUBSIDENCE.

frost: Frost occurs when the temperature of the air in contact with the ground, or at thermometer screen level, is below the freezing-point of water ('ground frost' or 'air frost', respectively). The term is also used of the icy deposits which may form on the ground and on objects in such temperature conditions (GLAZE, HOAR FROST).

Since the sensation of cold depends not only on air temperature but on the accompanying wind speed, the fourfold classification of frosts used in forecasts of this condition in the British Isles is varied with wind speed. Thus, frost is classified as 'slight', 'moderate', 'severe', or 'very severe' for screen temperature ranges of

−0·1° to −3·5°C, −3·6° to −6·4°C, −6·5° to 11·5°C, or below −11·5°C, respectively, if the accompanying wind speed is less than 10 knots; and for screen temperature ranges of −0·1° to −0·4°C, −0·5° to −2·4°C, −2·5° to −5·5°C, or below −5·5°C, respectively, if the wind speed exceeds 10 knots. See also GROUND FROST, FROST DAY.

frost day: A day of frost is defined as a period of 24 hours ending at i.e. 09 GMT, or at 21 GMT, where observations are made at 21 GMT, in which the minimum air temperature in the screen is below 0°C (32°F). See also GROUND FROST.

Table V gives average numbers of frost days at selected stations in the United Kingdom, based on the period 1946–60.

frost heaving: The uneven lifting and distortion of the ground close to the surface. It results from the expansion of water within the soil on freezing associated with the local formation of ice crystals, which accumulate into ice 'lenses'. The phenomenon may result in damage to road surfaces and loosening of root-hold of plants.

frost hollow: A local hollow-shaped region in which, in suitable conditions, cold air accumulates by night as the result of KATABATIC flow. Such regions are subject to a greater incidence of FROSTS, and to more severe frosts, than are the surrounding areas of non-concave shape.

frost-point: The frost-point (T_f) of a moist air sample is that temperature to which the air must be cooled in order that it shall be saturated with respect to ice at its existing pressure and HUMIDITY MIXING RATIO.

T_f is that temperature for which the saturation VAPOUR PRESSURE with respect to ice (e'_i) is identical with the existing vapour pressure (e') of the air, i.e.

$$e' = e'_i \text{ at } T_f.$$

Frost-point may be measured indirectly from wet- and dry-bulb temperature readings with the aid of humidity tables (see PSYCHROMETER), or directly with a FROST-POINT HYGROMETER.

frost-point hygrometer: A development by G. M. B. Dobson and A. W. Brewer of the well-known dew-point principle, for use in aircraft flying in air at sub-freezing temperatures. The thimble is cooled from below by a controlled flow of liquid air and is surrounded by a heating coil. Cooling and heating rates are adjusted till a deposit of ice on the surface is observed, either visually or photo-electrically, to remain constant; the temperature of the surface, measured by a resistance thermometer, is then the FROST-POINT of the air.

frost, protection against: Various methods are practised of affording protection to crops in orchards, vineyards, etc., against damage by frost. Protective measures may be effective only on 'radiation nights' characterized by calm or light wind conditions; no method is effective on occasions of frost in which there is substantial natural air movement.

The methods include: direct heating of the air near the ground; the production of a smoke screen over the crop (see SMUDGING); the flooding or sprinkling of crops, thus adding the thermal content of the water and increasing the effective specific heat of the soil; and the use of large fans designed to mix cold air near the surface with warmer air aloft.

frost smoke: An alternative for ARCTIC SEA SMOKE.

TABLE V—*Average number of frost days at selected stations in the United Kingdom, 1946–60.*

Station	Height above MSL	Latitude	Longitude	Jan.	Feb.	Mar.	Apr.	May	June	July	Aug.	Sept.	Oct.	Nov.	Dec.	Year
	metres															
Lerwick	82	60° 08′ N	01° 11′ W	10·8	11·5	7·9	4·9	1·3	0·1				0·8	2·7	5·9	45·9
Stornoway	3	58° 13′ N	06° 20′ W	9·1	10·0	6·7	3·3	0·6					1·0	3·9	6·1	40·7
Craibstone	91	57° 11′ N	02° 12′ W	15·7	14·7	8·4	5·3	1·4					1·0	4·5	10·2	61·4
Renfrew	8	55° 52′ N	04° 24′ W	12·8	12·8	8·1	4·0	0·9				0·2	2·8	5·9	9·9	57·5
Eskdalemuir	239	55° 19′ N	03° 12′ W	19·2	19·0	15·7	13·0	4·9	0·9	0·2	0·1	0·3	4·8	10·4	15·3	104·8
Douglas	87	54° 10′ N	04° 29′ W	6·3	8·7	3·5	0·3					1·3	0·1	0·7	2·9	22·5
Cockle Park	99	55° 13′ N	01° 41′ W	17·2	17·5	12·5	7·4	1·8					1·5	6·4	12·1	76·6
Gorleston	2	52° 35′ N	01° 43′ E	7·8	9·3	5·0	0·7	0·1				0·1	0·3	1·7	5·0	29·9
Cambridge	12	52° 12′ N	00° 08′ E	13·7	14·0	10·2	4·5	0·7					2·5	6·1	10·5	62·3
Birmingham (Edgbaston)	163	52° 28′ N	01° 56′ W	11·1	12·5	8·3	0·9						0·6	2·5	6·2	42·1
Kew	5	51° 28′ N	00° 19′ W	10·5	10·3	7·2	1·1						1·3	3·5	7·0	40·9
Southampton	20	50° 55′ N	01° 24′ W	12·1	10·8	7·3	1·1						1·1	3·5	7·3	43·2
Stonyhurst	115	53° 51′ N	02° 28′ W	13·7	14·1	9·4	2·3	0·2					1·6	4·5	9·2	55·0
Holyhead (Valley)	8	53° 15′ N	04° 32′ W	6·3	7·7	4·7	1·0							0·8	2·4	22·9
Falmouth	51	50° 06′ N	05° 03′ W	4·0	5·7	2·4	1·0							0·2	1·3	14·6
Aldergrove	69	54° 39′ N	06° 13′ W	11·5	12·5	7·6	4·3	1·3	0·2			0·3	1·1	5·1	7·2	51·1

TABLE VI—*Average number of days with gale, 1946–60*

Station	Latitude	Longitude	Jan.	Feb.	Mar.	Apr.	May	June	July	Aug.	Sept.	Oct.	Nov.	Dec.	Year
Lerwick	60° 08′ N	01° 11′ W	11·2	7·5	5·7	4·1	1·8	1·3	0·5	0·7	2·9	5·9	6·6	9·5	57·7
Tiree	56° 30′ N	06° 53′ W	6·1	3·1	2·1	1·5	0·3	0·0	0·2	0·5	1·5	2·5	3·4	7·3	28·7
Holyhead (Valley)	53° 15′ N	04° 32′ W	3·2	2·3	0·7	1·5	0·3	0·2	0·1	0·5	1·7	1·3	2·5	4·2	18·5
Isles of Scilly (St Mary's)	49° 56′ N	06° 18′ W	3·5	1·9	1·5	0·9	0·5	0·4	0·3	0·7	1·0	0·8	1·7	4·7	17·9
Shoeburyness	51° 32′ N	00° 49′ E	0·4	0·4	0·4	0·2	0·2	0·0	0·1	0·4	0·3	0·4	0·3	0·8	3·9
Spurn Head	53° 35′ N	00° 07′ E	3·5	2·1	1·4	1·5	0·4	0·7	0·3	0·1	0·8	1·4	2·2	2·9	17·3
Aberdeen Airport	57° 12′ N	02° 12′ W	1·9	1·3	0·6	1·1	0·0	0·1	0·0	0·1	0·8	0·5	0·4	0·9	7·7

Froude number: In fluid flow, a non-dimensional parameter (Fr) defined by the relationship $(Fr) = \bar{u}/\sqrt{(\delta g)}$, where $\bar{u}$ is the free stream velocity, δ the thickness of the BOUNDARY LAYER, and g the gravitational acceleration.

F-test: Also known as the 'variance ratio test'. If two independent estimates of the same VARIANCE are available, based respectively on n_1 and n_2 DEGREES OF FREEDOM, then the CHANCE EXPECTATION of their ratio (arranged to have the larger variance in the numerator) may be found from the F-table published, for example, by Brooks and Carruthers.* A value exceeding the chance expectation will be regarded with more or less confidence as implying that the two measured variances are not estimates of the same thing; for example, the smaller may be an estimate of NOISE while the larger estimates noise plus an organized contribution.

fundamental units: The units of mass (M), length (L), time (T) on which the less fundamental, or 'derived' units are based, e.g. the units of pressure or viscosity.

funnel cloud: The cloud formed at the core of a WATERSPOUT or TORNADO, sometimes extending right down to the earth's surface, attributed to the reduction of pressure at the centre of the vortex. Similar cloud formations are sometimes seen without a waterspout or tornado at the ground (cloud pendants). See also TUBA.

funnelling: The term 'funnelling' (sometimes 'canalization') is applied to the phenomenon in which the surface wind is constrained by topographical features to blow along a valley and is thereby increased in speed. Quantitative data on the phenomenon are lacking, mainly because of the difficulties inherent in expressing natural land features in a numerical form.

further outlook: A statement in brief and general terms appended to a detailed forecast and giving the conditions likely to be experienced in the 24 hours or more following the period covered by the actual forecast.

fusion: An alternative for 'melting'. See MELTING-POINT.

---

* BROOKS, C. E. P. and CARRUTHERS, N.; Handbook of statistical methods in meteorology. London, HMSO, 1953, p. 384.

G

gale: A WIND of a speed in the range 34 to 40 knots (force 8 on the BEAUFORT SCALE of wind force, where it was originally described as 'fresh gale'), at a free exposure 10 metres (33 feet) above ground. In general, a mean speed over a period of 10 minutes, as reported in synoptic code, is implied by the term 'gale'; where this is not intended, a phrase such as 'gusts to gale force' is used.

While the term 'gale' applies strictly to the speed limits given above, and higher winds are referred to in other terms, e.g. strong gale, storm, etc., statistics of gales refer to the attainment of mean speeds of 34 knots or over. See also GALE, DAY OF.

gale, day of: A day on which the wind speed at the standard height of 10 metres attains a mean value of 34 knots or more over any period of 10 consecutive minutes during the day. Table VI (p. 118) gives the average numbers of days with gale at selected coastal anemometer stations in Great Britain and is based on data for the period 1946–60.

gale warning: The Meteorological Office issues notice of the probability of gales (force 8 or more on the BEAUFORT SCALE of wind force in exposed situations or in the open sea) by broadcast warnings and by telegrams to ports and fishing stations recommended by the responsible local authorities.

Receipt of a gale warning notice by a station is made known by the hoisting of a black cone, three feet high (1 ft $\approx$ 30 cm) and three feet wide at base. Two cones are used, the south cone (point downwards) and the north cone (point upwards). For the detailed significance of these cones reference should be made to the instructions issued to all gale-warning stations or to the *Admiralty weather manual*.*

gamma: An obsolescent measure of magnetic field strength, equal to 10^{-5} GAUSS. The gamma is equal to the nanoTESLA (10^{-9}T) which now supplants it.

gamma (or γ) radiation: ELECTROMAGNETIC RADIATION of wavelength less than that of X-rays and of great penetrative power. Gamma radiation (also called 'gamma rays') is produced during the disintegration of many radioactive elements. Emitted by radioactive material in the ground, it is responsible for part of the total IONIZATION of the air at lower levels over land. It also constitutes the chief hazard in 'close-in' RADIOACTIVE FALLOUT. See also ALPHA PARTICLE, BETA PARTICLE.

gas: A fluid of unlimited capacity for expansion under diminishing pressure. The term is applied to any substance which obeys approximately the 'gas laws' of BOYLE and of CHARLES, and the combination of these two laws in the GAS EQUATION.

gas constant: See GAS EQUATION.

gas equation: The pressure (p), specific volume (α), density (ρ), and temperature (T) of a PERFECT GAS are related by the 'gas equation' ('equation of state'):

---

* London, Admiralty, Hydrographic Department. Admiralty weather manual 1938. London, HMSO, 1938, p. 334.

$$p\alpha = RT \text{ or } p = R\rho T$$

where R = 'specific gas constant' = R^*/M.

Values quoted in Appendix C of WMO publication *Technical regulations*† are:
R^* = 'universal gas constant' (gas constant per gramme mole of perfect gas)
= 8·31432 joules/mol degK
M = MOLECULAR WEIGHT of gas concerned.

For dry air up to about 25 km,
M = 28·9644
and R = 0·28705 joules/g degK.

For water vapour,
M = 18·0153
and R = 0·46151 joules/g degK.

For moist air the value of M is smaller, and that of R is greater, than the respective values for dry air by amounts which depend on the percentage weight of water vapour in the air—see MOLECULAR WEIGHT.

gathering ground: An area from which water is obtained by way of rainfall, drainage or percolation. See also CATCHMENT AREA and DRAINAGE AREA.

gauss: The unit of magnetic flux density, or magnetic induction, in the C.G.S. SYSTEM of units. See also GAMMA, TESLA.

Gaussian (frequency) distribution: See NORMAL (FREQUENCY) DISTRIBUTION.

gegenschein: A faintly luminous patch of light on the line of the ZODIACAL BAND at the point of the sky opposite to the sun. It is sometimes termed COUNTERGLOW, which is, however, also used of a different phenomenon.

general circulation: The term 'general circulation' has different meanings in different contexts and there is no unique definition. In its widest sense it is used to imply all aspects of the three-dimensional global flow and energetics of the whole atmosphere. In this sense the general circulation is exceedingly complex, involving fluctuations on all time and space scales as well as many other considerations. For many purposes, therefore, study of the general circulation is rendered more specific by the application of some form of averaging, in time or in space, or in both. This gives rise to various kinds of average circulation in which the temporal variations of high frequency and/or the spatial variations of small scale are filtered out so as to reveal the character of the longer-period/larger-scale events. The term 'general circulation' is often used to denote one or other of the various kinds of average circulation which can be so generated, usually supported by appropriate statistical information concerning the temporal or spatial variability about that average.

As an example, a systematic annual variation of the circulation is revealed by averaging the flow or pressure fields, which are intimately related, over individual calendar months or seasons. The number of years' data required for adequate suppression of the fluctuations is small in low and high latitudes but is much greater in middle latitudes, especially in the northern hemisphere where occurrences of strikingly abnormal mean circulation over large areas for a single month are not uncommon.

An idealized mean surface circulation with associated pressure distribution, appropriate to equinox and a uniform surface of the earth, is considered to be:
(i) A narrow belt of light or variable winds converging in a shallow belt of low pressure on the equator (DOLDRUMS).

---

† Geneva, World Meteorological Organization. Technical regulations, Vol. 1, General. WMO–No. 49. BD.2. 3rd edn. Geneva, WMO, 1968, p. 58.

(ii) TRADE WINDS (or 'tropical easterlies'), north-east in northern hemisphere and south-east in southern hemisphere, between about latitudes 30° N and 30° S and the doldrums.
(iii) Light, variable winds associated with high-pressure belts (subtropical anticyclones) in latitudes 30°–40° N and S.
(iv) Belts of 'westerlies', south-west in northern hemisphere and north-west in southern hemisphere, between about latitudes 40° and 60°.
(v) Variable winds converging in a low-pressure belt at about 60° N and 60° S ('temperate storm belts').
(vi) Regions of outflowing winds with an easterly component, diverging from high pressure near the poles.

The actual mean surface circulations in January and July are shown in Figures 17 and 18. The appreciable departures from the idealized flow, more especially in the northern hemisphere, are due mainly to the non-uniform character of the earth's surface, with associated continental winter anticyclones and summer depressions.

The general circulation is maintained against dissipative processes (mainly friction with the earth's surface) by the heat energy from the sun. The earth and its atmosphere are continuously losing heat by radiation, mostly in the infra-red region of the spectrum and in the long term there must be a global balance between this loss and the incoming solar energy. Because of the special radiative properties of its constituents, the earth's atmosphere prevents much of the long-wave radiation from being lost to space whilst at the same time being nearly transparent to solar radiation. In the absence of the atmosphere the average surface temperature of the globe would be around freezing instead of +15°C and the diurnal and annual ranges of temperature would be very much larger than they are.

Although there is a long-term global balance between incoming and outgoing radiation, considerable imbalances exist locally. Indeed, there is a substantial excess of effective incoming radiation in low latitudes up to about 40° N and S and a deficit from there to the poles. Alternatively, one may say that the mean atmospheric temperature in low latitudes is lower, and in high latitudes higher, than those appropriate to local radiative balance between incoming solar and outgoing terrestrial radiation. This situation is possible because of the global-scale mixing operation performed by the general circulation. In particular the circulation mixes the air of higher and lower latitudes. In middle latitudes the mixing is mainly effected by the cyclones and anticyclones which *inter alia* transform into kinetic energy the available potential energy established by differential heating.

The existence of a latitudinal mixing of air implies a three-dimensional mean circulation. At any given latitude the continuity demands that average mass transport of air polewards, integrated over all heights, equals that equatorwards; mean meridional flow in the upper air (of about 1 m/s) is polewards in low (0°–30°) and high (60°–90°) latitudes and equatorwards in middle (40°–60°) latitudes, balancing the opposite meridional components of the lowest layers. The (very small) mean vertical components of motion which are implied act upwards in the low-pressure belts near the equator and 60° N and S and downwards in the high-pressure belts at 30°–40° N and S and near the poles.

At any given level above the surface the mean flow is more nearly along a parallel of latitude than is the case at the surface. The main feature of the upper air mean circulation is an increase of the westerly wind component with height within the troposphere (increasing THERMAL WIND component) except in equatorial latitudes where easterly winds generally prevail. The strongest mean winds are the westerlies near the tropopause in about latitudes 30°–40° N and S; in winter their mean speed is in excess of 80 knots, in summer some 30–40 knots. Above the tropopause the latitudinal gradient of temperature is reversed and the westerly wind

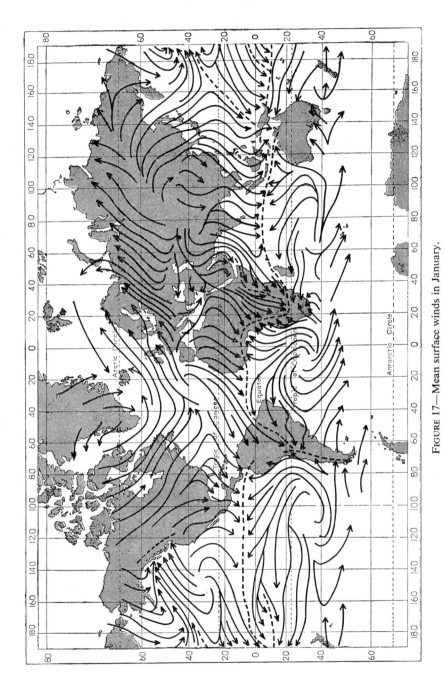

FIGURE 17—Mean surface winds in January.

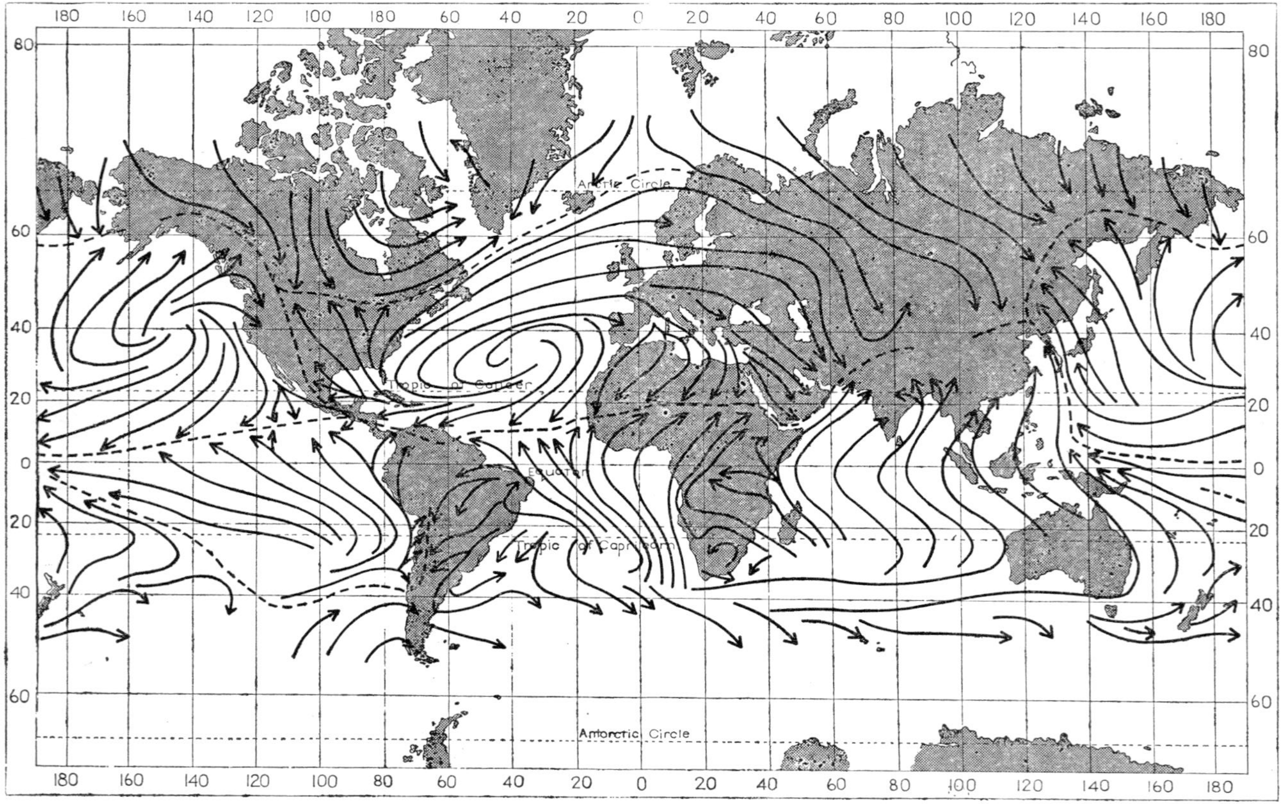

FIGURE 18—Mean surface winds in July.

component decreases with height except in high latitudes in winter—the 'polar-night JET STREAM'.

Apart from diagnostic analysis of the observations the large-scale motions of the atmosphere may be studied to some extent by means of laboratory experiments in which a fluid contained in a rotating vessel is differentially heated. Clearly the extent to which such a model can simulate realistically the behaviour of the actual atmosphere is severely limited; nevertheless useful insight into some of the properties of atmospheric motions can be gained. See DISHPAN EXPERIMENT.

An alternative line of study is to set up the mathematical equations governing the motion, heat balance, etc., of the atmosphere and to solve the system numerically on an electronic computer. Essentially one attempts to study the time evolution of the atmosphere starting from prescribed initial conditions. Such studies have come to be known as numerical simulation experiments because in a sense the computer is analogous to a laboratory with the system of equations taking the place of a physical model and indeed often being described as a mathematical model of the atmosphere. Such experiments may be conducted for a variety of reasons. Thus one may integrate forward in time for a long period in order to monitor the behaviour of the model and, if this closely resembles that of the real atmosphere, to exploit events within the model to gain a deeper understanding of the way the circulation operates and fulfils its various functions. Alternatively one may wish to appraise the effect on the circulation and climate of variations in the external parameters or of possible methods of climate modification. Numerical simulation is thus a powerful method for studying the atmospheric general circulation. See also POLAR VORTEX, QUASI-BIENNIAL OSCILLATION, SLANTWISE CONVECTION.

general inference: A term used in weather forecasting for a description of the general pressure distribution and the changes of pressure distribution which are in progress, together with a general statement of the type of weather likely to be experienced. It usually precedes a series of more detailed forecasts for individual districts and gives the framework on which these forecasts are based.

genitus: See CLOUD CLASSIFICATION.

geodesy: The science concerned with the size and shape of the earth.

geodynamic metre: See GEOPOTENTIAL.

geomagnetism: The study of the nature and causes of the earth's magnetic field and its variations.

The uniform dipole field which best fits the distribution of the earth's magnetism has 'geomagnetic' (or 'axis') poles at about 79° N, 70° W, and 79° S, 110° E, which define a system of 'geomagnetic co-ordinates' (latitude Φ, east longitude Λ); the 'geomagnetic axis' joining these poles is inclined at 11° to the geographical axis. The two 'magnetic' (or 'dip') poles, where a freely suspended magnet is vertical, are some 1000 to 2000 km distant from the respective geomagnetic poles. Except in higher latitudes, lines of equal geomagnetic latitude are almost parallel to lines of equal angle of DIP (angle of inclination of magnet with respect to horizontal), and angles between geomagnetic and geographic meridians approximate to measured angles of DECLINATION (azimuth setting of magnet with respect to geographic north). The field intensity is about 65μT (0.65 gauss) near the poles and 25μT near the equator. Decrease of intensity with height above ground proceeds approximately as the cube of distance from earth's centre.

Measurements of the magnetic field vector are made at a world-wide network of some 100 observatories. Continuous photographic recordings of the three independent components, horizontal force (H), angle of declination (D), vertical force (V), are made, and the variations recorded on such MAGNETOGRAMS are reduced to absolute measure by regular calibration. The high standard of accuracy of about 1 nT (1 gamma), is attained in these measurements.

More than 99 per cent of the earth's magnetic field is of internal origin. Observations since the 17th century have revealed the presence of slow, but cumulatively large, changes of strength and orientation of the field. Their cause is now thought to be large-scale vortices in the conducting molten core of the earth, slowly circulating across the earth's main field and so producing, by dynamo action, slowly changing regional magnetic fields. The origin of the main field itself is not yet satisfactorily explained.

Recent studies in PALEOMAGNETISM indicate some very large changes of field orientation on the geological time-scale. World-wide extension of these studies has yielded evidence of CONTINENTAL DRIFT on this time-scale.

The small part of the earth's field which is of external origin is produced by the entry into the high atmosphere of electromagnetic waves and particles emitted by the sun. Each of these produces characteristic effects on the magnetograms which thus provide a valuable continuous measure of solar activity.

The action of the solar wave-radiation is to ionize the high atmosphere and so make it electrically conducting. The sun and moon cause in the atmosphere tidal movements which, in the presence of the earth's magnetic field, induce, by dynamo action, electric currents in the conducting region. The magnetic field of these currents is observed at the ground, superposed on the earth's main field. The varying field produced by the thermal and tidal actions of the sun is, in general, clearly visible on the magnetograms as a characteristic local-time (S) variation of the magnetic elements. The purely tidal action of the moon on the atmosphere gives rise to a smaller varying field the existence of which is demonstrated by arranging the magnetic element values according to lunar-time (L) variation. The amplitude of the solar daily variation of the elements, on other than highly 'disturbed' days, is found to be a good relative measure of the sun's ionizing wave radiation. The lower IONOSPHERE, at 60 to 100 km, is the region mainly responsible for these magnetic effects.

The magnetic 'disturbance' produced by solar particles is a highly complex phenomenon which is most frequent in the 'auroral zones', at about 70° geomagnetic latitude, where it is never entirely absent. At higher levels of disturbance, especially beyond the rather arbitrary lower level of a 'magnetic storm', rapid field variations are world-wide and are accompanied by ionospheric disturbances, which produce anomalous radioreception, and by large earth currents, which adversely affect cable telegraphy. At such times AURORA is visible far equatorwards of its normal position. Field-strength fluctuations during large storms amount to about 3–8 per cent of the undisturbed value, depending on latitude. Field direction changes are in general a few degrees, but are much larger in high latitudes.

Measurement of disturbance intensity is usually made in terms of the range of the magnetic elements. For each Greenwich day, a subjective estimate of the degree of magnetogram disturbance, on the character (C) scale 0, 1, 2, is made for each observatory and averaged for all observatories to give an international character (Ci) figure on the scale 0·0, 0·1, etc. to 2·0. Such daily disturbance character figures are available from 1890, but are now largely replaced in practical use by the 'K index'. This index measures for each observatory, on the scale 0, 1, etc. to 9, the range, in excess of the characteristic wave radiation effect, of the most disturbed of the three magnetic elements in each 3-hour period, 0–3, 3–6, etc. to 21–24 GMT. The K index scale is semi-logarithmic in absolute force units and is fixed for each observatory, once for always, to suit the relative intensity of disturbance at the observatory. A 'planetary' disturbance index K_p for each 3-hour period is derived from the individual K indices of selected observatories. During the period of the INTERNATIONAL GEOPHYSICAL YEAR further disturbance indices Q, relating to each 15-minute period, were prepared at some observatories.

Equatorwards of the auroral zone, the diurnal maximum of disturbance occurs near local midnight (except near the geomagnetic equator) and the seasonal maximum occurs at the equinoxes; polewards of the zone, these maxima occur during the local day and summer, respectively. Although these and other systematic effects in the space and time variations of magnetic disturbance are reasonably well explained, there is yet no unifying theory of the storm phenomenon or of the closely associated aurora. Fundamental difficulties persist concerning the precise nature of the solar stream of charged particles which approaches the earth, and the interaction of this stream with the earth's main magnetic field. Rapidly varying electric currents, concentrated mainly in the auroral zones, together with a 'ring current' in the plane of the equator at a distance of several earth radii, explain qualitatively many of the observed features of disturbance. There is evidence that dynamo action in the high atmosphere is important in the production of the former currents. The existence of the equatorial ring current is not yet definitely confirmed.

The connection between magnetic disturbance and the physical state of the sun is shown, for example, by a 27-day recurrence tendency (solar rotation period) and an approximate 11-year (solar-cycle) variation of disturbance frequency and intensity. The correlation between individual magnetic disturbances and visual, radio or photographic evidence of solar activity is, however, relatively weak. Dependable prediction of magnetic storms has therefore not yet been achieved.

geometric mean: If $x_1, x_2, \ldots x_n$ are n measurements of the same kind, then $(x_1 \times x_2 \times \ldots x_n)^{1/n}$ is the geometric mean. See also MEAN.

geophysics: That branch of physics concerned with the earth and its atmosphere. The seven international Associations which at present constitute the International Union of Geodesy and Geophysics (IUGG) are Geodesy, Geomagnetism and Aeronomy, Scientific Hydrology, Meteorology and Atmospheric Physics, Physical Oceanography, Seismology and Physics of the Earth's Interior, and Vulcanology; other participating disciplines in the INTERNATIONAL GEOPHYSICAL YEAR project were Cosmic Rays, Ionosphere, and Solar Activity.

geopotential: The potential energy per unit mass of a body due to the earth's gravitational field referred to an arbitrary zero. The dimensions are L^2T^{-2}.

Geopotential (Φ) depends on geometric height (z) and GRAVITY (g) in accordance with the equation $\Phi = \int_0^z g\,dz$, mean sea level being the selected level of zero potential. In general, a geometrically level surface is not one of constant geopotential because of changes of the value of g along it.

The unit of geopotential is the potential energy acquired by unit mass on being raised through unit distance in a field of gravitational force of unit strength.

Geopotential is, from the dynamical point of view, a better measure of height in the atmosphere than is geometric height; energy is in general lost or gained when air moves along a geometrically level surface but not when it moves along an equigeopotential surface. A 'dynamic height' unit, as defined above, may be used but, by international agreement, a 'geopotential height' unit is preferred because it has the advantage of giving an even better fit, on average over the world, with geometric height, while retaining the dimensions and physical significance of geopotential. The two units are related by the equation: 1 geopotential metre = 0·98 dynamic metre.

The equation which defines the relationship between geopotential height (Z) and geometric height (z) is $Z = gz/9{\cdot}8$. Thus, where g has its near average value of 9·8 m/s², heights in geopotential metres and geometric metres are the same; for $g < 9{\cdot}8$ m/s² the height in geopotential metres is the smaller, for $g > 9{\cdot}8$ m/s² it is the bigger.

geopotential metre: See GEOPOTENTIAL.

geosphere: That part of the earth which is either solid or is composed of water, i.e. the LITHOSPHERE and HYDROSPHERE combined.

geostrophic approximation: See QUASI-GEOSTROPHIC MOTION.

geostrophic departure (or deviation): See AGEOSTROPHIC WIND.

geostrophic vorticity: Vorticity evaluated on the assumption of GEOSTROPHIC WIND conditions, as from the contours of a ISOBARIC SURFACE.

If $\nabla^2 z$ is the two-dimensional Laplacian of height at a point on an isobaric surface then the 'relative geostrophic vorticity' is given by

$$\zeta_G = (g/f)\nabla^2 z$$

where g is the gravitational acceleration and f the CORIOLIS PARAMETER. See also VORTICITY, LAPLACIAN.

geostrophic wind: That horizontal equilibrium wind (V_G), blowing parallel to the isobars, which represents an exact balance between the horizontal PRESSURE GRADIENT FORCE ($-(1/\rho)\nabla_H p$) and the horizontal component of the CORIOLIS FORCE (fV_G), where $\nabla_H p$ is the horizontal pressure gradient, ρ the air density, and f the CORIOLIS PARAMETER. Low pressure is to the left of the wind vector in the northern hemisphere, to the right in the southern hemisphere (see Figure 19). The magnitude of V_G is given by

$$V_G = |(1/f\rho)\nabla_H p|.$$

In terms of the height gradient of an isobaric surface ($\nabla_p z$), the magnitude of V_G is

$$V_G = |(g/f)\nabla_p z|$$

where g is the gravitational acceleration.

V_G blows parallel to the contours with the low values of height to the left in the northern hemisphere, and to the right in the southern hemisphere.

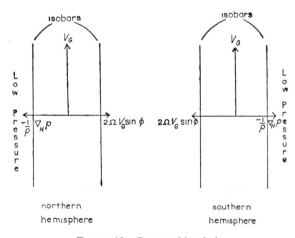

FIGURE 19—Geostrophic wind.

Geostrophic wind scales, based on the above relationships, are used and give a good approximation to the actual wind in the free atmosphere. A scale based on the isobaric surface relationship has the advantage of being independent of ρ (air density) and so may be used at any isobaric level.

glacial phase: A period during an ICE AGE (alternating with 'interglacial phases') when there is marked extension of ICE SHEETS equatorwards from polar regions; for the northern hemisphere average extension south of 75° N has been proposed (by G. C. Simpson) as a suitable definition of such a period.

glaciation: A term applied to the sudden transformation, near the top of a developing shower cloud, of supercooled cloud droplets into ice crystals; the transformation is often marked by the formation of ANVIL CLOUD.

glacier: An extensive mass of ice which is formed over land where there is net accumulation of snow. It is considered that growth of glaciers is favoured by snowy winters and cool summers, while recession or disappearance of glaciers is favoured by relatively dry winters and warm summers. Observed changes in thickness and areal extent of glaciers are used as important climatic indicators.

glacier wind: A gravitational (downhill) flow which develops above a GLACIER in day-time, especially in summer, because of the low temperatures then attained close to the glacier, relative to the surroundings. Maximum wind speed occurs some 2 m (6 ft) above ground.

glaciology: The study of the distribution and behaviour of snow and ice on the earth's surface. There is appreciable meteorological significance in this study, particularly in relation to climatic changes—see GLACIER. See also EUSTASY.

Glaisher stand: A form of stand devised by James Glaisher for the exposure of thermometers. The stand consists of a vertical portion, partially roofed, on which the thermometers are mounted, with a doubly roofed sloping rear portion designed to prevent the front portion becoming heated from the rear. The whole is capable of rotation about a vertical axis so that direct sunshine may be prevented from affecting the thermometers at all times. The stand has been superseded by the THERMOMETER SCREEN.

glaze (clear ice): A generally homogeneous and transparent deposit of ice formed by the freezing of supercooled drizzle droplets or raindrops on objects the surface temperature of which is below or slightly above 0°C.
It may be produced by the freezing of non-supercooled drizzle droplets or raindrops immediately after impact with surfaces the temperature of which is well below 0°C.
Glaze is observed at the earth's surface when raindrops fall through a layer of subfreezing temperature of sufficient depth.
On rare occasions glaze may attain sufficient thickness to bring down telegraph wires or present a serious hazard to ships. See BLACK ICE, ICE FORMATION ON AIRCRAFT.
Glaze on the ground must not be confused with 'ground ice', which is formed when:
    (i) water from a precipitation of non-supercooled drizzle droplets or raindrops later freezes on the ground,
    (ii) snow on the ground freezes again after having completely or partly melted, or
    (iii) snow on the ground is made compact and hard by traffic.

global radiation: The sum of direct and diffuse radiation received by unit horizontal surface. See RADIATION.

glory: The system of coloured rings similar to those of a CORONA round sun or moon, surrounding the shadow of an observer's head on a bank of cloud or mist. The phenomenon is also termed 'anticorona'. A several-fold effect is sometimes observed, while a FOGBOW may be seen to surround a glory.

When light passes through circular holes in an opaque screen, colours are produced by DIFFRACTION. If little mirrors all facing the sun could be substituted for the droplets in a cloud, the light from each mirror would behave as if it came through a hole from the reflection of the sun and similar diffraction colours would occur. The action of the drops is probably analogous. The mathematical theory developed by B. Ray is on these lines. Earlier writers had supposed that the phenomena were produced by the diffraction, by particles comparatively near the surface, of light reflected from deeper portions of the fog or cloud. See Plate 20.

Gold slide: An attachment, devised by E. Gold, for a marine mercury barometer to allow of the rapid correction and reduction to sea level of the reading of such a barometer with sufficient accuracy.

The ATTACHED THERMOMETER is mounted on a brass stock and the corresponding 'barometer correction scale' of millibars (1 mb per 6 degK difference from STANDARD TEMPERATURE) is mounted on a vertical slide. The position of the zero of this scale is altered according to the (closely approximate) corrections required for index error, latitude difference from 45°, and height of barometer above sea level; the required correction is then read from that part of the scale, in its adjusted position, opposite the end of the thermometer column.

goodness of fit: A term sometimes used for the measure of agreement between a set of observed frequencies and the frequencies expected according to some hypothesis which forms the subject of the CHI-SQUARE TEST.

gradient: The word 'gradient' is used in surveying and in common practice to indicate the slope of a hill, i.e. the change in height per unit distance along the hill. In mathematics, the gradient of a function φ is a vector, written grad φ or $\nabla \varphi$ or del φ, whose direction is that in which φ increases most rapidly and whose magnitude is the rate of increase of φ with distance in this 'up-gradient' or 'ascendant' direction. In meteorology, the PRESSURE GRADIENT FORCE acts from high to low values of pressure, i.e. in the 'down-gradient' or 'descendant' direction.

For synoptic and other purposes, attention is often confined to gradients in the horizontal plane, as for example the pressure gradient at mean sea level, as defined by mean-sea-level isobars. In common but not universal usage, the term 'temperature gradient' is reserved for temperature change with horizontal distance, change of temperature with height being referred to as 'temperature lapse rate' (or simply 'LAPSE rate'). The same is true of humidity. Similar considerations apply to gradients in an isobaric surface.

gradient wind: That equilibrium horizontal wind (V), blowing parallel to curved isobars of radius of curvature r, whose centripetal acceleration (V^2/r) represents the net inward horizontal force acting per unit mass of air. The only forces considered to be acting are the horizontal components of the PRESSURE GRADIENT ($-(1/\rho)\nabla_H p$) and CORIOLIS ($2\Omega V \sin \varphi$) forces (see Figure 20). The equations, for cyclonic and anticyclonic curvature of isobars, are:

PLATE 20   Glory: Tower Ridge, Ben Nevis.

PLATE 21 Halo.

cyclonic (acceleration in direction of pressure gradient force)

$$-(1/\rho)\nabla_H p + 2\Omega V \sin \varphi = -\frac{V^2}{r}$$

anticyclonic (acceleration opposite to pressure gradient force)

$$-(1/\rho)\nabla_H p + 2\Omega V \sin \varphi = +\frac{V^2}{r}.$$

In middle latitudes the gradient wind speed (V) is normally a rather closer approximation to the actual wind speed than is the GEOSTROPHIC WIND speed (V_G). V and V_G are related by the equation

$$V = V_G \pm \frac{V^2}{2r\Omega \sin \varphi}$$

(+ for anticyclonic, − for cyclonic curvature)

where the term $V^2/2r\Omega \sin \varphi$ is the so-called 'cyclostrophic component' of the wind. This latter is the AGEOSTROPHIC component of the gradient wind: it is normal to and to the left of the acceleration.

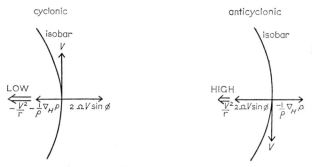

FIGURE 20—Gradient wind.

gramme: The unit of mass in the C.G.S. SYSTEM of units. It is one-thousandth part of the standard kilogramme which was originally supposed to represent the weight of a cubic decimetre of pure water at 4°C, but subsequent research has shown that the relationship is not exact. One gramme = 15·4 grains, or rather more than 1/30 ounce.

gramme-calorie: See CALORIE.

Grashof number: In heat transfer by free CONVECTION, a non-dimensional parameter ((Gr)) defined by the relationship $(Gr) = gl^3 \triangle T/\nu^2 T$ where l is a characteristic length, g the gravitational acceleration, ν the kinematic VISCOSITY, and $\triangle T$ the difference between the temperatures of the surface and the air in contact with it.

grass minimum temperature: The minimum temperature indicated by a thermometer freely exposed in an open situation at night with its bulb in contact with the tips of the grass blades on an area covered with short turf. See also GROUND FROST, CONCRETE MINIMUM TEMPERATURE.

graupel: The German word for 'soft hail'. See HAIL.

gravitational equilibrium: An alternative for DIFFUSIVE EQUILIBRIUM.

gravitational separation: The separation of particles, which are free to fall to earth, by virtue of their different TERMINAL VELOCITIES. For particles of a given substance, for example raindrops, the terminal velocities acquired increase with particle size.

Gravitational separation of the gases which form the atmosphere—usually then termed 'diffusive separation'—in accordance with the different weights of their molecules or (at high levels) atoms, is negligible below a level of about 80 km, because of the efficient mixing of the air at the lower levels. See also DIFFUSIVE EQUILIBRIUM.

gravity: The force of attraction between material bodies. The law of universal gravitation is that every mass attracts every other mass with a force which varies directly as the product of the attracting masses and inversely as the square of the distance between their centres of mass. The gravitational acceleration (g), i.e. the acceleration produced in a body which is free to move by the downward pull of the earth, is the force acting per unit mass of the body and is about 9·8 m/s² near the earth's surface. It decreases with increasing height in the atmosphere, being about 3 per cent less at a height of 100 km.

Because of the earth's rotation about its axis the observed force at the earth's surface is the vector resultant of the universal gravitational force and the centrifugal force which arises from the earth's rotation (force $\Omega^2 r$, where Ω is the earth's angular velocity and r the distance from the earth's axis, acting at right angles to the earth's axis). Since the latter force varies with latitude, being zero at the poles ($r = 0$) and a maximum at the equator (but of opposite sense there to the gravitational force), the force of apparent gravity also varies with latitude and has a maximum value at the poles and minimum value at the equator. The direction of the force of apparent gravity defines the direction of the local vertical, being normal to a *level* surface at the corresponding point. Because of the spheroidal shape of the earth (equatorial radius about 21·5 km longer than polar radius)—also the result of the earth's rotation—the force of gravity is not, in general, directed exactly towards the earth's centre. Small local anomalies of gravity occur and are associated with local topographical features and variations of mass distribution.

The formula for the variation of gravity over the earth's surface which has been used by the Meteorological Office since 1 January 1955 in applying corrections to barometer readings is:

$$g_{\varphi, h} = 980 \cdot 616(1 - 0 \cdot 0026373 \cos 2\varphi + 0 \cdot 0000059 \cos^2 2\varphi) - 0 \cdot 00009406 \, h$$

where $g_{\varphi, h}$ is the gravitational acceleration in centimetres per second² at latitude φ and at station height h in feet. However, the formula recommended by the World Meteorological Organization in *Technical regulations** contains additional terms which allow for the effect of the mean height of land or depth of water within 150 km of the point of observation. See also STANDARD GRAVITY.

gravity wave: A type of wave, also referred to as a 'gravitational wave', in which the controlling forces are GRAVITY and BUOYANCY. Such waves may be generated at a free surface of a single layer, as at an ocean surface, or at the boundary between adjacent layers or in a stably stratified medium. In the atmosphere additional forces may control the particle movement, as in a SHEAR-GRAVITY WAVE.

great circle: Any plane which passes through the centre of a sphere cuts the surface of the sphere in a 'great circle'.

---

* Geneva, World Meteorological Organization. Technical regulations, Vol. 1, General. WMO–No. 49, BD.2, 3rd edn. Geneva, WMO, 1968, Appendix B.

green flash: On some occasions the last glimpse obtained of the sun at sunset, or the first glimpse at sunrise, is a brilliant green—the 'green segment'—lasting a few seconds. On still rarer occasions a 'green flash' or 'green ray', also lasting a few seconds, shoots above the horizon from the upper limb. The explanation is the greater REFRACTION of the short waves (violet, blue, green) than of the long waves (red) of white sunlight, coupled with the greater degree of RAYLEIGH SCATTERING experienced by the violet and blue rays. In a hazy atmosphere such differential scattering may not be appreciable and the flash may then appear blue or violet. It is probable, though not yet confirmed, that an unusual degree of refraction, such as occurs with a low-level inversion of temperature, is required for the phenomenon.

Differential refraction of white light is also the cause of the analogous very rare phenomenon of the 'red flash' which may occur when the sun's disc appears just below a bank of clouds near the horizon.

The green flash has been observed in association with the moon and planets on rare occasions.

greenhouse effect: The effect, analogous to that which has been supposed to operate in a greenhouse, whereby the earth's surface is maintained at a much higher temperature than the temperature (about 250 K) appropriate to balance conditions with the solar radiation reaching the earth's surface. The atmospheric gases are almost transparent to incoming solar radiation, but water vapour and carbon dioxide in the atmosphere strongly absorb TERRESTRIAL RADIATION emitted from the earth's surface and re-emit the radiation, in part downwards. See ABSORPTION.

gregale: A strong north-east wind occurring chiefly in the cool season in the south central Mediterranean, but used also of a strong north-east wind in other parts of the Mediterranean, for example the south coast of France (grégal) and in the Tyrrhenian Sea (grecale).

grenades sounding: A method of atmospheric SOUNDING in which wind velocities and temperatures are inferred, as functions of height up to about 90 km, from an analysis of the speed of travel of sound waves originating from a series of grenades fired at height intervals of a few kilometres from an ascending METEOROLOGICAL ROCKET. The sounds of detonation are received by an array of microphones on the ground. The times and positions of grenade bursts are determined by optical or radio means.

grey body: A 'grey-body radiator' is defined as one which emits, for every wavelength, the same proportion of the maximum or 'black-body' radiation at a given temperature. Treatment of the atmosphere as a grey-body radiator has been found to give results which are too inaccurate to be useful; consideration of the detailed ABSORPTION spectrum is required. See RADIATION.

gridding: In synoptic meteorology, a system of graphical addition or subtraction of two fields represented by isopleths, used for example in upper air work and in obtaining 'anomaly patterns'.

Gridding may be performed on a single chart, as illustrated in Figure 21. Alternatively, two superimposed charts may be used, each with a set of isopleths, the points of intersection of which are made visible by illuminating the charts from below.

Grosswetter: A German term which is used to indicate, with special reference to long-range weather forecasting, the main features of weather over a specified region and period of time. The use of this term involves the concept that certain variable but non-random influences (not yet definable) govern the large-scale pattern of weather development. 'Grosswetterlagen' are the synoptic situations corresponding to the types of Grosswetter.

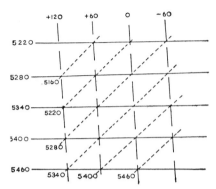

FIGURE 21—Example of gridding. Isopleths of heights of 500-mb surface (Z_{500} ——), 1000-mb surface (Z_{1000} — —) and 1000–500-mb thickness (T - - -), are drawn at 60m intervals.

Since $T = Z_{500} - Z_{1000}$, isopleths of Z_{500} may, for example, be obtained by graphical addition of T and Z_{1000}, or isopleths of Z_{1000} by graphical subtraction of T from Z_{500}. It is to be noted that the 1000–500-mb thickness isopleths pass through the intersections of the 1000-mb and 500-mb isopleths.

ground discharge: Lightning flash from cloud to ground. See LIGHTNING.

ground frost: From 1906 to 1960, inclusive, the Meteorological Office practice was to record a 'ground frost' when the grass minimum thermometer reached 30°F (-1°C) or below (30·4°F for thermometers read to tenths) and to base the 'number of days of ground frost' on this criterion. The reason for this choice is rather obscure; it may have been based on the belief that a temperature appreciably below 32°F (0°C) is required before damage is caused to the tissues of growing plants.

From 1 January 1961, the statistics issued have referred to the 'number of days with grass minimum temperature below 0°C' and no statistics have referred to 'ground frost'. The use of the term 'ground frost' in forecasts signifies a grass minimum temperature below 0°C (32°F).

ground, state of: Observations of the state of ground are made at selected stations. The code provides for 10 different conditions of the ground in the vicinity of a station, namely: dry; moist; wet (standing water in pools etc.); frozen; covered with glazed frost without snow; less than half covered with snow; more than half, but incompletely, covered with snow; completely covered with snow; half, but incompletely covered with loose, dry snow, dust or sand; and completely covered with loose, dry snow, dust or sand.

ground water: In HYDROLOGY, the water which is retained at all levels below the WATER TABLE.

growing season: That period of the year during which the growth of vegetation proceeds. For the common vegetation of north-west Europe a mean screen temperature exceeding 42°F (6°C) is regarded as an approximate critical value; the approximate length of growing season for any locality may therefore be obtained from the corresponding curve of annual variation of mean temperature.

Gulf Stream: Originating in the eastern area of the Gulf of Mexico, the ocean current known as the Gulf Stream flows through the Straits of Florida and up the eastern coast of the United States, following the edge of the continental shelf. Leaving the coast of America in about 40°N, it proceeds as a weaker and broader current across the Atlantic and reaches the British Isles in about 50°N. Its mean speed across the Atlantic is 5–8 km per day; near the American coast mean speeds of about 50 km per day are reached in spring and summer. Though one of the strongest and most constant of ocean currents, the Gulf Stream is subject to variability and even to reversals. While the term 'Gulf Stream' is popularly applied to the entire current system described above, more precise technical definition subdivides the system into the Florida Current to about 40°N, the Gulf Stream eastwards to 45°W, the North Atlantic Drift farther eastwards and northwards.

It is popularly supposed that the temperate climate of the British Isles is due to the warmth conveyed by the Gulf Stream; a more accurate view is that the temperate, maritime nature of the climate is due to the prevalence of south-west to west winds, which also cause the extension of the Gulf Stream towards the British Isles.

gust: A rapid increase in the strength of the wind relative to the mean strength obtaining at the time. It is a much shorter-lived feature than a SQUALL and is also different in nature, being due mainly to mechanical interference with the steady flow of air; it is, therefore, a pronounced feature of airflow near the ground, especially where there are large obstructions. Other factors, notably temperature lapse rate and vertical wind shear, are important, however, in determining the existence and magnitude of gusts. Such factors may produce gusts in circumstances where mechanical interference with the flow appears to be insignificant, as in CLEAR-AIR TURBULENCE.

The definition of the gust implies the existence of 'negative gusts' (lulls) of wind. The fluctuations of either sign are involved in the definition of 'gustiness'. The range between gusts and lulls increases with the mean wind strength; there is, however, also a dependence on anemometer exposure and, at most stations, on wind direction. See also GUSTINESS, TURBULENCE.

The strongest recorded gust at each of the seven stations of Table VI is listed in Table VII; the length of period of record is given for each station.

TABLE VII—*Strongest gust recorded on anemographs of selected Meteorological Office stations*

Station	Period of record years	Strongest gust kt	mile/h	Date
Lerwick	46	95	109	27 Jan. 1961
Tiree	31	101	117	26 Feb. 1961
Aberdeen Airport	17	88	101	31 Jan. 1953
Holyhead (Valley)	16	84	97	17 Jan. 1965
Isles of Scilly (St Mary's)	54	96	111	6 Dec. 1929
Shoeburyness	53	75	86	11 Jan. 1962
Spurn Head	29	79	91	7 Apr. 1943

gustiness: The important characteristics of a fluctuating wind, characterized by GUSTS and lulls about a mean level, are the frequency and strength of the gusts. The former is usually expressed by the number of wind maxima occurring within a specified period of time. The strength is usually defined in one of the following ways:

(i) In normal surface observations by a 'gustiness factor', i.e. by the percentage ratio of the difference between the maximum and minimum horizontal wind speeds to the mean wind speed recorded in a given period. In a selection of eight

British stations, values of the gustiness factor ranging from 25 to 100 per cent were obtained, large dependence of the factor on mean wind direction being found at some of the stations. A gustiness factor in ordinary observations may also be defined in terms of wind direction; the angular width (in radians) is a measure of lateral gustiness which for small values is nearly equivalent to the speed ratio.

(ii) In micrometeorology, if u', v', w' denote the fluctuations of the velocity components in three mutually perpendicular directions from the corresponding mean components, then the gustiness components may be represented by the standard deviations $\sigma_u = \sqrt{\overline{(u')^2}}$ etc. If the resultant mean wind $\overline{\mathbf{V}}$ is horizontal and in the direction of the x-axis, then the longitudinal, lateral and vertical 'intensities' of gustiness (or turbulence) are defined by $\sigma_u/\overline{\mathbf{V}}$, $\sigma_v/\overline{\mathbf{V}}$, $\sigma_w/\overline{\mathbf{V}}$, respectively.

For some purposes, especially the study of the effect of gusts on aircraft, the structure of individual gusts—as represented by the shape of the profile of gust velocity plotted against distance (or time)—is important. The most important 'gust shapes' are (a) the 'flat-topped' gust, in which the velocity increases uniformly, and (b) the 'sinusoidal' gust. In theoretical work, these are used as the standard gust in the U.K. and the U.S.A., respectively. Other gust-shape classifications are the 'sharp-edged' gust and the 'triangular' gust. The 'gust-length' is the distance occupied by the ascending part of the gust profile; it is typically about 30 m. See also TURBULENCE, EDDY, EDDY SPECTRUM.

H

haar: A local name in eastern Scotland and parts of eastern England for a sea fog which at times invades coastal districts. Haars occur most frequently in spring and early summer months. See ADVECTION FOG, SEA FRET.

haboob: The name, derived from the Arabian *habb* meaning to blow, applied to a DUSTSTORM in the Sudan north of about 13°N. Such storms occur from about May to September and are most frequent in the afternoon and evening.

Hadley cell: A simple thermal circulation first suggested by George Hadley in the 18th century as part explanation of the TRADE WINDS and still thought to be approximated to in the troposphere between latitudes 0° and 30°. If the effects of the earth's rotation are neglected, the circulation comprises a high-level poleward flow from heat source to heat sink in response to a horizontal pressure gradient (at a high level, pressure is greatest above the heat source since pressure decreases less rapidly with height in the warmer air column), and a compensating low-level flow towards the heat source; upward and downward motions at the heat source and sink, respectively, complete the circulation—see Figure 22.

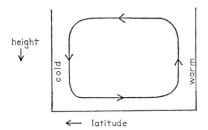

FIGURE 22—Hadley cell.

Haidinger's brush: A faint and transitory figure, in the shape of a yellowish brush with a small blue 'cloud' on either side, which appears when a source of strongly polarized light, such as the blue zenith sky with the sun close to the horizon, is closely observed. Because of physiological differences, the phenomenon is apparently not seen by all observers in quite the same form.

hail: Solid PRECIPITATION in the form of balls or pieces of ice (hailstones) with diameter ranging from 5 mm to 50 mm or even more. The stones fall from CUMULONIMBUS clouds and are commonly spherical or conical in shape although they sometimes form irregular lumps by agglomeration. The structure often consists of concentric layers of alternately clear and opaque ice, and the density varies between 0·1 and 0·9 g/cm³ according to structure.

Two other varieties of hail are recognised: (i) 'soft hail' (or 'graupel')—crisp, opaque and easily compressible pellets and (ii) 'small hail'—pellets with a soft hail nucleus and outer coating of clear ice. In either case the diameter rarely exceeds a few millimetres; they are classed as ICE PELLETS, but as hail if their diameter should exceed 5 mm. The density of hail varies between about 0·1 and 0·9 g/cm³, depending on structure.

Because of the high liquid-water concentration of cumulonimbus clouds, the growth of ice particles, following their initial formation on ice nuclei at temperatures well below freezing and at saturation with respect to water, is mainly by collision and coalescence with supercooled water drops. The particles are supported within the cloud by strong updraughts (of about 10 m/s) and so are able to attain an appreciable size.

The opacity of the ice is caused by the trapping of air bubbles within the ice on freezing of supercooled water drops on the ice particles. Since this occurs only at temperatures lower than about $-5°C$ it has been suggested that alternate clear and opaque rings occur when the stone is fluctuating at about this temperature level in the cloud, supported by a (varying) updraught nearly equal to the TERMINAL VELOCITY of the stone. It is, however, also suggested that the clear (opaque) rings reflect slow (rapid) freezing of supercooled water drops on the stone, associated with a high (low) water concentration in the cloud, and that the concentric rings result, therefore, from variations of liquid-water content. (The freezing of the drops involves the release of latent heat which must be dispersed before the freezing can be complete.)

hail, day of: Up to and including 1959 a day on which HAIL, of whatever variety, was observed even though in so small a quantity as to yield no measurable amount in the rain-gauge.

Since 1959, a day of hail has been counted as one on which hail or soft hail pellets, 5 mm or more in diameter, have fallen. Days with ice or snow pellets up to 5 mm in diameter, and also with ICE PELLETS and ICE PRISMS, are counted separately.

half-life: In RADIOACTIVITY, the time taken for the activity of a given quantity of a radioactive element to decrease to one-half of its original value. The half-lives of radioactive elements range from a small fraction of a second to many thousands of years.

half-width: See JET STREAM.

halo phenomena: The term 'halo', which might be applied to any circle of light round a luminous body, is restricted by meteorologists to a circle produced by REFRACTION through ice crystals, in contrast to CORONAE which are produced by DIFFRACTION. All the optical phenomena produced by REFLECTION and refraction of light by ice crystals are sometimes grouped together as halo phenomena.

The most common halo is a luminous ring of 22° radius surrounding the sun or moon, the space within the ring appearing less bright than that just outside. The ring, if faint, is white but if more strongly developed the inner edge is a pure red, outside which yellow may be detected. The halo of 22° is very common. In England it can be seen by an assiduous observer about one day in three. See Plate 21.

The angle of 22° is the angle of MINIMUM DEVIATION for light passing through a prism of ice (refractive index 1·31) with faces inclined at 60°. Thus the occurrence of the halo of 22° radius indicates the presence of ice crystals with faces inclined at 60°. Alternate faces of a hexagonal prism are inclined at this angle, and as hexagonal prisms are frequently found amongst ice crystals the halo is probably due to the refraction of light through such prisms.

A halo of 46° is to be seen occasionally, though seldom complete. This halo requires crystals with faces at right angles.

The halo of 22° is sometimes within a circumscribed nearly elliptical halo, the points of contact being at the highest and lowest points. The complete circumscribed halo is only seen when the elevation of the sun is 40° or more. With lower elevations separate tangent arcs are seen. These phenomena are explained by the presence of prismatic ice crystals floating with their axes horizontal.

Another group of phenomena requires prismatic crystals with their axes vertical. In this group are PARHELIA (mock suns) and the CIRCUMZENITHAL ARC.

In weather lore, haloes are often spoken of as presaging storms. Haloes are, however, too common to be reliable signs of impending exceptional weather.

harmattan: A dry wind blowing from a north-east or sometimes easterly direction over north-west Africa. Its average southern limit is about 5° N latitude in January and 18° N in July. Beyond its surface limit it continues southwards as an upper current above the south-west monsoon. Being both dry and relatively cool, it forms a welcome relief from the steady damp heat of the tropics, and from its health-giving powers it is known locally as 'the doctor' in spite of the fact that it carries with it from the desert great quantities of dust. This dust is often carried in sufficient quantity to form a thick haze, which impedes navigation on the rivers.

harmonic analysis: The resolution of a series of measurements (for example a TIME SERIES) into harmonic components, or sine waves, of which the periods are fixed in advance but the PHASE and AMPLITUDE are determined from the data. The periods chosen are usually sub-multiples of a particular period, the fundamental. Since the analysis involves the use of FOURIER SERIES it is also termed 'Fourier analysis'.

The reverse process, that of reconstituting the original function by addition of the harmonic components, is termed 'harmonic synthesis'.

Harmonic analysis is perhaps the simplest example of the general method of representing data by means of ORTHOGONAL FUNCTIONS. When the fundamental period can be determined from independent evidence, harmonic analysis is a powerful tool for separating harmonic components of variation from NOISE. It has been used effectively, for example, to investigate the semi-diurnal variation of pressure and the annual variation of temperature. If, however, the fundamental period is chosen by guesswork, or by inspection of the data, it is often impossible to determine the statistical significance of the results. In these circumstances harmonic analysis is likely to be less fruitful than methods of analysis which start from the CORRELOGRAM.

harmonic dial: A representation, on a polar diagram, of the results of HARMONIC ANALYSIS, in which periodic components of a given frequency are compared. The components may, for example, refer to different stations at a given epoch, or to results for different seasons at a given station. Points are plotted on the diagram at positions corresponding to the amplitude, measured on a linear scale from the origin, and to the epoch of maximum of the component. The scale of angular measure employed in the analysis and the corresponding scale of time are generally both shown on such diagrams.

harmonic mean: If $x_1, x_2, \ldots x_n$ are n measurements of the same kind, then the harmonic mean $\bar{x}_h$ is given by

$$\frac{1}{\bar{x}_h} = \frac{1}{n} \sum_{j=1}^{n} \frac{1}{x_j}.$$

See also MEAN.

haze: A suspension in the air of extremely small, dry particles invisible to the naked eye and sufficiently numerous to give the air an opalescent appearance. There is no upper or lower limit to the horizontal visibility in the presence of which haze may be reported.

In most cases the particles composing haze are small enough (less than about 1 μm) to cause differential scattering of sunlight and to contribute, for example, to sunrise and sunset colours.

For the reduction of visibility by water droplets see FOG and MIST. See also DUST HAZE.

hazemeter: A term sometimes used synonymously with VISIBILITY METER. The 'loofah hazemeter' is an instrument in which the intensity of the light scattered at a particular angle to the original beam passing through an enclosed air sample is used as a measure of the SCATTERING coefficient of the air and hence its visibility (neglecting direct ABSORPTION of light).

head-wind: See EQUIVALENT HEAD-WIND.

health-resort station: A CLIMATOLOGICAL STATION which participates in the Meteorological Office health-resort scheme whereby participating stations send daily a coded report or reports relating to observed temperature, rainfall, sunshine and weather for issue to the Press. See '*Observer's handbook*'* for details.

heap clouds: CLOUDS of appreciable vertical development (CUMULUS and CUMULONIMBUS), as opposed to LAYER CLOUDS.

heat: A form of energy, normally measured in CALORIES or JOULES. The dimensions are ML^2T^{-2}.

The transfer of heat to or from a substance is effected by one or more of the processes CONDUCTION, CONVECTION, and RADIATION. The common effect of such a transfer is to alter either the temperature or the state of the substance (or both). Thus, a heated body may acquire a higher temperature ('sensible' heat) or may change to a higher state (thus acquiring latent or 'hidden' heat).

The relation between the joule (J) and other heat units is:

1 15 degC calorie (cal_{15}) = 4·1855 J
1 International Steam Table calorie (IT cal) = 4·1868 J
1 60 degF British thermal unit = 1054·54 J
1 International Steam Table British thermal unit = 1055·06 J
See also SPECIFIC HEAT and LATENT HEAT.

heat capacity: An alternative for THERMAL CAPACITY.

Heaviside layer: A layer of the IONOSPHERE at about 100 km height, now usually termed the E-layer; it has also been termed the 'Kennelly–Heaviside layer'.

heiligenschein: A diffuse, white ring of light surrounding the shadow of the head of an observer which is cast on dewy grass. The phenomenon occurs mainly when the sun's elevation is low and the observer's shadow long.

heliostat: An instrument mounting designed to provide automatic orientation of the instrument towards the sun or automatic direction of the light from the sun on

---

* London, Meteorological Office; Observer's handbook, 3rd edn. London, HMSO, 1969, p. 6.

to the instrument. A familiar form is that used, in conjunction with a PYRHELIOMETER, to obtain continuous measurement of direct solar radiation.

helium: So named because of its original discovery in the sun's atmosphere, helium is one of the INERT GASES. It occurs in very low concentration in the atmosphere, $5 \cdot 2 \times 10^{-6}$ and $7 \cdot 2 \times 10^{-7}$ part per part of dry air by volume and weight, respectively. Being very light, with a molecular weight of only 4·003, it escapes continuously from the top of the atmosphere at a rate which is in approximate balance with the rate of production of ALPHA PARTICLES (helium nuclei) near the earth's surface.

helm wind: A strong, cold, north-easterly wind which blows down the western slope of the Crossfell Range in Westmorland and Cumberland. Its greatest frequency is in late winter and spring. When the helm wind blows, a heavy bank of cloud (the 'helm') rests along or just above the Crossfell Range and a slender, nearly stationary roll of whirling cloud (the 'helm bar'), parallel with the 'helm', appears above a point one to six kilometres from the foot of the fell. The helm wind is very gusty and often violent as it blows down the steep fell sides but ceases under the helm bar cloud. To the west of this point a light westerly may prevail over a short distance. The helm wind has a marked drying effect. The phenomenon is an example of a LEE WAVE.

The term 'helm wind' is applied to a similar wind with associated cloud elsewhere, e.g. in the Lake District, but the full development is confined to the strip of country east of the River Eden, particularly near Crossfell itself.

hertz: The unit of frequency of a periodic function, equal to one cycle per second.

heterosphere: Term proposed for that region of the ATMOSPHERE, upwards of about 80 km, in which the composition of the atmosphere changes, hence also the mean molecular weight of the gases, mainly because of the partial DISSOCIATION of oxygen and DIFFUSION. It forms a contrast with the underlying HOMOSPHERE.

high: A term sometimes used in synoptic meteorology to indicate a high-pressure system, for which the term ANTICYCLONE was coined by Sir Francis Galton.

hill fog: A term generally used of low cloud which envelops high ground. The production of saturation and condensation by forced uplift of the air is not necessarily implied in the use of this term as it is in the case of UPSLOPE FOG.

histogram: A graphical representation of the frequencies of occurrence of a variable in certain ranges. The variable is taken as abscissa, and on each range is drawn a rectangle of area proportional to the number of cases in that range. If the ranges are all made equal then the heights of the rectangles are proportional to the frequencies.

The histogram allows the general character of a frequency distribution to be taken in at a glance; it is especially useful when, as for winter temperatures in Alaska, a description in terms of the mean and standard deviation would be misleading. See also FREQUENCY DISTRIBUTION.

hoar-frost: Ice crystals in the form of scales, needles, feathers or fans deposited on surfaces cooled by radiation or otherwise. The deposit is frequently composed in part of drops of DEW frozen after deposition and in part of ice formed directly from water vapour at a temperature below 0°C (sublimation). The presence of fog, in so far as it checks the radiational cooling of surfaces, tends to prevent the formation of hoar-frost. See also RIME.

hodogram: See HODOGRAPH ANALYSIS.

hodograph analysis: A method of analysis of a wind sounding at a station. By a recent convention which is contrary both to previous practice and to normal mathematical convention, the individual wind vectors at selected levels are plotted on a polar co-ordinate diagram from the origin of the diagram (representing station position) towards the direction from which the wind blows (see Figure 23). This new method has the advantages of being easier to apply in practice and of being in accordance with the normal method of representing wind direction at a station. The lengths of the vectors are proportional to the corresponding wind speeds. The sense of each vector is towards the origin.

The lines joining the 'starting points' of airflow at successive levels form a hodograph (or hodogram). Each such line corresponds to the WIND SHEAR vector in the layer concerned. When taken in the appropriate sense (higher to lower level) it represents, on the assumption of geostrophic flow, the corresponding THERMAL WIND.

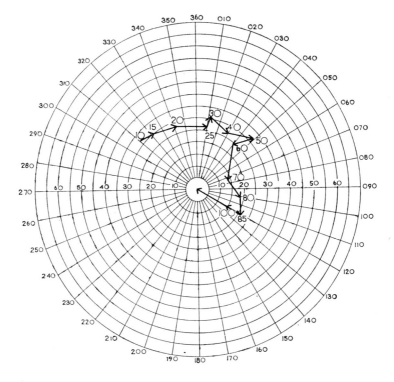

FIGURE 23—Hodograph relating to winds observed at Shanwell (56° 26′ N, 2° 52′ W) at 2230 GMT, 22 December 1961.
Successive arrows represent the vector difference between the wind at the top of a layer (denoted by figures, in tens of millibars at arrow tail) and the wind at the bottom of a layer (figures at arrow head). The bottom of the lowest layer is the earth's surface.

Hollerith system: A mechanical system used extensively for the processing of climatological data which are represented by the position of holes punched into cards of standard size.

homogeneous atmosphere: A hypothetical atmosphere in which air density is constant with height; the lapse rate is, by definition, the AUTOCONVECTIVE LAPSE RATE. The height (z) of such an atmosphere is from the hydrostatic and gas equations given by

$$z = p_0/g\rho_0 = RT_0/g,$$

where p_0, ρ_0 and T_0 are the surface pressure, density and temperature, R the gas constant for air, and g the gravitational acceleration. By ignoring the small variation of g with height, z is seen to be proportional to the absolute air temperature at the surface. For $T_0 = 273$ K, z is about 8 km. See also SCALE HEIGHT.

homogeneous condensation, freezing: See HOMOGENEOUS NUCLEATION.

homogeneous nucleation: Homogeneous or 'spontaneous' nucleation signifies the initiation of either CONDENSATION or FREEZING in the absence of condensation or freezing nuclei, respectively; the processes are also termed homogeneous or spontaneous condensation, and freezing, respectively.

Aggregates of water molecules continuously form and evaporate in supersaturated air which is free from condensation nuclei. The probability that such aggregates will attain a critical size big enough for them to become more stable than the vapour and so act as centres on which further rapid growth (spontaneous condensation) will occur is small for degrees of SUPERSATURATION less than about 400 per cent; such an order of supersaturation does not occur in the atmosphere.

In air which is deficient in freezing nuclei, lowering the temperature below 0°C increases the probability that aggregates of water molecules may take up an ice-like configuration and grow to a size sufficient for them to act as centres on which ice crystals may rapidly form. The variation with temperature of the probability of homogeneous nucleation is such that, in practice, there appears to be a critical temperature below which it occurs and above which it is absent. For water it is about $-40°C$.

homosphere: That region of the ATMOSPHERE, extending from the earth's surface to about 80 km, in which, neglecting water vapour, the composition of the atmosphere is constant (apart from some gases in very small concentration, e.g. carbon dioxide, ozone) and in which, therefore, the mean molecular weight of dry air is effectively constant. It forms a contrast with the overlying HETEROSPHERE.

horizon: In meteorology, this term signifies the line where the earth's surface and the sky apparently join. Neglecting REFRACTION, the distance of the horizon from an observer at height h is $\sqrt{(2ha)}$, where a is the earth's radius. For a height of 30 m the corresponding distance is 20 km; the actual distance is about 4 km greater on account of refraction. A level canopy of clouds 3 km high is visible from a point on the earth's surface for a distance of about 200 km, or the visible canopy has a width of 400 km. Distances corresponding to other heights may be obtained from Figure 24.

horse latitudes: The belts of variable, light winds and fine weather associated with the subtropical anticyclones at about 30°–40° latitudes. The belts move slightly north and south after the sun. The name arose from the old practice of throwing overboard horses, which were being transported to America or the West Indies, when the ship's passage was unduly prolonged.

hot-wire anemometer: See ANEMOMETER, ANEMOGRAPH.

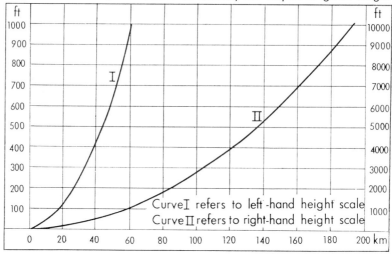

FIGURE 24—Relation between the height of an observing point in feet and the distance of the visible horizon in kilometres (neglecting refraction), or the height in feet of a cloud or other distant object and the distance in kilometres at which it is visible on the horizon.
Note: 1000 ft = 305 m; 80 km = 50 miles.

hour angle: The hour angle (H) of a heavenly body at any instant is the angle (usually expressed as a time on the scale 24 hours = 360°) between the observer's MERIDIAN and the meridian through the body. The angle is measured westwards from the observer's meridian.

humidity: The condition of the atmosphere in respect of its WATER VAPOUR content. The word 'humidity' used alone generally signifies RELATIVE HUMIDITY, but various other measures are employed such as HUMIDITY MIXING RATIO, VAPOUR CONCENTRATION, VAPOUR PRESSURE, SPECIFIC HUMIDITY, DEW-POINT, etc.

humidity mixing ratio: The humidity mixing ratio (r)—or, more generally, simply the 'mixing ratio'—of moist air is the ratio of the mass (m_v) of water vapour to the mass (m_a) of dry air with which the water vapour is associated, i.e.

$$r = m_v/m_a.$$

If e' is the vapour pressure, p the total pressure and ε the ratio of the densities of water vapour and dry air ($\varepsilon = 0\cdot 62197$), then

$$r = \varepsilon e'/(p-e').$$

Since e' is small compared with p, r (kg/kg) is a small quantity of about 0·01 kg/kg. For convenience, therefore, r is usually expressed in g/kg.

humidity slide-rule, tables: The slide-rule or tables, based on the psychrometric formula (with an assumed value of the product Ap in this formula appropriate to the conditions of exposure of the instruments), from which the DEW-POINT, VAPOUR PRESSURE, and RELATIVE HUMIDITY of an air sample may be obtained from readings of the dry- and wet-bulb hygrometer. See PSYCHROMETER.

humilis (hum): A CLOUD SPECIES (Latin, *humilis* low).
'CUMULUS clouds of only a slight vertical extent; they generally appear flattened'.*
See also CLOUD CLASSIFICATION.

hurricane: A name (of Spanish or Portuguese origin) applied to the intense TROPICAL CYCLONES which occur in the West Indies and Gulf of Mexico regions and off the Queensland coast. They are essentially of the same type as the western Pacific 'typhoon' and Bay of Bengal 'cyclone'.

In the BEAUFORT SCALE of wind force the name hurricane is given to a surface wind of force 12, corresponding to a mean speed of 64 knots or more. Mean speeds of this magnitude are very rarely attained in the British Isles, but the speed is frequently exceeded in gusts.

hurricane wave: The raising of the level of the sea by some 3–6 m in a restricted region near the centre of an intense TROPICAL CYCLONE.

hydrodynamic stability: An alternative for DYNAMIC STABILITY.

hydrogen: A gas, of molecular weight 2·02, which comprises $5·0 \times 10^{-7}$ and $4·0 \times 10^{-8}$ part per part of dry air by volume and weight, respectively.

Hydrogen is very abundant in the universe but occurs in only minute concentration throughout most of the atmosphere because its extreme lightness causes it to diffuse upwards to high levels. See also ATMOSPHERE.

hydrography: The study of water, more especially in the open seas and oceans, both from the physical standpoint and from that affecting the safety of navigation. In so far as tidal heights in estuaries and rivers are affected by river water, the science overlaps with that of HYDROLOGY.

hydrological balance: The hydrological balance, or hydrological budget, is the relationship between the EVAPORATION (E), PRECIPITATION (P), RUN-OFF (R), and the change in water storage ($\triangle S$), for a specified land area and period of time, and is expressed by the hydrological balance equation

$$P = E + R + \triangle S,$$

in which $\triangle S$ may be positive or negative.

hydrological budget: An alternative for HYDROLOGICAL BALANCE.

hydrological cycle: The full cycle of events through which water passes in the earth–atmosphere system, comprising EVAPORATION from land and water surfaces, CONDENSATION to form clouds, PRECIPITATION on to the earth's surface, movement and accumulation in the soil and bodies of water, and finally re-evaporation. 'Short-circuiting' of the complete cycle occurs in the form of evaporation of products of condensation and precipitation within the atmosphere.

hydrology: The study of the incidence and properties of water on and within the ground, including that held in rivers and lakes. It comprises studies of RAINFALL, EVAPORATION, RUN-OFF, GROUND WATER, SOIL MOISTURE, the HYDROLOGICAL BALANCE, snow and ice accumulation, and the chemistry of natural waters.

The practical applications of hydrology are extremely wide and the number of local and government agencies and other bodies concerned with the subject in

* Geneva, World Meteorological Organization. International cloud atlas, Vol. 1. Geneva, WMO, 1956, p. 13.

Great Britain is correspondingly large. The most comprehensive regular publications are the *Surface water year book of Great Britain** and the *Ground water year book.*†

hydrometeor: A generic term for products of CONDENSATION or SUBLIMATION of atmospheric water vapour. Hydrometeors include: ensembles of falling particles which may either reach the earth's surface (rain, snow, etc.) or evaporate during their fall (virga); ensembles of particles suspended in the air (cloud, fog, etc.); particles lifted from the earth's surface (drifting or blowing snow, spray); particles deposited on the earth or on exposed objects (dew, hoar-frost, etc.).

hydrometer: An instrument for measuring the density of liquids. In marine meteorology, hydrometers are used for determining the density of sea water.

hydrosphere: That part of the EARTH's surface which is covered by water substance.

hydrostatic equation: In an atmosphere at rest with respect to the earth, the variation of pressure (p) with height (z) is given by

$$\frac{\partial p}{\partial z} = -\rho g$$

where g is gravitational acceleration and ρ is air density.
In terms of GEOPOTENTIAL metres (Z) the equation is

$$\frac{\partial p}{\partial Z} = -9 \cdot 8 \rho,$$

the various quantities being expressed in SI UNITS.

The hydrostatic equation is very closely approximated when the air has horizontal motion relative to the earth and is therefore used as the basis for computations of height from vertical soundings.

hydrostatic equilibrium: The state of balance between the force of GRAVITY and the vertical component of the PRESSURE GRADIENT FORCE in the atmosphere. It is a state in which there is no vertical acceleration and the HYDROSTATIC EQUATION holds. See also QUASI-HYDROSTATIC APPROXIMATION.

hydroxyl: Hydroxyl molecules, of chemical formula OH, consist of one oxygen and one hydrogen atom. Hydroxyl has been identified in the spectrum of the AIRGLOW. It is considered to be formed by the DISSOCIATION of water vapour in the high atmosphere to form hydrogen atoms and hydroxyl.

hyetograph: A proprietary pattern of self-recording RAIN-GAUGE in which the recording pen is actuated by a series of stops attached to a vertical float rod.

hygrogram: The record made by a HYGROGRAPH.

hygrograph: A recording HYGROMETER. The type most familiar is the 'hair hygrograph' which uses the fact that human hair increases in length with increasing RELATIVE HUMIDITY. The changes in hair length are not linear, being proportionately

---

* Water resources board and Scottish development department. Surface water year book of Great Britain. London, HMSO.
† Water resources board and Scottish development department. Ground water year book. London, HMSO. (In the press.)

less at high relative humidity than at low, but the recorded changes are generally made linear by the design of the mechanism. Hair hygrographs are not very precise, being subject in particular to gradual changes, but have the advantage of operating above or below the freezing-point.

hygrometer: An instrument for measuring the humidity of the air. Among the many types of hygrometer are:

(i) The 'dry- and wet-bulb hygrometer' which is used in routine surface observations (see PSYCHROMETER).
(ii) Those hygrometers which use the property of expansion and contraction of certain materials with changing RELATIVE HUMIDITY as, for example, the 'hair HYGROGRAPH' in surface observations and gold-beater's skin in upper air observations.
(iii) 'DEW-POINT hygrometers' of the type designed by Daniell, Regnault and others, in which artificial cooling of a polished surface, whose temperature is measured, is continued until dew is seen to condense on it (see also FROST-POINT HYGROMETER).
(iv) Chemical hygrometers in which the quantity of moisture in a given mass of air is determined by direct weighing.
(v) Resistive hygrometers in which the effect of humidity on the electrical resistance of films of hygroscopic substances is determined.

hygrometric formula: An alternative for 'psychrometric formula'. See PSYCHROMETER.

hygroscope: An instrument for showing whether the air is dry or damp, usually by the change in appearance or dimensions of some substance. Hygroscopes are frequently sold in the form of weather predictors, e.g. 'weather houses' in which the appearance of the 'old man' or the 'old woman' is determined by the twisting and untwisting of a piece of catgut in response to changes of humidity.

hygroscopic: A hygroscopic substance is one which tends to absorb moisture by accelerating the condensation of water vapour.

hygrothermograph: A combined HYGROGRAPH and THERMOGRAPH, i.e. an instrument in which variations of atmospheric humidity and temperature are continuously recorded by separate traces on a single sheet. The instrument is sometimes termed a 'thermohygrograph'.

hypsography: The configuration of ISOHYPSES, i.e. lines of constant height of an ISOBARIC SURFACE.

hypsometer: Literally, an instrument for measuring height (Greek *hypsos*). In meteorology, however, the term is applied exclusively to an instrument in which the boiling-point of water is measured and the corresponding atmospheric pressure deduced from the boiling-point–pressure relationship (see Table VIII); if, as is normal, the instrument is used on a mountain and the pressure is known at some known (low) level, the pressures deduced at higher levels may be converted to heights by means of the ALTIMETER equation. An accuracy of one-hundredth of a degree (kelvin) in temperature measurement is required to obtain height to within 3 m.

TABLE VIII—*Dependence of boiling-point of water on atmospheric pressure*

Boiling-point K	Pressure*	
	mm of mercury	*mb*
374	787·67	1050·12
373	760·00	1013·23
372	733·16	977·45
371	707·13	942·74
370	681·88	909·08
369	657·40	876·44
368	633·66	844·79
367	610·64	814·10
366	588·33	784·36
365	566·71	755·54
364	545·77	727·62

* Pressure values are at 0°C, sea level, latitude 45°.

hypsometric formula: An alternative for altimeter equation. See ALTIMETER.

hythergraph: A CLIMAGRAM in which the selected meteorological elements are temperature and humidity or temperature and rainfall.

I

ICAO standard atmosphere: The STANDARD ATMOSPHERE adopted by the International Civil Aviation Organization. Its chief specifications are: mean sea level (MSL) pressure 1013·25 mb; MSL temperature 288·15 K (15°C); temperature lapse rate −6·5 degK/km up to 11 km where the temperature is 216·65 K (−56·5°C) and with an isothermal lower stratosphere above to 20 km. The extension, agreed in 1964,* of this atmosphere to 32 km specifies a lapse rate of −1·0 degK/km which gives a temperature, at 32 km, of 228·65 K (−44·5°C). Differences between the ICAO Atmosphere and the ICAN (International Commission for Air Navigation) Atmosphere which was previously adopted by many countries are generally negligibly small.

ice: Water substance in solid form. It occurs in the atmosphere and/or on the earth's surface in many forms such as ICE CRYSTALS, SNOW, HAIL, HOAR-FROST, RIME, GLAZE, GLACIER, etc.

The density of ice is generally about 0·917 g/cm³ (917 kg/m³). At 0°C the latent heat of fusion (pure ice to water), L_f, is 334 × 10³ J/kg and the latent heat of sublimation (pure ice to vapour), L_s, is 2834 × 10³ J/kg; these large values imply that the formation and disappearance of large masses of ice are important items in the heat budget of the earth–atmosphere system. The coefficient of linear expansion of pure ice at −10 to 0°C lies between 0·000050 and 0·000054; its specific heat at 0°C is 0·5. The physical properties of sea ice—in particular the values of latent and specific heats—may be very different from those of pure ice and vary greatly with temperature and SALINITY.

When a natural water surface cools, the water cooled at the surface is continually replaced by warmer water from below until the whole mass has fallen to 4°C (39°F), at which temperature water has its greatest density. The surface water then cools undisturbed until the freezing-point is reached and ice begins to form.

Three kinds of ice are produced in rivers: (i) sheet ice, which forms on the surface of the water, first of all near the banks and extending gradually towards the centre, and (ii) FRAZIL ICE and (iii) ground ice, both of which form in the rapidly moving stream in the centre of the river in very cold weather. Ground ice (or anchor ice) forms at the bottom, adhering to rocks and other substances in the river bed. It often rises to the surface, dragging with it masses of rock, and may destroy river structures.

When the sea freezes, the crystals formed contain no salt, but cannot easily be separated from the brine which is mixed up with them. When, however, the ice forms hummocks under the action of pressure, the brine drains out and leaves pure ice.

The *International meteorological vocabulary*† contains the Abridged International Ice Nomenclature which defines the terms used for sea ice.

---

* Montreal, International Civil Aviation Organization. Manual of the ICAO standard atmosphere, 2nd edn. Montreal, 1964.
† Geneva, World Meteorological Organization. International meteorological vocabulary. Geneva, WMO, 1966, Appendix A, pp. 249–251.

ice accretion: The formation and building up of a layer of ice on terrestrial objects or on aircraft in flight which are at sub-freezing temperatures. The ice forms by direct sublimation of water vapour or by the impingement of supercooled drops of precipitation, fog or cloud. See also ICE FORMATION ON AIRCRAFT.

ice age: A geological period during which great glaciers and ice sheets extended from the polar regions to as far equatorwards as about latitude 55°. Individual ice ages wax and wane in strength in 'glacial phases' and 'interglacial phases', respectively. Present-day climate is considered to be that appropriate to an interglacial phase of the Quaternary Ice Age which began about 1 million years ago and in which four main glacial periods are recognized, the last ending some 8000 to 40 000 years ago, depending on locality. At the peak of a glacial phase, land ice is considered to have covered an area about twice as large as at present. The term 'Little Ice Age' is applied to the period from about 1550 to 1850 which was conspicuous for its low mean temperature and equatorward extension of glaciers.

Other great ice ages occurred in the Permo-Carboniferous, some 250 million years ago and in the late Proterozoic and early Cambrian, some 500 million years ago; and there was more than one in the early Proterozoic, 700 to 1000 million years ago. The Permo-Carboniferous Ice Age is thought to have developed mainly within the present tropics, in South America, Africa, India and Australia. See also CLIMATIC CHANGES.

iceberg: A mass of ice, broken from a glacier ('glacier berg') or an ice-shelf ('barrier berg' or 'tabular berg'), which floats in the sea; the World Meteorological Organization nomenclature specifies a minimum height of five metres above sea level. The former are greenish in colour and irregular in shape and are typical of the Arctic; the latter are whitish in colour, more regular in shape and typical of the Antarctic.

Erosion of icebergs proceeds most rapidly at and below the water line. Favourable winds sometimes carry icebergs to latitudes 40°–50° or even below.

Iceberg ice is as pure as distilled water, containing only four parts of solid per million. Icebergs also contain a considerable quantity of air, ranging in various specimens examined from 7 to 20 or 30 per cent.

iceblink: A typical whitish glare on low clouds above an accumulation of distant ice. See also WATER SKY.

ice-bulb temperature: That temperature (T_i) at which pure ice must be evaporated into a given sample of air, adiabatically and at constant pressure, in order to saturate the air at temperature T_i under steady-state conditions. The temperature recorded by an ice-covered 'wet bulb' of a PSYCHROMETER may not exactly accord with this definition. See also THERMODYNAMIC TEMPERATURES.

ice-crystal cloud: A cloud which is composed (almost) exclusively of ice crystals, as opposed to water droplets. The CLOUD GENERA Ci, Cc and Cs are normally ice-crystal clouds.

ice crystals: Ice crystals form in the atmosphere on ICE NUCLEI at temperatures appreciably below freezing-point and, it is generally thought, at about saturation with respect to water. They exist in simple form, notably needles and hexagonal columns and plates, and in more complex star-shaped and branched forms. They multiply by SPLINTERING and grow, by diffusion of water vapour on to them, into forms and at rates which depend on a variety of conditions, for example the internal and surface structure of the crystals and the prevailing conditions of temperature and supersaturation. The crystals collide and coalesce with other crystals or supercooled water droplets to form SNOWFLAKES, GRAUPEL or HAIL.

ice-crystal theory: An alternative term for the BERGERON (–FINDEISEN) THEORY of PRECIPITATION.

ice day: An ice day is defined as a period of 24 hours, beginning normally at 09 GMT, on which the maximum temperature is less than 0 degC (32 degF).

ice formation on aircraft: Ice may form on aircraft either on the ground or in flight. ICE ACCRETION in flight may constitute a danger by affecting the aerodynamic characteristics of the engine performance, or in other ways.
There are four types of airframe icing:

(i) HOAR-FROST. A white crystalline coating of ice which forms in clear air, by sublimation of water vapour, when an aircraft surface is at a temperature below 0°C and is colder than the frost-point of the air with which it is in contact. This may occur when an aircraft moves rapidly (normally by descent) from a level at which the temperature is below 0°C into a layer of warmer and relatively moist air.

(ii) RIME. A light, white, opaque deposit which forms in clouds of low water content consisting of small, supercooled water drops.

(iii) CLEAR ICE or GLAZE. A coating of clear ice which forms in clouds of high water content consisting mainly of large, supercooled water drops, or which forms when rain (usually supercooled) falls on an aircraft whose temperature is below 0°C.

(iv) Cloudy ice or mixed ice: In flight, rime and clear ice are often mixed because a cloud contains a large range of drop sizes and because ice crystals, liquid droplets and snowflakes may all be present in a cloud. These conditions affect the formation of ice and result in a rough, cloudy deposit called 'cloudy ice' or 'mixed ice'.

When a supercooled water drop impinges on an aircraft, part of the drop freezes immediately at the leading edges, part may be lost before freezing, while the remainder freezes subsequently. In the case of a small drop the final freezing occurs quickly, giving opaque ice on and near the leading edge; a light deposit of this kind may, however, be lost by the action of wind. For a large drop the final freezing is less rapid and the drop has the opportunity to spread over the wing, leading to the more dangerous type of clear ice.

Observations indicate that icing risks are greatest at temperatures not too far below freezing-point, from, say, 0°C to $-12°C$. No lower temperature limit can, however, be set for possible ice formation which has been reported, for example, at a temperature of $-60°C$. Icing rate tends to increase with liquid water content of clouds and with cloud-particle size (greater for cumuliform clouds than for layer clouds). Clear ice is associated with the passage of fronts, chiefly warm fronts in winter, and is comparatively rare in the British Isles.

The engine and air-intake systems are liable to icing in a variety of conditions which may include ambient temperature substantially above 0°C and clear air.

Icelandic low: The 'Icelandic low' has a value of about 994 mb on the January chart of mean surface pressure. This mean depression is centred between Greenland and Iceland and represents the aggregate of many deep depressions. In summer, depressions in this region are less intense.

ice nucleus: A generic term which includes both 'freezing nucleus' and 'sublimation nucleus'. See NUCLEUS.

ice pellets: Precipitation of transparent or translucent pellets of ice, which are spherical or irregular, rarely conical, and which have a diameter of 5 mm or less.

They are recorded in two types, and the appropriate type is always specified in the present-weather code:

(a) Frozen raindrops or largely melted and refrozen snowflakes (formerly 'grains of ice').
(b) Pellets of snow encased in a thin layer of ice (formerly 'small hail').

ice prisms: A fall of unbranched ice crystals, in the form of needles, columns or plates, often so tiny that they seem to be suspended in the air. These crystals may fall from a cloud or from a cloudless sky. (Rarely observed in the United Kingdom.)

Ice Saints: St Mamertus on 11 May, St Pancras on 12 May and St Gervais on 13 May known on the continent as the 'cold Saints' days'. It is said in France that these three days do not pass without a frost. A Buchan cold spell from 9–14 May covers this period—see BUCHAN SPELLS.

ice sheet: A large area of land ice with a dome-shaped, almost smooth surface. The largest ice sheets now existing are those of the Antarctic and Greenland.

icing: See ICE ACCRETION and ICE FORMATION ON AIRCRAFT.

IGC: Abbreviation for INTERNATIONAL GEOPHYSICAL CO-OPERATION.

IGY: Abbreviation for INTERNATIONAL GEOPHYSICAL YEAR.

illuminometer: The common name for a DAYLIGHT recorder.

inclination, magnetic: An alternative for DIP, MAGNETIC.

incus (inc): A supplementary cloud feature (Latin, *incus* anvil).
  'The upper portion of a CUMULONIMBUS spread out in the shape of an anvil with a smooth, fibrous or striated appearance.'* See also CLOUD CLASSIFICATION.

independence: A term used in its natural sense in statistical work, but often with the implication that the COVARIANCE or CORRELATION between corresponding values of the elements concerned does not differ significantly from zero.

index: The pointer or other feature in an instrument whose position with regard to the scale determines the reading. The term is also sometimes applied to the fixed mark which constitutes the zero; thus the 'index error' of a barometer is the error which is due to faulty positioning of the scale.

index correction: The quantity which (with proper sign) has to be added to an instrumental reading to correct for index error. It is of the same magnitude as the INDEX error, but of the opposite sign.

index cycle: A term often applied to alternating periods of predominantly ZONAL and MERIDIONAL FLOW. The corresponding variations of the ZONAL INDEX are, however, only very approximately cyclic in nature. See CYCLE.

index error: See INDEX, INDEX CORRECTION.

---

* Geneva, World Meteorological Organization. International cloud atlas, Vol. 1. Geneva, WMO, 1956, p. 16.

Indian summer: A warm, calm spell of weather occurring in autumn, especially in October and November. The earliest record of the use of the term is at the end of the eighteenth century, in America, and it was introduced into the British Isles at the beginning of the nineteenth century. There is no statistical evidence to show that such a warm spell tends to recur each year.

indirect circulation: See DIRECT CIRCULATION.

inert gases: The elements HELIUM, NEON, ARGON, KRYPTON, XENON, RADON, THORON, and ACTINON, termed 'inert gases' (sometimes also 'noble gases' or 'rare gases') because of their chemical inactivity. Apart from argon (0·93 per cent by volume) these gases occur in the air in only minute quantities.

inertial flow: Flow in the absence of external forces. See CIRCLE OF INERTIA.

inertia stability: A type of DYNAMIC STABILITY, associated with the earth's rotation, in which an air particle, embedded in a wind flow along a circle of latitude, tends to return to this latitude on being subjected to a small displacement from it. Inertia instability arises when such a displacement results in an acceleration of the particle away from its original latitude.

The condition for stability is that the CORIOLIS PARAMETER should exceed the northward increase of the geostrophic west-wind component, i.e. $f > \partial u_G / \partial y$. Similarly, instability arises when $f < \partial u_G / \partial y$.

Similar conditions apply to air currents which are not east–west and may be extended to displacements along sloping isentropic surfaces which are most likely to give rise to instability. The criterion for stability in such currents is $f > (\partial v / \partial n)_\theta$ where v is the velocity and differentiation is normal to the flow along the isentropic surface.

Strictly, these criteria apply only in the simple case of straight flow with no variation of velocity along the flow. In the more general case of curved flow with velocity variations the criteria become much more complicated.

inertia wave: A type of stable atmospheric wave which arises because of the inertia of a mass moving over the earth's surface and the associated CORIOLIS FORCE arising from the earth's rotation.

infra-red radiation: ELECTROMAGNETIC RADIATION in the approximate wavelength range from about 0·7 to 1000 μm; 52 per cent of the total solar RADIATION intensity is contained within this range of wavelengths, the amount at wavelengths greater than 4 μm being very small.

In meteorology, the term is often used loosely as an alternative for LONG-WAVE RADIATION.

insolation: A term which is used in various senses:

(i) The intensity at a specified time, or the amount in a specified period, of direct solar RADIATION incident on unit area of a horizontal surface on or above the earth's surface.

(ii) The intensity at a specified time, or the amount in a specified period, of total (direct and diffuse) radiation incident on unit area of a specified surface of arbitrary slope and aspect.

The insolation depends on (*a*) SOLAR CONSTANT, (*b*) calendar date, (*c*) latitude (involving variations of length of day and degree of obliquity of sun's rays), (*d*) slope and aspect of surface, and (*e*) degree of transparency of atmosphere.

The purely astronomical factors (including variations of length of day but omitting all effects of atmospheric attenuation which depend in part on latitude) are accounted for in the tables and chart contained in the *Smithsonian meteorological tables*.* These show the daily solar radiation totals on a horizontal surface at various latitudes and dates, on the assumption of a perfectly transparent atmosphere; seasonal and annual totals are also given. Their main features are a primary daily maximum at the summer pole, a secondary daily maximum at about 45° latitude in the summer hemisphere, a minimum (zero) at and near the winter pole, and an annual total at the equator about 2·4 times that at either pole.

These theoretical results have only partial relevance to what is found in practice. If, for example, account is taken of attenuation of the radiation by the atmosphere, the maximum at the summer pole disappears because of the oblique nature of the radiation received in high latitudes. If, further, account is taken of systematic latitudinal variations of atmospheric attenuation of radiation by clouds, the secondary maximum at latitude 45° moves to the semi-arid climatic zone at about 35°.

instability: See STABILITY.

instability index: A numerical measure of the static instability of the atmosphere, particularly in relation to the likelihood of thunderstorm development—see STABILITY.

Various such indices have been proposed, most of them based on empirical studies relating the distribution of temperature, usually also of humidity, at selected levels within the troposphere, to thunderstorm incidence. The examples quoted in this text are BOYDEN INDEX and SHOWALTER INDEX.

instability line: In synoptic meteorology, a line or belt generally some hundreds of kilometres in length, along which slight or moderate instability (convective) phenomena exist but which is not marked by a surface front; such a line may, however, be marked by an UPPER-LEVEL TROUGH. When the instability phenomena are of a violent nature the term SQUALL LINE is generally used.

intensification: In synoptic meteorology, 'intensification' of a pressure system signifies an increase with time of the pressure gradient around the centre of the system. The converse term is 'weakening'.

inter-diurnal variation: Day-to-day change (of a specified element).

interglacial phase: A period of an ICE AGE during which (as now) permanent ice at low levels is confined to latitudes polewards of about 75°.

internal energy: For PERFECT GASES internal energy is a function only of temperature; the change of internal energy per unit mass for a temperature increment dT is

$$dE = c_v dT.$$

For an atmospheric column of unit cross-section, extending from the surface to height h, the internal energy is given by

$$E = c_v \int_0^h \rho T \, dz$$

where c_v is the specific heat at constant volume, ρ the density, T the temperature and z the height. See also POTENTIAL ENERGY.

---

* LIST, R. J.; Smithsonian meteorological tables. *Smithson Misc Collns, Washington*, **114**, 1968, pp. 418, 419.

International Geophysical Co-operation (IGC): Period from 1 January to 31 December 1959, during which the major part of the observational programme of the INTERNATIONAL GEOPHYSICAL YEAR was, by international agreement, continued.

International Geophysical Year (IGY): An international programme of observation in nearly all branches of GEOPHYSICS, on a world-wide scale, from 1 July 1957 to 31 December 1958, near sunspot maximum.

International Polar Year: See POLAR YEAR.

International Quiet Sun Years (IQSY): An international programme of observation in nearly all branches of GEOPHYSICS, on a world-wide scale, from 1 January 1964 to 31 December 1965, near sunspot minimum.

inter-quartile range: See QUARTILE.

intertropical convergence zone: The intertropical convergence zone (ITCZ) is a relatively narrow, low-latitude zone in which air masses originating in the two hemispheres converge. Since characteristics distinct from a FRONT of higher latitudes are involved, the term ITCZ is generally preferred to the alternative 'intertropical front' (ITF). The term 'equatorial front' is sometimes also used.

Over the Atlantic and Pacific Oceans, where it is closely related to the DOLDRUMS, the ITCZ is the boundary between the north-easterly and south-easterly trade winds. Over the continents it is replaced by the boundary between other wind systems with components directed towards the equator, for example in Africa between the HARMATTAN and the south-west monsoon. The ITCZ moves northwards in the northern summer and southwards in the southern summer, its mean position being somewhat north of the equator. Over the oceans the range of movement is small, but over the continents it may be very large.

The horizontal convergence associated with the ITCZ implies generally upward motion in the lower troposphere and cloudy, showery weather. The zone is subject to day-to-day oscillations of position and variations of activity. Shallow low-latitude depressions, which may on occasion greatly intensify, are often associated with the zone.

intertropical front: See INTERTROPICAL CONVERGENCE ZONE.

intortus (in): One of the CLOUD VARIETIES (Latin, *intortus* twisted).

'CIRRUS, the filaments of which are very irregularly curved and often seemingly entangled in a capricious manner.'* See also CLOUD CLASSIFICATION.

inversion: An inversion (of temperature) is said to occur at a point, or through a layer, at which or through which temperature increases with increasing height; such a feature is an inversion of the condition of a positive LAPSE of temperature which normally obtains in the atmosphere.

An inversion is marked on a normal temperature–height diagram (height the vertical ordinate) by a line which slopes upwards to the right. It implies great static STABILITY and absence of turbulence at the level concerned. See also INVERSION LAYER.

inversion layer: An atmospheric layer in which there is an INVERSION of temperature. Vertical motion of air through such a layer is inhibited by considerations of static

---

* Geneva, World Meteorological Organization. International cloud atlas, Vol. 1. Geneva, WMO, 1956, p. 14.

STABILITY. If the air under the inversion is relatively moist, the layer itself often lies above a layer of stratified cloud. An inversion layer near the earth's surface ('surface inversion') occurs on a RADIATION NIGHT; inversion layers at higher levels are often associated with the tropopause and anticyclones, and sometimes with fronts.

ion: The name (selected by Faraday) for an electrically charged atom or molecule; the presence of ions in a gas or electrolyte renders it an electrically conducting medium.

Ions are formed in the low atmosphere mainly by radiations from radioactive material in the air and by cosmic radiation; on a local scale, the rupture of water drops or the friction of wind-driven snow or sand may be important in producing atmospheric ions. Cosmic rays are an increasingly important ionizing source with increase of height above sea level. Solar ultra-violet radiation and X-rays are mainly responsible for the intense ionization of the atmosphere above about 50 km.

In the process of ion formation, an electron is ejected from a neutral particle (molecule or atom) which thus acquires a positive charge, while the electron attaches itself to another neutral particle which acquires a negative charge; an 'ion pair' is thus formed. These so-called 'small ions' are highly mobile; they may disappear by recombination or, near the earth's surface, by becoming attached to electrically neutral (e.g. pollution) particles, forming 'large ions' of much lower mobility, with a resulting decrease in air conductivity.

ionization: The process of ION formation.

ionization potential: The minimum energy (ELECTRON-VOLTS) required to remove an electron from an atom or molecule. It is also termed 'ionization energy'.

ionogram: In radio echo sounding of the IONOSPHERE an automatic record of corresponding values of wave frequency and VIRTUAL HEIGHT.

ionosphere: That portion of the earth's ATMOSPHERE, extending upwards from about 60 km to an indefinite height, which is characterized by a concentration of IONS and free ELECTRONS high enough to cause reflection of radio waves.

The IONIZATION, which is caused at these levels mainly by solar ultra-violet and X-ray radiation, reaches peak values, separated by shallow troughs, in rather well-defined regions of the atmosphere, giving rise to the ionospheric E-, F_1- and F_2-layers situated at about 110, 160 and 300 km, respectively. There is sometimes no peak value but only a 'ledge' in which the height gradient of electron concentration is small. Radio reflections are sometimes also obtained from heights of 65–80 km—the so-called D-layer which is not, however, characterized by a maximum ionization concentration.

The E- and F_1-layers are, to a first approximation, 'regular' in that their peak values of ion and electron densities and corresponding heights have systematic latitudinal, diurnal, seasonal and sunspot-cycle variations in reasonable accord with CHAPMAN LAYER theory (electron density highest near the equator, at midday, in summer, and at sunspot maximum, respectively, when the intensity of the ionizing radiation is greatest). The corresponding features of the F_2-layer, on the other hand, show many anomalies due to solar and lunar tidal effects and the influence of the earth's magnetic field. The large, short-period changes of electron height distribution and concentration in this layer are closely correlated with geomagnetic storms and are termed 'ionospheric storms'.

The properties of the ionosphere are regularly investigated at a world-wide network of stations by an automatic radio-echo technique ('ionospheric echo sounding') in which short bursts of radio waves of different frequencies are emitted in rapid succession at vertical incidence and the resulting reflected waves registered on an 'ionogram'. Since the electron concentration required for reflection of radio waves increases with the frequency of the waves the technique yields a CRITICAL FREQUENCY (f_0) of penetration of the regions of electron peak concentration and the corresponding VIRTUAL HEIGHT. Rockets and satellites are also used to sound the ionosphere and have yielded information, in particular, of the electron concentrations in the troughs separating the various layers and at levels above the F_2-layer.

Observations show that the ionosphere is in a state of constant and complex motion which is in part systematic (including tidal), and in part random, in nature. In the lower ionosphere the charged particles are generally moved bodily with the uncharged particles; at higher levels the motion of the charged particles is mainly independent of that of the uncharged particles and is much affected by the forces exerted by the earth's magnetic field.

IQSY: Abbreviation for INTERNATIONAL QUIET SUN YEARS.

iridescence: A word formed from the name of Iris, the rainbow goddess, to indicate rainbow-like colours; an alternative is 'irisation'.

Iridescence in the form of tinted patches of red and green, sometimes of blue and yellow, is occasionally observed on high clouds, generally within about 30° of the sun. The boundaries of the tints are not circles with the sun as centre but tend to follow the outlines of the cloud. Iridescent clouds are considered to be parts of CORONAE, the coloration being caused by DIFFRACTION of sunlight by very small cloud particles.

irisation: An alternative for IRIDESCENCE.

irrotational motion: Motion in which there is no VORTICITY. It is defined by the equation curl $\mathbf{V} = 0$ where $\mathbf{V}$ is the three-dimensional velocity vector which defines the motion.

isallo-: A prefix used, in conjunction with another word, to denote lines drawn on a map or chart to display the tendency (time rate of change) of any element, each line being drawn through places at which the element has the same tendency, e.g. ISALLOBAR, isallotherm.

isallobar: A line of constant BAROMETRIC TENDENCY. Such lines are sometimes drawn on synoptic charts, mainly as an aid to forecasting the movement of features of the pressure distribution.

isallobaric wind: That part of the AGEOSTROPHIC WIND which is associated with local time change of the PRESSURE GRADIENT.

One of the conditions for the application of the GEOSTROPHIC WIND equation is that the pressure distribution should be steady. In the treatment by D. Brunt and C. K. M. Douglas of the effect on the wind of a changing horizontal pressure gradient, it was shown that the main term additive to the geostrophic wind corresponds to a wind directed normally to the ISALLOBARS towards the isallobaric low and of magnitude $\dfrac{1}{\rho f^2} \operatorname{grad}_H \dfrac{\partial p}{\partial t}$, where ρ is air density, f the Coriolis parameter $= 2\Omega \sin \varphi$, and $\operatorname{grad}_H \dfrac{\partial p}{\partial t}$ the horizontal gradient of the isallobars. This, now generally known as the 'isallobaric wind', rarely exceeds 5 m/s, but is at times important in its effects on horizontal convergence and rainfall.

isanomaly: A line of constant anomaly (i.e. difference from normal) of a meteorological variable. See ISO-.

isentropic: Without change of ENTROPY, generally equivalent in meaning to ADIABATIC. Isentropic surfaces in the atmosphere are surfaces of constant POTENTIAL TEMPERATURE.

isentropic analysis: An analysis of the physical and dynamical processes taking place in the free atmosphere on the basis of the location and configuration of the various ISENTROPIC surfaces and distribution of atmospheric properties (e.g. temperature, humidity mixing ratio) and motion on them. 'Isentropic charts' are the main tool of this work.

iso-: A prefix meaning 'equal', extensively used in meteorology in conjunction with another word, to denote lines drawn on a map or chart to display the geographical distribution of any element, each line being drawn through places at which the element has the same value, e.g. ISOBAR, ISOTHERM. The words 'isogram' or 'isopleth' are sometimes used as generic names of this type.

isobar: A line of constant (atmospheric) pressure.

'Surface isobars' are based on pressure readings corrected to a common level (see REDUCTION to sea level). Such isobars, drawn at intervals of one millibar or more (depending on scale of chart), produce characteristic patterns such as are illustrated in the articles on ANTICYCLONE, COL, DEPRESSION, RIDGE, SECONDARY, and TROUGH.

isobaric analysis: Analysis along a surface of constant pressure; the analysis usually includes variations of GEOPOTENTIAL, temperature, humidity, wind.

For study of conditions in the upper air there are theoretical and practical advantages in employing isobaric analysis rather than analysis along surfaces of constant geopotential. The GEOSTROPHIC WIND speed is related to contour spacing in a way that is independent of air density and permits the use of the same wind scale at different isobaric levels.

isobaric surface: Surface of constant (atmospheric) pressure.

isoceraunic line: A line of constant percentage frequency of days in a year on which thunder is heard. See ISO-.

isochasm: A line of constant frequency of visible AURORA. See ISO-.

isochrone: A line drawn on a map in such a way as to join places at which a phenomenon is observed at the same time, e.g. lines indicating the places at which rain commences at a specified time.

isoclinic (or isoclinal) line: A line of constant magnetic DIP. See ISO-.

isodynamic line: A line of constant total magnetic intensity. See ISO-.

isogon: A line of constant wind direction. See ISO-.

isogonic (or isogonal) line: A line of constant magnetic DECLINATION. See ISO-.

isogram: See ISO-.

isohaline: A line of constant SALINITY. See ISO-.

isohel: A line of constant sunshine duration. See ISO-.

isohyet: A line of constant rainfall amount. See ISO-.

isohypse: An alternative for CONTOUR.

isokinetic sampling: Air sampling which takes place at a rate such that the flow velocity in the sampling tube is equal to that of the environment being sampled. To ensure that no error occurs in the sampling of particular matter, isokinetic conditions should hold.

isomeric line: The distribution of monthly rainfall over an area may be represented by a plot of percentage of average annual rainfall attained in the month at each station (isomeric values). A line of constant percentage on such a chart is an isomeric line or isomer.

isoneph: A line of constant cloud amount. See ISO-.

isopleth: See ISO-.

isopycnic surface: A surface of constant (atmospheric) density. Such surfaces intersect ISOBARIC SURFACES in a BAROCLINIC, but not in a BAROTROPIC, atmosphere.

isosteric surface: A surface of constant (atmospheric) SPECIFIC VOLUME. Since specific volume is the inverse of density, an isosteric surface is also ISOPYCNIC.

isotach: A line of constant wind speed. See ISO-.

isotherm: A line of constant temperature. See ISO-.

isothermal: Of equal temperature. The term 'isothermal layer' was originally applied to the region now known as the STRATOSPHERE. The term is, in this restricted sense, now obsolete but is still used to denote any layer with ZERO LAPSE rate of temperature, revealed by vertical sounding of the atmosphere.

isovel: An alternative for ISOTACH.

J

Jacob's ladder: See CREPUSCULAR RAYS.

jet stream: A fast narrow current of air, generally near the tropopause, characterized by strong vertical and lateral wind shears. A jet stream is usually some thousands of kilometres in length, hundreds of kilometres in width and some kilometres in depth. The vertical and horizontal shears are as a rule of the order 5–10 m/s (10–20 kt) per kilometre and 5 m/s (10 kt) per 100 kilometres, respectively, but much stronger shears can occur with more intense jet streams. For certain purposes, arbitrary minimum speeds varying from about 25 to 50 m/s (50 to 100 kt) have been used to define a jet stream.

The distance from a jet-stream core to the nearest point at which the wind velocity falls to one half of the peak value is termed the 'half-width' of the jet stream.

Two main types of jet stream are recognized, (i) subtropical (westerly) and (ii) polar front (westerly). The subtropical jet stream is relatively constant in position in a given season and dominates mean seasonal wind charts; in contrast, the polar-front jet stream is highly variable in position from day to day over a wide range of temperate latitudes and is therefore masked on such charts. In addition, a westerly 'polar-night jet stream' (high latitudes in winter) occurs at times within the stratosphere above 50 mb, and an easterly jet stream occurs in the stratosphere in the equatorial belt of the eastern hemisphere.

joule: The unit of work or energy in the SI and M.K.S. SYSTEM of units. This unit was also recommended at the Ninth General Conference of Weights and Measures (1948) for use as the unit of HEAT.

The joule is the work done by a force of one NEWTON in moving its point of application one metre in the direction of the force. The dimensions are ML^2T^{-2}.

$$1 \text{ J} = 1 \text{ W s} = 10^7 \text{ ERG}$$
$$1 \text{ } 15° \text{ CAL} = 4\cdot1855 \text{ J}$$
$$1 \text{ IT CAL} = 4\cdot1868 \text{ J}.$$

K

Kármán's constant: A non-dimensional quantity (k) in the equation which defines the nature of the wind structure in the low atmosphere in adiabatic lapse rate conditions—see LOGARITHMIC VELOCITY PROFILE. k is found experimentally to have a value close to 0·4.

katabatic wind: On a 'radiation night' of clear skies and slack pressure gradient, TERRESTRIAL RADIATION from the earth's surface causes a layer of cold air to form near the ground, with an associated INVERSION of temperature. If the ground is sloping, the air close to the ground is colder than air at the same level but at some horizontal distance. Downslope gravitational flow of the colder, denser air beneath the warmer, lighter air results and comprises the 'katabatic wind'. It is known also as the 'drainage wind' or 'mountain breeze'. This downslope wind, and the associated formation of FROST HOLLOWS, have important agricultural effects. On a larger scale it is experienced, for example, as the fiord winds of Norway; on a still larger and more violent scale, it largely comprises the outflowing winds from the Greenland and Antarctic continents.

katafront: As defined by T. Bergeron, a FRONT (warm or cold) along which the warm air is descending relatively to the cold air. Downward motion of the warm air at most levels is generally implied, with frontal activity feeble or absent.

katathermometer: A large-bulb spirit thermometer with graduations at 35°C and 38°C; the bulb is heated to a temperature above 38°C, the time is taken for the spirit to fall from the 38°C mark to that at 35°C, and the wind speed producing the cooling is computed.

In another form of the instrument used in relation to human bioclimatology, two bulbs, one wet, the other dry, are employed and the cooling of the bulbs (at about human body temperature) effected by the wind is measured.

Kelvin effect: An effect, relating to the saturation vapour pressure over liquid surfaces, discovered by Lord Kelvin.

The saturation VAPOUR PRESSURE (p) at a curved liquid surface (e.g. spherical water droplet) exceeds that over a similar plane surface (p_∞) by an amount which increases with decreasing radius of curvature (r) of the surface. The relationship between them is given by

$$\log_e \frac{p}{p_\infty} = \frac{2M\sigma}{\rho R^* T r}$$

where M is the molecular weight of the vapour, σ the surface tension of the liquid, ρ the liquid density, R^* the universal gas constant, and T the temperature.

The effect of surface curvature is important for droplets of diameter less than about 1 μm; at 10°C, for example, the ratio p/p_∞ is 1·023 for a water droplet of diameter 0·1 μm, and 1·002 for a diameter of 1 μm. The effect is of fundamental importance in the initiation of CONDENSATION in the atmosphere. See also NUCLEUS.

Kern's arc: A faint and very rare HALO which is centred on a point opposite to, and on the same circle as, a CIRCUMZENITHAL ARC.

Kew-pattern barometer: A portable marine BAROMETER designed by P. Adie in 1854 for the Kew Committee of the British Association. The scale is graduated to take account of changes of the level of the mercury in the steel cistern so that it is only necessary to read the top of the mercury column. The tube is constricted to minimize PUMPING when the barometer is used at sea. Similar barometers, known as 'station' barometers, but without the constriction, were subsequently adopted for use on land and are regularly employed at British meteorological stations.

khamsin: A southerly wind blowing over Egypt in front of depressions passing eastwards along the Mediterranean or North Africa, while pressure is high to the east of the Nile. As this wind blows from the interior of the continent it is hot and dry, and often carries much dust. It is most frequent from April to June.
    Gales from the south or south-west in the Red Sea are also known as khamsin.

kilocalorie: The metric thermal unit, being the quantity of heat required to raise the temperature of one kilogramme of water from 14·5°C to 15·5°C. See also CALORIE, JOULE.

kilogramme: The unit of mass in the M.K.S. SYSTEM of units. See also GRAMME, SI UNITS.

kilomole: The kilogramme-molecular weight of a substance, i.e. the weight of it in KILOGRAMMES which is numerically equal to its MOLECULAR WEIGHT.

K-index, geomagnetic: See GEOMAGNETISM.

kinematical analysis: Analysis of the field of atmospheric wind flow.

kinetic energy: The ENERGY possessed by a body by virtue of its motion. It is a scalar quantity of magnitude $\frac{1}{2}mv^2$, where m is the mass and v the velocity of the particle.
    D. Brunt estimated that the total kinetic energy of the large-scale motion of the atmosphere is of the order 3×10^{20} joules (8×10^{13} kilowatt-hours). He also estimated that, in the absence of solar radiant energy, dissipation of the atmosphere's kinetic energy by turbulence, generated mainly at the earth's surface, would be almost complete after six days.

Kirchoff's law: See RADIATION.

kite: In the now out-moded use of kites carrying aloft instruments for sounding the upper atmosphere, heights of over 20 000 feet (6 km) were attained.

kite balloon: A captive BALLOON, of such a form as to have aerodynamic lift, sometimes used for carrying METEOROGRAPHS and other types of sounding instrument.

knot: A speed of one nautical mile per hour and approximately 0·5 m/s. See MILE.

kona storm: A storm over the Hawaiian Islands, associated with a depression passing north of the islands, and characterized by strong or gale southerly winds and heavy rain.

konimeter: An instrument for counting the dust particles in a known volume of air. See DUST COUNTER.

kosava: A RAVINE WIND which occurs on the Danube, south-east of Belgrade.

Koschmieder's law: A law which states that the apparent brightness, B_s, of a black object at distance d, due entirely to scattered light, is related to the brightness of the horizon at the same azimuth, B_h, by the equation
$$B_s = B_h (1 - e^{-\beta d})$$
where β is the SCATTERING coefficient, assumed constant throughout the part of the atmosphere concerned.
See also VISIBILITY, AIR-LIGHT.

krypton: One of the INERT GASES, comprising 1.0×10^{-6} and 3.0×10^{-6} part per part of dry air by volume and weight, respectively. Its molecular weight is 83·80.

K-theory: The K-theory, also termed the EXCHANGE COEFFICIENT theory, of turbulent mixing is based on the assumption that the vertical flux of a conservative air-mass property ($\bar{E}$) which is effected by atmospheric eddies may be expressed as the product of an exchange coefficient (K) and the vertical gradient of the property concerned, i.e.
$$\text{flux} = K\, \partial \bar{E}/\partial z.$$
The value of the theory is limited by the extent to which K is dependent on height and atmospheric stability and on the atmospheric property concerned (heat, water vapour, momentum, etc.). A typical value of K is 1 m²/s. See also TURBULENCE, DIFFUSIVITY, EDDY CONDUCTIVITY.

Kuro Shio: Variously translated 'blue salt' or 'black' stream, a warm-water current of a characteristic dark blue colour, the main branch of which flows north-eastwards along the south coast of Japan before merging in the general drift of the North Pacific. This current is analogous to the Gulf Stream of the Atlantic in that it carries warmth to higher latitudes. A branch of the Kuro Shio sets into the Japan Sea.

kurtosis: A statistic describing the peakedness of a symmetrical FREQUENCY DISTRIBUTION, defined in terms of the MOMENTS as $\beta_2 = \mu_4/\mu_2^2$. For the NORMAL FREQUENCY DISTRIBUTION $\beta_2 = 3$.
The measure ($\beta_2 - 3$) is sometimes called the 'excess of kurtosis'. A distribution for which $\beta_2 > 3$ is termed 'leptokurtic' and has a high peak, deficient shoulders and long tails, compared with the normal distribution; while one for which $\beta_2 < 3$ is 'platykurtic' and has comparatively large shoulders and deficiencies in peak and tails.

L

labile: A term sometimes used to denote a condition of static instability or of neutral equilibrium in the atmosphere, i.e. a LAPSE rate equal to or exceeding the ADIABATIC. In German, the word is used to denote instability only.

lacunosus (la): One of the CLOUD VARIETIES (Latin, *lacunosus* having holes).
'Cloud patches, sheets or layers, usually rather thin, marked by more or less regularly distributed round holes, many of them with fringed edges. Cloud elements and clear spaces are often arranged in a manner suggesting a net or a honeycomb.
This term applies mainly to CIRROCUMULUS and ALTOCUMULUS; it may also apply, though very rarely, to STRATOCUMULUS.'* See also CLOUD CLASSIFICATION.

lag: A delay between a change in conditions and its indication by an instrument. While instrumental construction is usually designed to minimize lag, it is in some cases deliberately introduced, e.g. in mercurial barometers (by narrowing of tube) to avoid PUMPING, and in Symons's earth thermometers (by immersion of bulb in micro-crystalline wax) to ensure negligible change of indicated temperature on withdrawal of the instrument from the earth for reading.
In statistics, the lag is the displacement in time between terms being compared in the same or different series.

lag correlation: An alternative for AUTOCORRELATION. See also CORRELOGRAM.

Lagrangian change: The Lagrangian change of an element (e.g. pressure) is the time rate of change of the value of the element possessed by an individual moving fluid parcel ('change following the motion' or 'substantial change' or 'individual change') and is usually written, for example, dp/dt.
Lagrangian change is related to EULERIAN CHANGE (e.g. $\partial p/\partial t$) by the relationship

$$\frac{d}{dt} = \frac{\partial}{\partial t} + u\frac{\partial}{\partial x} + v\frac{\partial}{\partial y} + w\frac{\partial}{\partial z}$$

where u, v, w are the velocity components in the x, y, z directions, respectively.

Lagrangian similarity: Dimensional treatment in which the rate of vertical spread of particles by TURBULENCE is related to the characteristic scales of velocity and length referred to under SIMILARITY THEORY.

Lambert's law: See ABSORPTION.

Lambert's projection: See PROJECTION.

laminar boundary layer: A very shallow layer of air, adjacent to a fixed boundary, in which the air velocity increases very rapidly but fairly regularly from the boundary (where it is zero) to the free airstream. The molecular viscous stress in such a layer greatly exceeds the REYNOLDS STRESS.

---

* Geneva, World Meteorological Organization. International cloud atlas, Vol. 1. Geneva, WMO, 1956, p. 15.

laminar flow: Fluid flow is described as 'laminar', or 'streamline', if each particle follows the precise path of its predecessors; an essential feature of such flow is that there is no mixing of adjacent layers of the fluid.

land- and sea-breezes: Local winds, caused by the unequal heating and cooling of adjacent land and water surfaces under the influence of solar radiation by day and radiation to the sky at night, which produces a gradient of pressure near the coast. During the day the land is warmer than the sea and a breeze, the sea-breeze, blows onshore; at night and in the early morning the land is cooler than the sea and the land-breeze blows offshore. The breezes are most developed when the general pressure gradient is slight and the skies are clear. In such circumstances the sea-breeze usually sets in during the forenoon and continues until early evening reaching its maximum strength during the afternoon; the land-breeze may set in about midnight or not until the early morning. The land- and sea-breezes are much influenced by topography and vary considerably from one part of the coast to another. In the British Isles the pure sea-breeze rarely exceeds Beaufort force 3, but in lower latitudes it may reach force 4 or 5. The land-breeze is usually less developed than the sea-breeze. The sea-breeze does not usually extend more than 30–40 km on either side of the coastline, and often its extent is considerably less. Its farthest penetration inland is often marked by a line of horizontal convergence of wind and vigorous convection, sometimes termed a 'sea-breeze front'. The current is shallow at its commencement, but in favourable circumstances the depth may exceed 0·5 km at the time of its maximum development.

langley: An obsolescent unit of energy per unit area, equivalent to 1 CALORIE/cm².

Laplacian: A mathematical operator, designated ∇^2 and defined in three-dimensional Cartesian co-ordinates by

$$\nabla^2 = \frac{\partial^2}{\partial x^2} + \frac{\partial^2}{\partial y^2} + \frac{\partial^2}{\partial z^2}.$$

The two-dimensional Laplacian is defined by

$$\nabla^2 = \frac{\partial^2}{\partial x^2} + \frac{\partial^2}{\partial y^2}.$$

For a scalar field φ the quantity $\nabla^2 \varphi$ at a point is a measure of the difference between the value of the scalar φ at that point and the average value of φ in the neighbourhood of the point (in either three or two dimensions). For example, the factor $\nabla^2 z$ in the expression for the GEOSTROPHIC VORTICITY at a point 0 is, to a close approximation, given by

$$\nabla_0^2 z = \frac{z_1 + z_2 + z_3 + z_4 - 4z_0}{b^2}$$

where z_0 is the height at point 0 and z_1, z_2, z_3 and z_4 are the heights at points each a (small) distance b from the point, the lines $z_1 z_0 z_3$ and $z_2 z_0 z_4$ being at right angles to each other.

lapse: 'Temperature lapse' (or often simply 'lapse') denotes decrease of temperature with increasing height in the atmosphere. 'Temperature lapse rate', or 'lapse rate', signifies decrease of temperature per unit increase of height, taken positive when temperature decreases with height.

The temperature lapse rate, as denoted for example by a 'lapse line' drawn on a temperature–height diagram, is of fundamental significance in atmospheric ADIABATIC vertical motion and static STABILITY. The lapse rate averages about 0·6 degC per 100 metres in the TROPOSPHERE, where, however, it is very variable in space and time; over a limited interval of height it may be negative, corresponding to an INVERSION. The lapse rate is very small in the lower STRATOSPHERE.

The term 'lapse' is also used, as in 'humidity lapse rate', to signify the rate of decrease with height of certain meteorological variables other than temperature.

latent heat: The quantity of heat absorbed or emitted, without change of temperature, during a change of state of unit mass of a material; it has normally been expressed in the unit calories/gramme. In future in SI UNITS it will be expressed in joules/kilogramme. The dimensions are L^2T^{-2}.

In meteorology, the energy transformations associated with changes of state of water are very important; heat is absorbed in the changes from ice to water to water vapour and is released in changes which are in the opposite sense.

The latent heat of fusion (ice to water) at 0°C is 334×10^3 J/kg (79·7 IT cal/g). The latent heat of sublimation (ice to water vapour) at 0°C is 2834×10^3 J/kg (677 IT cal/g), varying with temperature by less than 0·2 per cent in the meteorological range. The latent heat of vaporization (water to water vapour) at 0°C is 2501×10^3 J/kg (597 IT cal/g) and at 40°C is 2406×10^3 J/kg (575 IT cal/g).

1 IT calorie = 4·1868 joules = $4·1868 \times 10^7$ ergs.

See also HEAT, ENTHALPY.

latent instability: See STABILITY.

latitude: The geographical latitude is defined as the angular elevation of the celestial POLE above the surface tangential to the spheroid which represents the earth. It is also equal to the angle between the normal to this surface and the plane of the equator. The geographical latitude φ at any point differs only slightly from the geocentric latitude φ', the latter being the angle between the radius of the earth passing through the point and the plane of the equator. The relation between φ and φ' is, approximately, $\varphi - \varphi' = 69·6'' \sin 2\varphi$. Geocentric latitude is found to be more useful than geographical latitude in seismology as the calculation of distance from place to place is facilitated.

Astronomical latitude is defined as the elevation of the celestial pole above the level of a mercury surface, or in other words the angle between the plumb-line and the plane of the equator.

For an explanation of geomagnetic latitude and magnetic latitude see GEOMAGNETISM.

law of storms: A nautical expression denoting the mariners' rules for avoiding the so-called 'dangerous half' of a TROPICAL CYCLONE (that part of the cyclone which is to the right of the cyclone's path in the northern hemisphere, to the left in the southern hemisphere).

layer clouds: CLOUDS of no marked vertical development, also termed 'sheet clouds', as opposed to 'heap clouds'. The layer cloud types comprise CIRRUS, CIRROCUMULUS, CIRROSTRATUS, ALTOCUMULUS (except altocumulus castellanus), ALTOSTRATUS, NIMBOSTRATUS, STRATUS, and STRATOCUMULUS.

leader stroke: See LIGHTNING.

least squares: When a set of observations or frequencies can be represented by values calculated from a formula containing adjustable parameters, the criterion of least squares (i.e. the least sum of the squared discrepancies) is generally used to determine the 'best' values for the parameters, usually with satisfactory results.

Other ways of measuring an overall discrepancy include the use of weighted least squares (if it is more important to fit some of the data than others) or the minimizing of the CHI-SQUARE STATISTIC.

lee depression: An alternative for OROGRAPHIC DEPRESSION.

leeward: Leeward of a point signifies the 'downwind' direction from the point, e.g. eastwards in the case of a west wind.

lee waves: A system of stationary air-waves ('standing waves') forming, under certain conditions, to the lee of a hill or mountain which presents a mechanical obstruction to the wind. The existence of such waves is sometimes shown by the presence of cloud near the wave crests ('lee-wave clouds'). The 'lee wavelength', i.e. the distance between adjacent wave crests, is usually between 4 and 40 km. The 'lee-wave amplitude', i.e. half the vertical distance from wave trough to crest, is very small at very low and high levels and reaches a maximum at an intermediate level.

Meteorological conditions suitable for lee-wave formation are (i) a stable layer of air between less stable layers at the ground and at a higher level, (ii) a wind of at least 15 knots across the top of the ridge, and (iii) constancy of wind direction up to the top of the stable layer. R. S. Scorer expressed the required conditions mathematically as an upward decrease of l^2 where

$$l^2 = \left(\frac{g\beta}{V^2} - \frac{1}{V}\frac{d^2V}{dz^2} \right)$$

where β is the static stability ($\theta^{-1}\partial\theta/\partial z$, where θ is potential temperature and z the height), g the gravitational acceleration, and V the wind speed component in the direction across the hill. The first of the terms is the more important.

See STANDING WAVE, MOUNTAIN WAVE.

Lenard effect: The separation of electric charges that accompanies the shattering of water drops. In pure water the larger drops are found to carry a positive charge to earth while the very fine droplets carry a negative charge into the surrounding air. Experiments show, however, that the magnitude of the effect and even the charge signs may be altered by various factors, such as the presence of impurities.

The Lenard effect forms the main basis of G. C. Simpson's theory of charge generation and distribution in a thundercloud—see THUNDERSTORM.

lenticularis (len): A CLOUD SPECIES (Latin derivation from *lens* lentil).

'Clouds having the shape of lenses or almonds, often very elongated and usually with well-defined outlines; they occasionally show irisation. Such clouds appear most often in cloud formations of orographic origin, but may also occur in regions without marked orography.

This term applies mainly to CIRROCUMULUS, ALTOCUMULUS and STRATOCUMULUS'.* See Plate 16. See also CLOUD CLASSIFICATION.

leptokurtic: See KURTOSIS.

leste: A hot, dry, southerly wind occurring in Madeira and northern Africa in front of an advancing depression.

---

* Geneva, World Meteorological Organization. International cloud atlas, Vol. 1. Geneva, WMO, 1956, p. 12.

levanter: A humid easterly wind in the Straits of Gibraltar. It is most frequent from June to October, but may occur in any month. At Gibraltar it causes standing waves and complex eddies which may cause dangerous conditions to the lee of the Rock. It is usually associated with stable air under an inversion and BANNER and ROTOR cloud to leeward. Strong winds and/or day heating cause the cloud base to lift and the cloud to thin or disperse.

leveche: A hot, dry, southerly wind which blows on the south-east coast of Spain in front of an advancing depression. It frequently carries much dust and sand, and its approach is indicated by a strip of brownish cloud on the southern horizon.

level: A surface is 'level' if it is everywhere at right angles to the direction of the force of gravity which is indicated by a plumb-line; along such a surface the GEOPOTENTIAL is constant. In general, a level surface in the atmosphere does not quite coincide with one of equal geometric height above mean sea level.

libeccio: A strong, squally south-westerly wind in the central Mediterranean, most common in winter.

life-cycle: A term used in synoptic meteorology to include the processes of formation, deepening, occluding and filling-up which are typical of a frontal DEPRESSION. The cycle usually occupies some three days but is subject to a wide range of variation of length.

lifting condensation level: See CONDENSATION LEVEL.

lightning: A visible electric discharge ('lightning flash') associated with a THUNDERSTORM. The various types include the 'cloud discharge' occurring within the THUNDERCLOUD, the 'air discharge' between a part of the cloud and the adjacent air, and the 'ground discharge' between cloud and ground. Centres of charge of opposite sign are neutralized by the occurrence of each flash, that in the air being the SPACE CHARGE.

High-speed recording of lightning discharges by BOYS'S CAMERA and also studies of associated electric field charges have shown that the ground discharge comprises at least two 'strokes' or 'streamers' which are generally as follows: first, a 'leader stroke' from cloud to ground in which the luminosity is relatively faint, the time of travel variable but often between 10^{-1} and 10^{-2} second, the progress sometimes in distinct steps ('stepped leader') and the path sometimes tortuous and with many branches of luminosity extending out from the main channel; second, an immediate 'return stroke' from ground to cloud which is very rapid and vivid and which lights up the main channel and any branches. Observations have shown, however, that leader strokes directed upwards from ground to cloud may predominate from very high structures. Sometimes (in the British Isles in a minority of cases) the ground discharge comprises a series of double strokes separated by a time interval of the order of microseconds; any subsequent leader stroke ('dart leader') is much more rapid than the first, since it follows the relatively ionized path to ground made by the first leader.

The typical lightning flash discharges a quantity of about 20 coulombs and involves a potential difference of some 10^8 or 10^9 volts. The sign of the charge carried by discharges to earth is predominantly negative.

See also BALL LIGHTNING, CHAIN LIGHTNING, FORKED LIGHTNING, PEARL-NECKLACE LIGHTNING, RIBBON LIGHTNING, ROCKET LIGHTNING, SHEET LIGHTNING, STREAK LIGHTNING.

light waves: ELECTROMAGNETIC RADIATION contained within the VISIBLE SPECTRUM, i.e. between about 0·4 and 0·7 μm.

line-squall: A phenomenon accompanying the passage, at any particular place, of a SQUALL LINE or active COLD FRONT. The characteristic features of a line-squall are (i) arch or line of low black cloud, (ii) rapid rise of wind speed and veer of wind, (iii) rapid drop in temperature, (iv) rapid rise in pressure, and (v) severe thunderstorm, often with hail.

lithometeor: A little-used generic term for non-aqueous solid particles suspended in the air or lifted from the earth's surface, e.g. haze, smoke, dust, drifting sand.

lithosphere: That part of the EARTH which is solid.

local time: Time reckoned from the epoch noon which at any place is the time of transit across the meridian of either the true sun or mean sun (according as local apparent time or local mean time is being used). A local time (LT) variation is one based on the mean sun, while a sunshine recorder or sundial indicates local apparent time. See TIME.

logarithmic velocity profile: The theoretical variation of mean wind speed near the earth's surface (within the 'surface BOUNDARY LAYER'), derived under various restrictive conditions. For aerodynamically rough flow—the meteorologically significant case—the wind profile is given by the equation

$$\frac{\bar{u}}{u_*} = \frac{1}{k} \ln \left(\frac{z}{z_0} \right)$$

where $\bar{u}$ is the mean wind velocity at height z, u_* the FRICTION VELOCITY, z_0 the ROUGHNESS LENGTH, and k is KÁRMÁN'S CONSTANT (about 0·4). The equation applies only for $z \geqslant z_0$. Observations show that the equation applies only in conditions of neutral stability.

longitude: The longitude of any place is the angle between the geographical MERIDIAN through that place and a standard or 'prime meridian', which is taken to be the meridian of Greenwich.

long-range forecast: See FORECAST.

long wave: In synoptic meteorology, a smooth, wave-shaped contour pattern on an isobaric chart with a wavelength of about 2000 km, relating more especially to the middle or high troposphere. Some four or five such waves (also termed 'Rossby waves') typically extend across a hemispherical chart.

Assuming a sinusoidal long-wave pattern of wavelength L in a BAROTROPIC atmosphere with no viscosity, and in which absolute vorticity is conserved, Rossby obtained the formula

$$c = U - \frac{\beta L^2}{4\pi^2}$$

where c is the wave speed (positive, eastward; negative, westward), U the mean zonal wind speed and β is the ROSSBY PARAMETER, i.e. the northward rate of change of the CORIOLIS PARAMETER. The formula accounts for progressive, stationary and retrograde waves ($c > 0$, $c = 0$, $c < 0$, respectively) but not for waves which may progress faster than the zonal current ($c > U$).

long-wave radiation: In its usual meteorological usage, an alternative for TERRESTRIAL RADIATION.

looming: An optical phenomenon, associated with a greater-than-normal rate of decrease of air density with height near the surface, in which objects which are normally below the HORIZON become visible. See MIRAGE.

Loschmidt's number: The number of gas molecules per unit volume at normal temperature and pressure. By AVOGADRO'S LAW this is the same for all gases and equals $2 \cdot 687 \times 10^{25}$ per m^3.

low: A term commonly used in synoptic meteorology to denote a region of low pressure, or DEPRESSION.

Lowitz, arcs of: On rare occasions arcs slightly concave towards the sun extend obliquely downwards and inwards from the parhelia of the 22° HALO. They are formed by REFRACTION through ice crystals oscillating about the vertical.

lumen: The derived unit of luminous flux in the SI UNITS system. The lumen (lm) is the flux emitted within unit solid angle (1 STERADIAN) by a source of unit luminous intensity (1 CANDELA) radiating equally in all directions.

luminance: The luminance of a surface is expressed by its luminous intensity per unit projected area, the plane of its projection being perpendicular to the direction of view. The unit is the CANDELA per square metre (sometimes known as 1 nit).

luminosity: For a given wavelength, the ratio of the luminous flux to the radiant energy flux. It varies throughout the VISIBLE SPECTRUM from zero at either end of the spectrum to a maximum at wavelength $0 \cdot 555$ μm. The ratio is also termed 'luminous efficiency', or 'visibility ratio'.

luminous efficiency: An alternative for LUMINOSITY.

luminous night clouds: An alternative for NOCTILUCENT CLOUDS.

lunar: Relating to *luna*, the moon; thus a lunar rainbow is a rainbow formed by the rays of the moon.
    The 'lunar month' or 'lunation' or 'synodic month' is the period of revolution of the moon round the earth relative to the line joining the centres of the sun and earth and averages about 29·53 days. The 'sidereal month' is the moon's revolution period relative to the fixed stars and averages about $27\frac{1}{3}$ days. These periods are to be distinguished from the civil 'calendar month' (January, February, etc.). A 'lunar day' signifies, in geophysics, the interval between successive lunar TRANSITS and averages about 24 h 51 min.

lustrum: A period of five consecutive years, which is sometimes used for grouping meteorological statistics which extend over a long period of years.

lux: The derived unit of illumination in SI UNITS, being 1 LUMEN per square metre (1 lm/m^{-2}). It is related to the obsolescent unit, the foot-candle, by
$$1 \text{ foot-candle} = 10 \cdot 76 \text{ lux}$$
$$1 \text{ kilolux} = 92 \cdot 9 \text{ foot-candles.}$$

lysimeter: An instrument for measuring the rate of PERCOLATION of rain through a soil.

M

Mach angle: See SHOCK WAVE.

Mach lines: See SHOCK WAVE.

Mach number: A pure number, significant in the movement of bodies through the air, defined as the ratio of the airspeed of a body to the speed of sound at the corresponding temperature. Neglecting a very small variation with water-vapour concentration, the speed of sound varies directly with the square root of the temperature (kelvin) of the air; in the ICAO STANDARD ATMOSPHERE it is, for example, 341 m/s (762 mile/h) at mean sea level and 295 m/s (660 mile/h) at 11 km.

mackerel sky: A sky covered with CIRROCUMULUS or, occasionally, high ALTO-CUMULUS clouds, arranged in a regular pattern of 'waves' and small gaps and resembling the scales of a mackerel; the cloud variety is VERTEBRATUS.

macroclimate: The general CLIMATE of a substantial part of the earth's surface, as for example all or most of a continent.

macroclimatology: The study of MACROCLIMATES.

macrometeorology: The study of large-scale processes in the atmosphere occurring over substantial regions of the earth's surface, up to and including the GENERAL CIRCULATION of the atmosphere itself.

macroviscosity: A quantity denoted N and defined by O. G. Sutton as $N = u_* Z_0$ where u_* is the FRICTION VELOCITY and Z_0 the ROUGHNESS LENGTH. Aerodynamically rough flow—the meteorologically important case—signifies a value of N greater than about 10^{-2} m²/s, i.e. a value some 10^3 times greater than the kinematic VISCOSITY (v). The parameter N plays a part in fully rough flow analogous to that of v in smooth flow.

maestro: A north-westerly wind in the Adriatic. It is most frequent on the western shore and in summer. North-westerly winds in other parts of the Mediterranean, notably in the Ionian Sea, and on the coasts of Sardinia and Corsica are also known as maestro.

magnetic character: The magnetic character of each day, from the point of view of disturbance, is classified 0, 1, 2 on the local scale and 0·0, 0·1, ..., 2·0 on the international scale—so-called C and Ci figures, respectively. See GEOMAGNETISM.

magnetic crochet: A short-lived disturbance (lasting generally less than one hour) of the earth's magnetic field, observed only in the sunlit hemisphere synchronously with an intense SOLAR FLARE. The disturbance is essentially an augmentation of the normal solar diurnal variation of the magnetic elements and is attributed to a sudden increase in ionospheric conductivity produced by solar flare wave radiation. It is also termed a (magnetic) 'solar flare effect' (sfe). See also GEOMAGNETISM.

magnetic storm: See GEOMAGNETISM.

magnetogram: The record from a MAGNETOGRAPH.

magnetograph: An instrument for obtaining continuous records of the earth's magnetic field. In the usual (La Cour) type, three magnets are suspended with their axes horizontal. One has its axis in the magnetic meridian, the second is perpendicular to the magnetic meridian, and the third can move in the vertical plane, the magnets responding to variations of DECLINATION (D), horizontal force (H), vertical force (V), respectively. To each magnet is attached a mirror which reflects light from a fixed lamp on to photographic paper which covers a rotating cylinder. A system of prism reflections enables the reflected traces to be contained within a single chart, even in the greatest magnetic storm. Regular time marks are made automatically and 'scale values' determined experimentally. Comparison with absolute (MAGNETOMETER) measurements of each element at known times enables the value of the reference or 'base line' for each element to be calculated and hence the absolute value of each element at any time. A fresh magnetograph record is obtained daily. A 'quick run' record with much expanded time-scale may also be operated.

Various electronic methods of measuring the absolute values of the earth's total field and standard components have been developed since about 1950. They have been introduced to a limited extent at some of the permanent observatories. Satisfactory recording of MICROPULSATIONS is possible only with this type of equipment which has the additional advantage of affording an output in digital form. See also GEOMAGNETISM.

magnetohydrodynamics: The study of the motion of an electrically conducting fluid in the presence of a magnetic field. Geophysical spheres in which such study is important include those concerned with motions in the earth's core and in the earth's high atmosphere.

magnetometer: An instrument for obtaining absolute measurements of the earth's magnetic field. The fundamental measured elements are generally the angles of DECLINATION (D) and DIP (I), and the value of the horizontal component (H); total force (F) and vertical force (V) are, however, sometimes measured. See GEOMAGNETISM.

magnetosphere: Term proposed for that composite region of the earth's high atmosphere, including the EXOSPHERE and much of the underlying IONOSPHERE, in which the earth's magnetic field exerts strong control over the motion of ionized particles.

mamma (mam): A supplementary cloud feature, previously termed 'mammatus' (Latin, *mamma* udder).

'Hanging protuberances, like udders, on the under surface of a cloud.

This supplementary feature occurs mostly with CIRRUS, CIRROCUMULUS, ALTOCUMULUS, ALTOSTRATUS, STRATOCUMULUS and CUMULONIMBUS.'* See Plate 14. See also CLOUD CLASSIFICATION.

manometer: An instrument for measuring differences of pressure. Ordinarily the weight of a column of liquid is balanced against the pressure to be measured. The mercury BAROMETER is, therefore, a form of manometer.

---

*Geneva, World Meteorological Organization. International cloud atlas, Vol. 1. Geneva, WMO, 1956, p. 16.

map projection: See PROJECTION.

mares' tails: The popular name for tufted CIRRUS clouds.

Margules's formula: See FRONT.

marine climate: An alternative for MARITIME CLIMATE.

maritime climate: A type of CLIMATE which is dominated by the near presence of the sea and is characterized by small diurnal and annual ranges of air temperature. Such a climate prevails over islands and windward parts of continents, e.g. the British Isles, more especially the extreme west. See CONTINENTAL CLIMATE.

Marsden square: Mainly used for identifying the geographical position of meteorological data over the oceans. System devised by Marsden in 1831 who divided a Mercator chart of the world into squares of 10° latitude by 10° longitude. Each square is numbered, and is again subdivided into 100 one-degree squares which are numbered from 00 to 99 so that, given the position of the square, the first figure of the sub-square denotes the latitude and the second figure the longitude.

mass concentration: See SPECIFIC HUMIDITY.

maximum: In meteorology, the highest value reached by a specified element in a given period. See EXTREMES.

maximum thermometer: Usually, a mercury-in-glass THERMOMETER which has a constriction in the bore, close to the bulb. If the thermometer is exposed horizontally the mercury is able, on expansion, to flow from the bulb past the constriction but not, on subsequent contraction, in the opposite direction. The end of the mercury column farther from the bulb indicates the maximum temperature attained since the last 'setting' of the thermometer; the setting is effected by shaking the mercury past the constriction.

mean: If $x_1, x_2, \ldots x_n$ are a set of measurements of the same kind, the most important mean values are defined as follows:

ARITHMETIC MEAN $\quad \bar{x} = \dfrac{1}{n} \sum\limits_{j=1}^{n} x_j$

GEOMETRIC MEAN $\quad \bar{x}_g = (x_1 \times x_2 \times, \ldots x_n)^{1/n}$

HARMONIC MEAN $\quad \dfrac{1}{\bar{x}_m} = \dfrac{1}{n} \sum\limits_{j=1}^{n} \dfrac{1}{x_j}.$

The term mean used without qualification generally implies the arithmetic mean. However, an expression such as 'the mean temperature of the atmosphere' is ambiguous unless it is accompanied by a statement of the positions at which the temperature has been measured or estimated. In a vertical air column, the 'mean temperature with respect to pressure' implies that the averaging process has been applied to a set of temperatures which are at equal intervals of pressure. The 'mean temperature with respect to height' is an average of another set of temperatures, namely those at equal height intervals.

mean deviation: In a series of n values, the mean deviation (e) is the mean of all the deviations (x) from the arithmetic mean, taken without regard to sign, i.e. $e = \Sigma|x|/n$. In a series with approximately NORMAL DISTRIBUTION, $e \approx 0{\cdot}8\sigma$, where σ is the STANDARD DEVIATION.

mean effective diameter: See CLOUDS, PARTICLE DISTRIBUTION IN.

mean free path: Mean distance travelled by the molecules or atoms of a gas between consecutive collisions with other molecules or atoms. Mean free path is a function of gas pressure. At low atmospheric levels it is of the order 10^{-9} m, increasing with height to about 10^{-2} m at 85 km, to 10^2 m at 180 km, and to still greater values at higher levels.

mean sea level: See SEA LEVEL.

median: When a series of n observations is arranged in order of magnitude, the central value (n odd), or the mean of the two central values (n even), is termed the 'median'. The median may differ appreciably from the MEAN, in which case the frequency distribution is said to be 'skew'.

median volume diameter: See CLOUDS, PARTICLE DISTRIBUTION IN.

mediocris (med): A CLOUD SPECIES (Latin, *mediocris* medium).
'CUMULUS clouds of moderate vertical extent, the tops of which show fairly small protuberances.'* See also CLOUD CLASSIFICATION.

Mediterranean-type climate: A distinctive type of subtropical CLIMATE, included in the Köppen classification, which is characterized by dry, hot, sunny summers and mild, moderately rainy winters. The type is found in the land regions bordering the Mediterranean, central and coastal southern California, central Chile, extreme south of South Africa, and parts of southern Australia.

medium-range forecast: See FORECAST.

megathermal climate: A CLIMATE of high temperature: more specifically, in the Köppen classification, one in which no month has a mean temperature below 18°C. Such conditions are found in the more humid of the tropical or subtropical regions.

melting band: A conspicuous horizontal band often observed in vertical cross-section radar displays of precipitation.
On reaching the melting (0°C) level, falling snowflakes begin to melt. Since an ice particle with a wet skin reflects almost as well as if it were an entirely liquid particle, the strength of the echo returned from the region below 0°C is greater than that above. Since, further, (i) the melting particles are spheroidal in shape and have a larger surface area than raindrops, and (ii) the rate of fall of the particles increases and their volume concentration decreases when completely melted, the radar reflectivity of the melting particles exceeds that of the raindrops below. The various effects combine to produce a bright band which has a maximum intensity a few hundred metres below the 0°C level. See also RADAR METEOROLOGY, BRIGHT BANDS.

---

* Geneva, World Meteorological Organization. International cloud atlas, Vol. 1. Geneva, WMO, 1956, p. 14.

melting-point: That temperature, characteristic of a given substance at a given pressure, at which the change of state from solid to liquid occurs, For pure ice at standard atmospheric pressure the melting-point is 0 degC.

Member: A Member of the WORLD METEOROLOGICAL ORGANIZATION, as defined in its Convention; primarily a State, territory or group of territories having or maintaining a Meteorological Service.

member: A person elected or designated and serving on the Executive Committee or other committee or on a Technical Commission or on a working group of the WORLD METEOROLOGICAL ORGANIZATION.

meniscus: The curved upper surface of liquid in a tube. The meniscus is concave for water, convex for mercury. The curvature effect arises from SURFACE TENSION. Scales and measures are graduated on the assumption that readings are taken at the centre of the meniscus in either case, i.e. lowest point of water meniscus, highest point of mercury meniscus.

Mercator's projection: See PROJECTION.

mercury: Mercury is a metallic element of great value in the construction of meteorological instruments. In the mercury barometer its great density enables the length of the instrument to be made moderate, while the low pressure of its vapour at ordinary temperatures makes possible a nearly perfect vacuum in the space above the top of the barometric column. In the mercury thermometer there is no risk of condensation in the upper end of the stem, as there is with the spirit thermometer.

Specific gravity	= 13·5951 at 0°C
Specific heat	= 0·0335 at 0°C
Vapour pressure	= 0·00021 mb at 0°C
	= 0·00343 mb at 30°C
Freezing-point	= 234·2 K
Coefficient of expansion	= 0·000182 per degC

meridian: The (geographical) meridian at any point of the earth's surface is the semi-GREAT CIRCLE which passes through the point and terminates at the geographical POLES. The 'prime meridian' is that which passes through Greenwich.

The meridian of an observer is that semi-great circle of the CELESTIAL SPHERE which passes through the observer's ZENITH and terminates at the celestial poles.

The magnetic meridian at any point of the earth's surface is the direction of the compass at that point. The 'geomagnetic meridian' or 'dipole meridian' at any point is the meridian which passes through the point and terminates at the geomagnetic poles—see GEOMAGNETISM.

meridional cell: See MERIDIONAL CIRCULATION.

meridional circulation: Generally, a closed circulation in a vertical plane oriented along a geographic MERIDIAN. It is also termed 'meridional cell'—see, for example, HADLEY CELL.

meridional extension: Marked elongation of upper ridges and troughs in a north–south direction.

meridional flow: Airflow in the direction of the geographic MERIDIAN, i.e. south–north or north–south flow.

mesoclimate: The CLIMATE of a moderately restricted region of the earth's surface; an urban district, as opposed to a neighbouring rural district, is near the lower end of the scale of areas encompassed by this term.

mesoclimatology: The study of MESOCLIMATES.

mesometeorology: The study of the atmosphere, pursued on a geographical scale between those employed in MICROMETEOROLOGY and SYNOPTIC METEOROLOGY. A several-fold increase in the number of synoptic stations over a restricted region of southern England, say, would meet the observational needs of mesometeorology, the object of which is a study of those substantial local variations of meteorological phenomena which are missed on the normal synoptic network.

mesopause: The top of the MESOSPHERE, at a height of about 80 km, marked by a temperature minimum and hence temperature inversion.

mesosphere: That part of the ATMOSPHERE, between the STRATOPAUSE at about 50 km and the MESOPAUSE at about 80 km, in which temperature generally falls with increasing height. An alternative definition, in which the mesosphere was considered to include also the layer from about 20 to 50 km in which temperature generally increases with height, is not now favoured.

mesothermal climate: A CLIMATE of moderate temperature; more specifically, in the Köppen classification, a climate in which the mean temperature of the coldest month lies between $-3°C$ and $+18°C$ (see MICROTHERMAL CLIMATE and MEGATHERMAL CLIMATE). Such conditions are found mainly between latitudes 30° and 45° but extend up to about 60° on the windward side of continents.

meteor: As defined in the *International cloud atlas*,* 'a phenomenon, other than a cloud, observed in the atmosphere or on the surface of the earth, which consists of a precipitation, a suspension or a deposit of aqueous or non-aqueous liquid or solid particles, or a phenomenon of the nature of an optical or electrical manifestation'. Meteors so defined are classified into four groups, namely HYDROMETEOR, LITHOMETEOR, PHOTOMETEOR and ELECTROMETEOR; the last three of these terms, in particular, are very little used.

In its more commonly used sense, a meteor, or 'shooting star', is a fragment of solid material (iron or stone) of undetermined origin, which enters the upper regions of the atmosphere and is there made visible by incandescence caused by compression of air in front of the meteor, the meteor itself evaporating in the air. (In another usage, the term 'meteor' is applied only to the visible trail, the particle itself being termed a 'meteoroid'.) In clear-sky conditions an individual observer normally sees a few meteors per hour; in large 'meteor showers', however, many more are to be seen. Most meteors are visible for only one or two seconds but a very large meteor may have a luminous trail that persists for half an hour or longer. The size of meteors ranges from about a centimetre downwards, most being very much smaller. Those too small to be detected visually or by the reflection of radio waves from the ionized trail which they leave in the high atmosphere are termed 'micrometeors' comprising, at very small particle size, 'meteoric dust'. The occasional bodies large enough to reach the ground are termed 'meteorites'. The large and very bright meteors are known as 'fireballs'; those whose destruction in the atmosphere is associated with an air explosion (audible up to a distance of about 50 miles) are termed 'bolides'.

---

* Geneva, World Meteorological Organization. International cloud atlas, Vol. 1. Geneva, WMO, 1956, p. 61.

Meteors are grouped into (i) 'meteor showers' which compose parts of great streams of particles orbiting the sun; they are mostly associated with comets and appear to enter the atmosphere from those points of the heavens and around those calendar dates given in Table IX, and (ii) 'sporadic meteors', of random occurrence, which account for the bulk of total meteor activity. Evidence advanced by E. G. Bowen to the effect that meteor-shower occurrence is a significant factor in the time distribution of world-wide rainfall, through the formation of condensation nuclei, is not generally accepted as convincing.

From synchronized meteor observations at a known distance apart (now obtained photographically through wide-angle telescopes) the velocity and brightness and the heights of appearance and disappearance of individual meteors may be measured. Such observations show that most visible meteors appear at about 110 km and disappear at about 80 km, with a secondary maximum frequency of disappearance at about 45 km and minimum frequency at 55 km. Since the rate of burn-out of a meteor is a function of air density, the density and hence the temperature of the air may be calculated from such data. The results indicate a temperature maximum at about 60 km at least as high as that near the ground and a rapid increase of temperature above about 85 km.

The IONIZATION of the air which is produced along the path of a meteor is significant in the sporadic E-LAYER. The radio-echo technique has been used with the ionized trails produced by meteors in order to determine density, temperature and winds at the atmospheric levels concerned.

TABLE IX. *Meteor-shower occurrence*

Shower	Date of maximum	Shower	Date of maximum
Quadrantids	3–4 January	Perseids	5–14 August
Lyrids	21 April	Giacobinids	10 October
η-Aquarids	4–6 May	Orionids	20–23 October
o-Cetids	14–23 May	Taurids	3–10 November
Arietids	29 May–18 June	Leonids	16–17 November
ζ. Perseids	1–16 June	Geminids	12–13 December
β-Taurids	26 June–5 July	Ursids	22 December
δ. Aquarids	28 July		

meteorite: A METEOR, or 'meteoroid', large enough to survive passage through the earth's atmosphere and so reach the earth's surface.

meteorograph: An instrument which gives an automatic record of two or more of the ordinary meteorological elements. The term has been used more especially of an instrument, now obsolete, attached to a kite or balloon to measure the pressure, temperature and humidity of the upper atmosphere.

meteorological office: In aeronautical terminology, an office designated to provide meteorological information for international air navigation. Sub-division is made in this terminology into:

(i) regional, (ii) main, (iii) dependent, (iv) supplementary, (v) watch offices, defined as follows:
- (i) Regional meteorological office: headquarters of a meteorological region which directs, controls and inspects the various stations of the region; it is qualified to issue directives, technical instructions, regional forecasts and warnings.
- (ii) Main meteorological office: a meteorological office competent to prepare forecasts, supply meteorological information and briefing to aeronautical personnel, supply meteorological information required by an associated dependent or supplementary office.

(iii) Dependent meteorological office: a meteorological office competent to prepare forecasts under the guidance of a main meteorological office, supply meteorological information and briefing to aeronautical personnel, supply meteorological information required by an associated supplementary office.

(iv) Supplementary meteorological office: a meteorological office competent to supply aeronautical personnel with meteorological information received from a main or a dependent office and with meteorological reports otherwise available.

(v) Meteorological watch office: a meteorological office competent to maintain watch over meteorological conditions within a defined area or along designated routes or portions thereof for the purpose of supplying meteorological information, in particular, meteorological warnings. A watch office may be an independent office or may be part of a main or dependent meteorological office.

meteorological optical range (MOR): As recommended by the World Meteorological Organization in 1957, a quantity which is identical for practical purposes with VISIBILITY, and is defined as the length of path in the atmosphere required to reduce the luminous flux in a collimated beam of light to 0·05 of its original value. (The 'light' is defined as that emanating from an incandescent lamp at a colour temperature of 2700 K.)

meteorological rocket: A small solid-fuel rocket designed to carry a payload of a few kilogrammes to a height of 60 to 90 km. This payload is often a ROCKETSONDE, CHAFF or grenades (see GRENADES SOUNDING). These rockets are usually unguided and the settings of the launcher have to be determined from BALLISTIC WIND measurements.

meteorological reconnaissance flight: An aircraft flight made for the specific purpose of obtaining information in a region inadequately served by surface observations (generally over the sea). These flights usually include vertical ascents at selected points on the route.

meteorology: The science of the atmosphere: from the Greek *meteoros*, lofty or elevated and *logos*, discourse. Meteorology embraces both WEATHER and CLIMATE and is concerned with the physical, dynamical and chemical state of the earth's atmosphere (and those of the planets), and with the interactions between the earth's atmosphere and the underlying surface.

methane: Gas, of chemical formula CH_4, which occurs in minute concentration in the atmosphere (about $2·0 \times 10^{-6}$ part by volume—see AIR).

Methane is released to the atmosphere mainly by the decay of biological products and is destroyed mainly by atmospheric ozone. Its mean lifetime in the atmosphere is considered to be not greater than 200 years.

metre: The unit of length in the M.K.S. SYSTEM of units and in SI UNITS. It was intended to be equal to one forty-millionth of the Paris meridian, but errors entered into the calculation and it must now be considered as an arbitrary length. For many years it was defined as the distance, at the melting-point of ice, between two lines engraved upon a platinum-iridium bar kept in Paris. In October 1960 it was decided at the International Bureau of Weights and Measures to abandon the ruled metre bar as a standard and to redefine the metre as being 1 650 763·73 wavelengths, in a vacuum, of the radiation corresponding to the transition between two specified energy levels

(2 P_{10} and 5 D_5) of the krypton atom of mass 86. This is a measurement whose accuracy can be maintained to one part in 100 million.

An Order in Council in 1898 defined the inch as 25·400 millimetres from which it may be deduced that one metre is 39·370113 inches.

micro-: A prefix meaning small, from the Greek *mikros:* for example, microbarograph, microseism. In units it signifies one-millionth and is designated μ, as in MICROMETRE, μm.

microbarograph: An instrument designed for recording small and rapid variations of atmospheric pressure. In the Shaw–Dines microbarograph, differences in pressure between a large and well-lagged reservoir of air and the external atmosphere are reflected in movements of a bell-shaped float in a bath of mercury and recorded on a clockwork-driven chart by a pen which is directly linked with the float. The reservoir is provided with a slow leak to atmosphere in the form of fine capillary tubing. In all but rapid changes of atmospheric pressure the leak equalizes the reservoir and atmospheric pressures, and the pen remains at the centre of the chart.

In the more modern and very sensitive Jones–Forbes microbarograph, flexible metal bellows constitute an ANEROID capsule and are made to drive a capacitance micrometer. The instrument is compensated for temperature changes and is protected against mechanical distortion and vibration. Its frequency response is made adjustable by the provision of variable leaks between the capsule and the atmosphere.

microclimate: The physical state of the atmosphere close to a very small area of the earth's surface, often in relation to living matter such as crops or insects. In contrast to CLIMATE, microclimate generally pertains to a short period of time.

microclimatology: The study of MICROCLIMATES.

micrometeorology: The study of the fine structure of the physical processes occurring in the atmosphere. This branch of meteorology is much concerned with atmospheric conditions close to limited regions of the earth's surface but embraces also the detailed structure of physical processes, more especially TURBULENCE, at higher levels of the atmosphere.

micrometre: A measure of length, equal to 10^{-6}m, denoted μm; previously termed 'micron'.

micron: See MICROMETRE.

micropulsations: In GEOMAGNETISM, rapid variations of the magnetic elements of a (nearly) periodic nature. The period ranges from a fraction of a second to some hundreds of seconds, classification being made in terms of the period range. Electronic equipment, as opposed to the La Cour types of MAGNETOGRAPH, is necessary for the measurement of pulsations of short period.

Natural pulsations have a wide variety of characteristics and have been shown to be closely related to conditions in the IONOSPHERE and MAGNETOSPHERE. Pulsations have also been observed to result from a high-altitude nuclear explosion.

microseisms: Term usually restricted, by convention, to a type of quasi-periodic motion of the ground which is unrelated to earthquakes, explosions, or such artificial agencies as industry and traffic.

The periods of microseisms range from a fraction of a second to several minutes. Atmospheric phenomena—notably the effect of wind on the oceans and resultant effect on the ocean bottom—play an important part in their production which is, however, not well understood. Microseisms have long been used as an early indication of tropical-storm development over the sea. Tripartite stations, spaced in a triangle at distances a few kilometres apart, have more recently been developed as a useful aid in the tracking of tropical storms across sea areas.

microthermal climate: A CLIMATE of low temperature: more specifically, in the Köppen classification, a climate of long, cold winters and short summers, the mean temperature of the coldest month being less than $-3°C$ and that of the warmest month being greater than $10°C$. Such conditions are found in the interior and eastward parts of continents between about latitudes $40°$ and $65°$.

microwaves: RADIO WAVES of short wavelength (generally in the range from a fraction to some tens of centimetres) employed, for example, in RADAR.

Mie scattering: SCATTERING of ELECTROMAGNETIC RADIATION by spherical particles. The theory of such scattering, developed by G. Mie in 1908, does not depend on the assumption that the scattering particle has a radius small compared with the wavelength of the radiation and thus has wider application than the theory appropriate to RAYLEIGH SCATTERING. Extension of the theory to particles of various types and shapes, together with formulae and numerical data required to apply the theory, are contained in the book *Light scattering by small particles.**

mil: A unit, used in the preparation of standard ballistic meteorological messages, defined by 6400 mils $= 360°$. See BALLISTICS.

mile: The 'statute mile' is defined as 1760 yd or 5280 ft. The 'geographical mile' is defined as the length of one minute of arc of longitude at the equator, i.e. 1/21 600 of the earth's equatorial circumference and equals $6087·2$ ft. The 'nautical mile' was originally taken, for navigational purposes, as the length of one minute of arc of latitude. Owing to the spheroidal shape of the EARTH this length varies with latitude, being given by $(6076·8 - 31·1 \cos 2\varphi)$ ft where φ is latitude. The precise length of 6080 ft was adopted by the British Admiralty as the nautical mile. The U.S. nautical mile is $6080·21$ ft, while the 'international nautical mile' is exactly 1852 metres ($6076·12$ ft).

millibar: The thousandth part of a BAR. The millibar (mb), equivalent to 10^3 c.g.s. units of pressure (DYNES/cm²), has replaced the inch or millimetre of mercury as the unit of pressure in the Meteorological Office since 1 May 1914. See also PRESSURE.

$$1 \text{ bar} = 10^3 \text{ millibars} = 10^6 \text{ dynes/cm}^2 = 10^5 \text{ NEWTONS/m}^2.$$

Mills period: A minimum period of leaf wetness in apple trees in spring which, expressed as a function of mean air temperature during the period, is favourable for the development of apple scab infection.

Some examples of critical period length at various temperatures are: 30 hours at $42°F$ ($5·6°C$); 14 hours at $50°F$ ($10°C$); 10 hours at $59°F$ ($15°C$).

minimum: In meteorology, the lowest value reached by a specified element in a given period. See EXTREMES.

---

* VAN DE HULST, H. C.; Light scattering by small particles. New York, John Wiley and Sons, 1957.

minimum deviation: In REFRACTION, the minimum change of direction suffered by radiation in passing through a refractive medium, as for example light rays through ice crystals. Since changes of direction of the refracted rays are least near the position of minimum deviation, rays emerging after total refraction at and near this position contribute most to the brightness seen by an observer. Minimum deviation is fundamental in the production of the simple refraction phenomenon of the HALO and in the refraction plus REFLECTION phenomenon of the RAINBOW.

The 'angle of minimum deviation' (D) in a plane normal to the faces of a prism intersecting at angle A is given by the formula:

$$n = \frac{\sin\{(A + D)/2\}}{\sin(A/2)}$$

where n is the REFRACTIVE INDEX of the medium. For the passage of light through hexagonal ice crystals ($A = 60°$, $n = 1\cdot31$), D is nearly $22°$, corresponding to the radius of the small halo.

minimum temperature: The lowest temperature attained, usually in the THERMOMETER SCREEN (screen minimum temperature) or on the ground (GRASS MINIMUM TEMPERATURE or CONCRETE MINIMUM TEMPERATURE), during a given period.

Because of their significance in the formation of frost and fog, much attention has been given to the problem of forecasting night minimum temperatures of either type. Various empirical formulae have been derived, using mainly the observed dry-bulb and wet-bulb temperatures and dew-point, and the observed values (or those estimated for the cooling period) of surface wind speed and cloud amount. Owing to variations of topographical influences and length of cooling period, such formulae may be applicable only to a specified location and time of year.

minimum thermometer: Usually, an alcohol-in-glass THERMOMETER which carries a small index within the bore, below the surface of the liquid. Initial 'setting' of the instrument is effected by raising the bulb of the thermometer higher than the stem; the index then falls till its lower end meets the end of the spirit column, being prevented from further fall by surface tension. Contraction of the spirit on cooling causes the index to be dragged back closer to the bulb by the end of the spirit column, while expansion of the spirit on heating does not affect the position of the index. The position of that end of the index farther from the bulb thus indicates the minimum temperature attained since the last setting of the thermometer. The thermometer is supported in the screen so that the stem has a slight slope downwards towards the bulb. With this arrangement the movement of the index towards the bulb is slightly assisted by gravity.

mintra: That temperature above which, at a given pressure, theory indicates that no CONDENSATION TRAILS will form. The variation of mintra temperature with pressure is shown by a line on the Meteorological Office TEPHIGRAM chart.

mirage: An atmospheric optical phenomenon produced by REFRACTION of light in the layers of air close to the earth's surface due to large temperature gradients in the vertical and associated changes of REFRACTIVE INDEX.

Two main classes of mirage occur, (i) 'inferior' and (ii) 'superior', in which the virtual image is below and above the object, respectively. The inferior mirage is seen over a flat, strongly heated surface (e.g. desert or road) and gives the illusion of an expanse of water; it is caused by the strong upward refraction of light from the clear sky towards the observer. The superior mirage is seen above a flat surface of much lower temperature than the air above it; light from an object is in this case bent downwards towards the observer, as in LOOMING (see Figure 25). In such physical conditions multiple reflections may give rise to various images, some displaced laterally with respect to the object, as in FATA MORGANA.

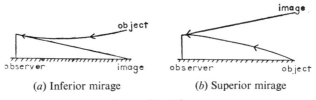

(a) Inferior mirage (b) Superior mirage

FIGURE 25—Mirages.

mist: A state of atmospheric obscurity produced by suspended microscopic water droplets or wet hygroscopic particles. The term is used for synoptic purposes when there is such obscurity and the associated visibility is equal to or exceeds 1 km; the corresponding relative humidity is greater than about 95 per cent. The particles contained in mist have diameters mainly of the order of a few tens of micrometres. See also HAZE, FOG.

mistral: A north-westerly or northerly wind which blows offshore along the north coast of the Mediterranean from the Ebro to Genoa. In the region of its chief development its characteristics are its frequency, its strength and its dry coldness. It is most intense on the coasts of Languedoc and Provence, especially in and off the Rhône delta. On the coast, speeds are about 40 knots but in the Rhône valley a speed of over 75 knots has been reached.

mixed cloud: A cloud which is composed of both ice crystals and water droplets. The CLOUD GENERA As, Ns and Cb are normally mixed clouds.

mixing condensation level: See CONDENSATION LEVEL.

mixing length: That distance (l) moved by a discrete EDDY in a turbulent fluid, carrying with it the momentum and other properties appropriate to the mean motion at the original point of the fluid, before mixing again with the general flow. This quantity, introduced in eddy diffusion by L. Prandtl (1925) on analogy with the 'mean free path' appropriate to the process of molecular diffusion, appears in expressions for the eddy velocity (u'), REYNOLDS STRESS (τ), and eddy VISCOSITY (K_M), as follows:

$$u' = l \frac{\partial \bar{u}}{\partial z},$$

where u' represents an instantaneous departure from the average velocity $\bar{u}$;

$$\tau = -\overline{\rho u'w'}$$
$$= \rho l^2 \left(\frac{\partial \bar{u}}{\partial z} \right) \left| \frac{\partial \bar{u}}{\partial z} \right|,$$

where the bar represents a time average;

$$K_M = l^2 \frac{\partial \bar{u}}{\partial z}.$$

mixing ratio: The ratio of the mass of a particular gaseous constituent (e.g. ozone) of the atmosphere to the mass of air with which the constituent is associated.

The term is most often used in respect of the admixture of water vapour with dry air in the atmosphere—see HUMIDITY MIXING RATIO.

mizzle: See SCOTCH MIST.

m.k.s. system: A system of units based on the metre, the kilogramme and the second as FUNDAMENTAL UNITS.

Formerly much less widely used in meteorology than the c.g.s. system, the m.k.s. units have recently been adopted, together with the ampere, the kelvin and the candela, as the basis of the International System (SI) of Units. See SI UNITS.

mock moon: An alternative for PARASELENE.

mock sun: An alternative for PARHELION.

mock sun ring: An alternative for PARHELIC CIRCLE.

mode: In a series of values, the value of most frequent occurrence. An approximate rule, when there is only one mode, is: mode = median -3 (mean $-$ median).

model atmosphere: A simplified representation of the atmosphere devised for studying atmospheric behaviour and often for the purposes of producing NUMERICAL WEATHER FORECASTS. The simplifications allow attention to be focused on those atmospheric properties which are thought to be of the greatest importance and usually permit these properties to be represented in the mathematical system of equations.

For many purposes the earth's atmosphere can be regarded as behaving as a stratified fluid and it can therefore be treated as consisting of a number of layers whose properties and interrelations can be expressed mathematically.

The simplest and most restrictive model is the BAROTROPIC model in which a single pressure surface, usually 500 mb, is considered, More sophisticated BAROCLINIC models involve consideration of many isobaric (or similar) surfaces. This enables vertical variations in winds and temperature to be represented more realistically. By considering the distribution of moisture, the effects of condensation and evaporation which are known to be important can also be incorporated.

modified refractive index: For convenience in radio meteorological work, in which the curvature of radio waves with respect to the curvature of the earth has to be considered, a 'modified refractive index' (M.R.I.) is employed. The M.R.I. is defined by the equation

$$M' = (n-1 + h/a) \times 10^6 \text{ M units}$$

where M' denotes the M.R.I., n is the refractive index of the air, h is height above earth's surface, and a is the earth's radius. For explanation of M units see also REFRACTIVE INDEX.

moist air: A term which in physical meteorology usually signifies simply a mixture of DRY AIR and water vapour. In synoptic meteorology and climatology the term is applied to air of high RELATIVE HUMIDITY.

moisture content: See SPECIFIC HUMIDITY.

mole (or mol): The gramme-molecular weight of a substance, i.e. the weight of it in grammes which is numerically equal to its MOLECULAR WEIGHT.

molecular weight: The weight of a molecule of an element, defined on a scale in which the molecular weight of carbon 12 is 12·0000.

The mean molecular weight (M) of moist air, as applied, for example, in the GAS EQUATION, is

$$M = \frac{m_1 + m_2}{m_1/M_1 + m_2/M_2}$$

where m_1, m_2 are relative weights of dry air and water vapour, respectively,

M_1 is mean molecular weight of dry air $= 28 \cdot 9644$,

and M_2 is molecular weight of water vapour $= 18 \cdot 0153$.

mole fraction: The mole fraction N_i of the ith component of a mixture of gases is defined by

$$N_i = \frac{m_i/M_i}{\Sigma (m_i/M_i)}$$

where m_i is the mass of the ith component in a given volume or mass of the mixture and M_i is its molecular weight, the summation indicated being made over all components. It is identical with FRACTIONAL VOLUME ABUNDANCE. See also AIR.

moment of inertia: The moment of inertia (I) of a body about an axis is the sum of the products of the mass (m) of each element of the body and the square of its perpendicular distance (r) from the axis, taken for all elements of the body, i.e. $I = \Sigma mr^2$.

moments: In statistics, if $x_1, x_2, \ldots x_n$ are measurements of the same kind then the first moment (μ_1), second moment (μ_2), third moment (μ_3), etc. about the mean $\bar{x}$ of their distribution are defined by

$$\mu_1 = \frac{1}{n} \sum_{j=1}^{n} (x_j - \bar{x}) \;,$$

$$\mu_2 = \frac{1}{n} \sum_{j=1}^{n} (x_j - \bar{x})^2 \;,$$

$$\mu_3 = \frac{1}{n} \sum_{j=1}^{n} (x_j - \bar{x})^3 \;,$$

and so on.

momentum: The (linear) momentum of a particle is the product of its mass and its velocity. It has dimensions MLT^{-1}.

monochromatic radiation: RADIATION of a single wavelength. 'Monochromatic flux', 'monochromatic intensity', etc. signify the flux, intensity, etc. per unit wavelength interval.

monsoon: Derived from the Arabic word 'mausim', meaning 'season', the term originally referred to the winds of the Arabian Sea which blow for about six months from the north-east and for six months from the south-west, but is now used also of other markedly seasonal winds. The essential cause is the differential heating of large land and sea areas, altering with season.

Monsoon conditions are best developed in the subtropics, as in east and south-east Asia (north-east and south-west monsoons in winter and summer, respectively).

The rainy season associated with the south-west monsoon is the outstanding feature of the climate of these regions and the term 'the monsoon' is there popularly used to denote the rains, without reference to the winds.

Monsoon conditions occur also, but to a lesser degree, in northern Australia, parts of western, southern and eastern Africa, and parts of North America and Chile. The term is also employed, for example, in north-west Germany. See also INTERTROPICAL CONVERGENCE ZONE.

Monte Carlo methods: A term used to describe various techniques involving the use of RANDOM NUMBERS, or random sequences, for objectives which are difficult or impossible to achieve by wholly systematic methods. For example, the distribution of the CHI-SQUARE STATISTIC can be determined by a Monte Carlo method when the expected frequencies are too small to validate a direct mathematical determination, while a multi-dimensional integral can be evaluated approximately by a Monte Carlo process when a systematic evaluation is impracticable.

month: See LUNAR, CALENDAR.

moon: The earth's only (natural) satellite. Its radius is 1738 km, mass about $7 \cdot 38 \times 10^{22}$ kg, mean density $3\frac{1}{3}$ times that of water, mean distance from the earth 384 400 km. Its SIDEREAL PERIOD of revolution averages about $27\frac{1}{3}$ days, but varies as much as three hours from this value on account of the eccentricity of its orbit and its 'perturbations'. The SYNODIC PERIOD of revolution (i.e. the interval from new moon to new moon) has a mean value of about $29 \cdot 53$ days, but varies by about 13 hours on account of the eccentricities of the orbits of the moon and earth.

The brightness of the moon is caused by its reflection of direct sunlight falling on it; while the faint illumination of the dark segment is caused by reflected sunlight which has been first reflected to the moon by the earth's surface and atmosphere ('earthlight')—see ALBEDO. The moon's albedo is about 10 per cent; this value and the considered absence of appreciable atmosphere imply a mean temperature of the moon's surface, in radiative equilibrium with absorbed solar radiation, of 267 K. The moon causes a measurable, though very small tidal movement of the earth's atmosphere but has no other substantiated meteorological effect. See also LUNAR.

moon, phases of: The appearance of the moon, by custom restricted to the particular phases of 'new moon' when nothing is visible, 'first quarter' when a semi-circle is visible with the illuminated bow on the west, 'full moon' when a full circle is visible and 'last quarter' when a semi-circle is visible with the bow on the east. The changes of phase are caused by changes in the relative positions of earth, moon and sun. The moon rotates on its axis once in each orbital revolution and so the same face of the moon is always turned towards the earth.

mother-cloud: A cloud from which another cloud develops. See CLOUD CLASSIFICATION.

mother-of-pearl clouds: An alternative term for NACREOUS CLOUDS. The name was given to them by H. Mohn (1893) because of their brilliant IRIDESCENCE, similar to that shown by mother-of-pearl.

motion, equations of: The equations of motion of the atmosphere are obtained by applying Newton's second law of motion, which equates the total force acting per unit mass, to the acceleration produced. For meteorological purposes the equation is required in terms of 'relative motion', i.e. in terms of accelerations and velocities measured on the rotating earth.

In vector notation the equation is

$$\dot{\mathbf{V}} = -\frac{1}{\rho}\nabla p - 2\mathbf{\Omega} \wedge \mathbf{V} + \mathbf{g} + \mathbf{F}$$ (see ACCELERATION for explanation of symbols).

In Cartesian co-ordinates in which x and y are in a horizontal plane to east and north, respectively, and z is positive vertically upwards, the equations of relative motion are:

$$\frac{du}{dt} = -\frac{1}{\rho}\frac{\partial p}{\partial x} - 2\Omega(w\cos\varphi - v\sin\varphi) + F_x,$$

$$\frac{dv}{dt} = -\frac{1}{\rho}\frac{\partial p}{\partial y} - 2\Omega u\sin\varphi + F_y,$$

$$\frac{dw}{dt} = -\frac{1}{\rho}\frac{\partial p}{\partial z} + 2\Omega u\cos\varphi - g + F_z,$$

where u, v, w are the velocities in the positive x, y, z directions respectively; F_x, F_y, F_z are the components of friction in these directions, due to viscosity or, in mean flow on the meteorological scale, to Reynolds stresses; φ is latitude; and d/dt is the change with time of an individual particle of air ('individual change' or 'change following the air motion').

mountain breeze: An alternative for KATABATIC WIND.

mountain wave: An air wave which is stationary, or almost stationary, with respect to the earth's surface. Such a wave sometimes has cloud in the wave crest. It is formed over and/or to leeward of a hill or mountain which obstructs the airflow. See STANDING WAVE, LEE WAVES.

mountain wind: See VALLEY WIND.

moving averages: In a series of numbers a_1, a_2, a_3, etc., the '3-term moving averages' (also termed 'running means') are $(a_1 + a_2 + a_3)/3$, $(a_2 + a_3 + a_4)/3$, etc. Similarly n-term moving averages may be formed where n is any integer less than the total of numbers in the series; the greater n is, the greater the SMOOTHING of the original series. A moving average differs from the corresponding moving sum only by a constant factor.

Where the original numbers form a TIME SERIES, moving averages may be used either (i) to eliminate from the original series a variation of known periodicity (e.g. by the formation of 12-term running means from a series of successive mean monthly temperatures), or (ii) to smooth out short-period fluctuations and so reveal to better effect any long-period fluctuation present in the original series. Theory and experiment show, however, that spurious PERIODICITY may be introduced by the use of data which have been smoothed by the formation of moving averages.

See also FILTERING for a discussion of weighted moving averages.

M-regions, solar: The hypothetical restricted solar regions which are thought to emit, usually during a few successive months, SOLAR CORPUSCULAR STREAMS. The existence of such M- (magnetically active) regions is inferred from a marked tendency for small and moderate magnetic storms to recur at intervals of about 27 days, which is the solar rotation period. M-regions have not yet been identified with any visible solar feature.

mutatus: See CLOUD CLASSIFICATION.

N

nacreous clouds: A rare type of stratospheric cloud, also termed 'mother-of-pearl clouds'. The reported occurrences of this cloud have been mainly in Norway and Scotland in winter on occasions of strong and deep west/north-west flow (deep depression over northern Scandinavia). The clouds are somewhat lenticular in form, very delicate in structure, and show brilliant IRIDESCENCE at angular distances up to about 40° from the sun's position. The colouring is most brilliant shortly after sunset (or before sunrise) and endures for a considerable time after sunset. The mean of C. Störmer's height measurements of the cloud is 24 km. The clouds show little or no movement; this fact, together with the circumstances of their occurrence strongly suggests that they are in the nature of MOUNTAIN WAVE clouds. The nature of the particles is not known but the associated optical effects suggest diffraction by spherical particles of diameter less than 250 μm.

natural co-ordinates: A system of co-ordinates in which the motion is referred to a right-handed set of axes **t, n, k**, where **t** is a unit vector in the direction of the flow, **n** is a unit vector at right angles and to the left of **t**, and **k** is a unit vector directed vertically upwards.

In this system $\partial/\partial s$ represents differentation along the STREAMLINES and $\partial/\partial n$ represents differentiation along the ORTHOGONAL LINES. Also

$$\mathrm{grad} \equiv \nabla = \mathbf{t}\frac{\partial}{\partial s} + \mathbf{n}\frac{\partial}{\partial n}$$

$$\mathrm{div}\,\mathbf{V} = \frac{\partial \mathbf{V}}{\partial s} + \mathbf{V}\frac{\partial \psi}{\partial n}$$

$$\mathrm{curl}\,\mathbf{V} = \mathbf{V}\frac{\partial \psi}{\partial s} - \frac{\partial \mathbf{V}}{\partial n}$$

where ψ is the angle measured positive counter-clockwise from east to the velocity vector; $\partial \psi/\partial s$ is the streamline curvature usually denoted by K_s, and $\partial \psi/\partial n$ is the ORTHOGONAL CURVATURE usually denoted by K_n.

Navier–Stokes equations: The equations of motion appropriate to a viscous fluid. Where the fluid is incompressible, the component of acceleration, with reference to axes fixed in space, which acts in the x direction of the Cartesian system of co-ordinates is related to the corresponding forces acting per unit mass by the equation

$$\frac{du}{dt} = -\frac{1}{\rho}\frac{\partial p}{\partial x} + v\left(\frac{\partial^2 u}{\partial x^2} + \frac{\partial^2 u}{\partial y^2} + \frac{\partial^2 u}{\partial z^2}\right) + X$$

where v is the kinematic VISCOSITY and X the component of the external forces per unit mass in the x direction. Analogous expressions obtain in the y and z directions. See also MOTION, EQUATIONS OF.

nebule: A measure of atmospheric opacity, defined by the statement that a screen of 100 nebules transmits the fraction 1/1000 of the incident light. The definition implies that a screen of opacity 1 nebule has a TRANSMITTANCY of 0·933 and that 1 nebule per kilometre is equivalent to an EXTINCTION COEFFICIENT of 0·069 per kilometre. The number of nebules per kilometre of air varies from about 1 in very good visibility to about 10 000 in thick fog.

Reference to the unit DECIBEL shows that an opacity of 100 nebules corresponds to a difference of flux density of $10 \log_{10} (1000/1) = 30$ decibels.

nebulosus (neb): A CLOUD SPECIES (Latin, *nebulosus* fog covered).
'A cloud like a nebulous veil or layer, showing no distinct details.
This term applies mainly to CIRROSTRATUS and STRATUS.'* See also CLOUD CLASSIFICATION.

neon: One of the INERT GASES, comprising $1·8 \times 10^{-5}$ and $1·2 \times 10^{-5}$ part per part of dry air by volume and weight, respectively. Its molecular weight is 20·183.

nephanalysis: The interpretation of SATELLITE cloud pictures in terms of cloud types and amounts. The analysis is transmitted either as a FACSIMILE picture or in the form of a five-figure code.

nephoscope: An instrument for determining the direction of motion of a cloud and its angular velocity about a point on the ground directly below it; the product of the angular velocity (radians/hour) and the cloud height, measured or estimated, in miles gives a measure of the cloud speed in miles per hour.

The types of nephoscope most commonly employed are the Fineman reflecting (or mirror) nephoscope and the Besson comb nephoscope—see '*Observer's handbook.*'† A camera obscura arranged to project a view of the clouds near the zenith on to a graduated board may be used as a nephoscope.

net radiation: An alternative for RADIATION BALANCE.

neutral atmosphere: A hypothetical atmosphere in neutral vertical equilibrium throughout, characterized by dry adiabatic lapse rate conditions. It is also termed 'adiabatic atmosphere' or 'convective atmosphere'. See ADIABATIC.

neutral point: See POLARIZATION.

newton: The M.K.S. and SI unit of force. It is the force that produces an acceleration of 1 m/s^2 when applied to a mass of 1 kg. The dimensions are MLT^{-2}.

$$1 \text{ newton} = 10^5 \text{ DYNES}.$$

nightglow: The night-time AIRGLOW emission, being the feeble light of the night sky emitted continuously by the upper atmosphere. It has also been termed 'night skylight' and 'permanent aurora'.

The light of the nightglow is measured by a photometer or a spectroscope. It is estimated to account, on average, for nearly half the intensity of light which is present on a clear, moonless night.

The 'auroral green line' 5577 Å ($\text{Å} = 10^{-10}\text{m}$) of atomic oxygen, the SODIUM D line (5893 Å), and HYDROXYL bands are among the prominent identified emissions in the complex nightglow spectrum. Photometer measurements reveal the presence of systematic solar-cycle and shorter-period time variations of intensity similar to

---

* Geneva, World Meteorological Organization. International cloud atlas, Vol. 1. Geneva, WMO, 1956, p. 12.
† London, Meteorological Office. Observer's handbook, 3rd edn. London, HMSO, 1969, pp. 36–40.

those of electron density in the IONOSPHERE. Measurements indicate that emissions are generally most intense at 80–100 km but occur also at 200–300 km.

Nightglow is caused by the radiation emitted in the night-time chemical reactions of the ionized and dissociated products which are formed during the day by solar ultra-violet radiation in the atmospheric gases of the ionosphere.

night skylight: An obsolescent alternative for NIGHTGLOW.

night-sky lights: The light received near the ground on a clear, moonless night is estimated to be composed, on average, of direct light from stars and nebulae (about 40 per cent), ZODIACAL LIGHT (15 per cent), NIGHTGLOW (40 per cent), and light from all these sources scattered by the earth's atmosphere (5 per cent).

night-sky recorder: A camera with long-focus lens which is mounted with its axis directed towards one of the celestial poles. The camera shutter is arranged, by a clockwork mechanism, to be open only when the sun is 10° or more below the horizon. As the earth rotates images of the stars trace arcs of concentric circles on the film. Quantitative assessment may be made of the incidence of clouds (or fog or thick haze) which cause interruptions of the record.

Use of this instrument, which is also termed a 'starshine recorder', is very limited.

nimbostratus (Ns): One of the CLOUD GENERA (Latin, *nimbus* rainy cloud and *stratus* spread out).

'Grey cloud layer, often dark, the appearance of which is rendered diffuse by more or less continuously falling rain or snow, which in most cases reaches the ground. It is thick enough throughout to blot out the sun.

Low, ragged clouds frequently occur below the layer, with which they may or may not merge.'* See Plate 10. See also CLOUD CLASSIFICATION.

Nipher shield: A form of screen, based on a suggestion by F. E. Nipher in 1897, which is fitted to a rain-gauge (or snow-gauge), for the purpose of eliminating, as far as possible, wind eddies at the mouth of the gauge and so enabling a truer catch of rain to be made. Such screens are not used in the Meteorological Office.

nitric oxide: A gas, of chemical formula NO, which occurs in minute quantities (of the order 10^{-9} by volume ratio) in the low atmosphere where it is of industrial origin. It is produced in the high atmosphere by DISSOCIATION and subsequent chemical reactions.

In various experiments nitric oxide has been injected into the high atmosphere and has been followed either visually (glow produced at night) or by radio.

nitrogen: A chemically inactive gas, of molecular weight 28·016, which comprises 78·09 and 75·54 parts per 100 parts of dry air by volume and weight, respectively.

Nitrogen exists only in the molecular form (formula N_2) in the lower atmosphere, but has been identified in atomic form in the high atmosphere where it suffers DISSOCIATION to a small extent.

nitrogen cycle: A complex circulation of nitrogen involving the soil, plants, animals and the atmosphere. In that part of the circulation which involves the atmosphere, nitrogen is released to the atmosphere from the soil (as NITROGEN and NITROUS OXIDE) by the action of nitrogen-fixing bacteria. R. M. Goody and C. D. Walshaw

---

* Geneva, World Meteorological Organization. International cloud atlas, Vol. 1. Geneva, WMO, 1956, p. 11.

estimate the average magnitude of the earth–atmosphere nitrogen cycle to be not less than 10^7 molecules/m²s.

nitrogen dioxide: A gas, of chemical formula NO_2, which occurs in minute quantities (of the order 10^{-9} by volume ratio) in the low atmosphere where it is of industrial origin. It is produced in the high atmosphere by DISSOCIATION and subsequent chemical reactions.

nitrous oxide: A gas, of chemical formula N_2O, which occurs in the approximately uniform volume ratio of about 4×10^{-7} throughout the troposphere and lower stratosphere. N_2O is considered to be supplied to the atmosphere from the soil as part of the NITROGEN CYCLE and to be destroyed by DISSOCIATION in the high atmosphere.

n-method: See SUPERPOSED-EPOCH METHOD.

noble gases: An alternative for INERT GASES.

noctilucent clouds: Tenuous but at times brilliant clouds in the very high atmosphere. These clouds, also termed 'luminous night clouds', have been observed during the midnight hours of the summer months in latitudes higher than about 50°. The clouds are to be seen in appropriate viewing conditions (direct illumination by sunlight against a dark sky and in the absence of lower clouds) more frequently than was previously thought. They resemble cirrostratus in appearance but have a bluish-white to yellow colour. In the British Isles they are usually seen towards the northern horizon, but extend on occasion to high elevations.

Measurements have shown the clouds to be at a height of 80–85 km and to have a movement from the north-east at speeds between about 100 and 300 knots. Pronounced wave formation is often visible and slower movement of the waves towards the north-east, contrary to that of the cloud material, has been reported. It is yet uncertain whether the clouds are composed of dust or of ice particles. A particle radius of the order 10^{-1} μm is indicated by measurements of the strong POLARIZATION of the light from the clouds. No clear association between the appearance of the clouds and such occurrences as volcanic eruptions or meteor showers has yet been shown. See Plate 18.

nocturnal radiation: The excess RADIATION emitted by the earth's surface at night relative to that received by the earth's surface from the atmosphere (mainly from clouds and from atmospheric water vapour and carbon dioxide). It is also termed 'effective radiation'.

Since the earth radiates as a black body at its own temperature while the atmosphere is transparent over an important range of wavelengths at terrestrial temperatures (see ATMOSPHERIC WINDOW) the excess is nearly always positive and results in a fall in temperature of the earth's surface at night. Nocturnal radiation is greatest when the air is cloudless and relatively dry.

The term is perhaps unfortunate because the same process occurs by day; the energy loss is then, however, generally small compared with the influx of solar radiation.

noise: In the description of space or time fields a term used for random errors of observation and other unwanted and unco-ordinated effects which are irrelevant to the purpose of the measurements forming the field.

normal: The name given to the average value over a period of years of any meteorological element such as pressure, temperature, rainfall or duration of sunshine.

Normal meteorological values are subject to 'uncertainty' owing to year-to-year variability of the observations; the computed STANDARD ERROR of the mean, decreasing with increasing length of period over which averages are taken, is a measure of this uncertainty. In the selection of a suitable length of period, a compromise must be struck in that the period must be long enough for the computed standard error of the mean to be small, but must not be so long that there is a risk that the period contains an appreciable SECULAR TREND (change of normal) of the observations. A period of about 30 years has in the past been thought to be a reasonable compromise and is generally used in deriving normals.

Climatological normals for Great Britain and Northern Ireland are contained in a number of Meteorological Office publications (temperature 1931–60*; sunshine, 1931–60†; rainfall, 1916–50‡). For the British Isles, monthly and annual charts of pressure, wind, temperature, rainfall, snow, thunder, sunshine, fog and cloud are given in the *Climatological atlas of the British Isles*.§ Tables of temperature, relative humidity and precipitation for the world are contained in another Meteorological Office publication.¶

normal (frequency) distribution: The normal (also termed the Gaussian) distribution corresponds to the distribution of random errors about a population mean which is indicated by the NORMAL LAW OF ERRORS. The equation on which the distribution is based is

$$y = \frac{\exp(-x^2/2\sigma^2)}{\sigma(2\pi)^{\frac{1}{2}}}$$

where σ is the standard deviation. There is maximum frequency (y) of small errors (x), the frequency decreasing rapidly with increase of error size. The curve is therefore bell-shaped and is symmetrical about the y-axis along which the mean, mode and median all coincide.

The normal distribution is specified entirely by the values of the MEAN and STANDARD DEVIATION; the particular one in which the mean is zero and the standard deviation unity is termed the 'unit normal distribution'.

Probabilities of error (departure) occurrence greater than or less than a specified size (expressed as the ratio x/σ) are given by the 'probability integral', or 'error function' (erf), which is tabulated in books on statistics. For example, probabilities of a departure, of unspecified sign, greater than a given multiple of σ are: σ, 0·138; 2σ, 0·046; 3σ, 0·003; 4σ, 0·00006. The probabilities of a departure of a specified sign are half those quoted.

The normal distribution occupies a central position in statistical theory because of the number and variety of distributions which tend to normality and because several of the main statistical SIGNIFICANCE tests (e.g. STUDENT'S t-TEST, the F-TEST and the test for the significance of a CORRELATION coefficient) are based on the assumed normality of the distributions of the data. The distributions of some of the meteorological elements, notably of temperature and pressure, often approximate to the normal distribution while those of other elements, e.g. daily rainfall, depart widely from it. The operations of averaging data, or converting them into departures

---

* London, Meteorological Office. Averages of temperature for Great Britain and Northern Ireland, 1931–60. London, HMSO, 1963 (reprinted with amendments 1969).

† London, Meteorological Office. Averages of bright sunshine for Great Britain and Northern Ireland, 1931–60. London, HMSO, 1963.

‡ London, Meteorological Office. Averages of rainfall for Great Britain and Northern Ireland, 1916–50. London, HMSO, 1958 (reprinted 1967).

§ London, Meteorological Office. Climatological atlas of the British Isles. London, HMSO, 1952.

¶ London, Meteorological Office. Tables of temperature, relative humidity and precipitation for the world. Parts I to VI. London, HMSO, 1958, 1959, 1967.

from normal, tend to produce distributions which are nearer to normal; a reasonably close approach to normality is required before the above tests may properly be applied.

normalized series: In statistics, series obtained by subtracting the MEAN from each term of a series and dividing the result by the STANDARD DEVIATION. The use of normalized series simplifies the comparison of data expressed in disparate units.

normal law of errors: When an observation is subject to a large number of errors of measurement which are independent and individually small, then the central limit theorem states that repeated measurements will fall into the NORMAL FREQUENCY DISTRIBUTION. Such a measurement is sometimes said to conform to the normal law of errors.

Raw meteorological measurements rarely conform to the normal law of errors, but averages taken over many terms may do so well enough to allow the standard statistical tests to be applied. PERSISTENCE in the data or extreme SKEWNESS in the distribution of the individual data, can prevent the appearance of normality in means taken over samples of attainable size.

Normand's theorem: Of various propositions enumerated by C. Normand in relation to the thermodynamics of the atmosphere, that generally termed 'Normand's theorem' is to the effect that, on an AEROLOGICAL DIAGRAM, the dry adiabatic through the dry-bulb temperature, the saturated adiabatic through the wet-bulb temperature, and the saturation mixing ratio line through the dew-point temperature of an air sample, all meet in a point.

norte, norther: A strong, cold, northerly wind which blows mainly in winter on the shores of the Gulf of Mexico. Here it is sometimes humid and accompanied by precipitation. The northers reach the Gulf of Tehuantepec as cold, dry winds, where they often set in suddenly and quickly raise a heavy sea.

North Atlantic Drift: See GULF STREAM.

nor'wester: A violent, convective type of storm, often accompanied by a LINE-SQUALL, which occurs in Bengal and Assam in the months March to May. The storms are so named because of their pronounced tendency to move from the north-west.

n.s.r.t.: Near surface reference temperature, measured by a thermistor probe mounted in the ship's water intake just inboard of the sea valve.

n.t.p.: An abbreviation for 'normal temperature and pressure', signifying a temperature of $0°C$ and a pressure of 760 mm of mercury (1013·2 mb), these being the selected standard conditions under which volumes of gases are compared. It is also termed S.T.P. (standard temperature and pressure).

nucleation: In meteorology, the initiation of either of the phase changes from water vapour to liquid water, or from liquid water to ice. The normal process in the atmosphere is one of 'heterogeneous nucleation' in which the phase change is initiated on minute foreign particles—see NUCLEUS. In the absence of nuclei, the phase change is one of HOMOGENEOUS NUCLEATION or 'spontaneous nucleation'.

nucleus: In meteorology, a minute solid particle suspended in the atmosphere.
Classification of nuclei is now generally made into 'Aitken nuclei' of radius less than $0·2$ μm, 'large nuclei' (radius from $0·2$ to 1 μm) and 'giant nuclei' (radius > 1

μm). Nuclei are most numerous in the Aitken range, at about 0·05 μm. At the earth's surface, concentrations as low as about $1000/cm^3$ are measured in country air and over the oceans but numbers may be up to nearly 1000 times greater than this in industrial areas. The particle mass is generally between 10^{-12} and 10^{-16} g. The concentration decreases with height, more rapidly in the case of the larger than of the smaller particles. The large and giant nuclei act as nuclei of CONDENSATION, the former being the more effective because they are much the more numerous; condensation nuclei are hygroscopic in nature, i.e. have an affinity for water. The high saturation VAPOUR PRESSURE associated with drops of very small radius (see KELVIN EFFECT) prevents the Aitken nuclei from acting as condensation nuclei.

Nuclei are dispersed into the atmosphere by such processes as duststorms, volcanic eruptions, formation of sea salt spray, and combustion. Chemical analysis shows that the nuclei contain the ions, SO_4, NH_4, NO_3, Na, and Cl.

A 'freezing nucleus' is one on which an ice crystal forms by the freezing of a water droplet. A 'sublimation nucleus' is one on which direct deposition of ice from water vapour occurs; the extent (if any) to which this sublimation process actually occurs in the atmosphere is doubtful. 'Ice nucleus' is a generic term which includes both freezing and sublimation nuclei. Such nuclei are much less common than condensation nuclei. Their number, or at least their effectiveness, increases as the temperature decreases below 0°C. Measurement indicates that their size distribution is of the order 0·1 to several micrometres and that they are composed of volcanic dust and clay and other soil particles of crystalline structure similar to that of ice. Ice crystals appear to form on such nuclei at saturation with respect to water and to grow by sublimation. Freezing of water droplets is generally considered to occur spontaneously (without the aid of nuclei) below a temperature of about −40°C—so-called 'homogeneous nucleation'.

numerical weather forecast: An OBJECTIVE FORECAST, in which the future state of the atmosphere is determined by the numerical solution of the basic theoretical equations involved. The calculations are so lengthy as to necessitate, in general, the use of an electronic computer.

Numerical forecasts are usually performed by one of two main methods. The first describes the horizontal motions by means of the VORTICITY EQUATION and assumes QUASI-GEOSTROPHIC ASSUMPTION. This method has the advantage of automatically eliminating all non-meteorological components of the motion so that a relatively long time interval can be used in the numerical solution without causing COMPUTATIONAL INSTABILITY; it also evades a practical difficulty caused by the almost exact compensation of the Coriolis and pressure-gradient terms in the PRIMITIVE EQUATIONS of motion which leaves the crucial acceleration terms as small residuals. The disadvantages are that the more intense synoptic developments (when the horizontal DIVERGENCE becomes large) are not properly accounted for, and it is difficult to include the effects of heating, moisture and friction in a satisfactory manner. The second method uses the primitive equations of motion directly and is becoming of growing practical importance as both the speed and power of computers and the sophistication of mathematical and numerical techniques increase.

Various atmospheric 'models' are used for numerical forecasts, varying chiefly in the degree of complexity with which the vertical structure of the atmosphere is accounted for. The simplest and most restrictive is the BAROTROPIC model in which a single pressure surface, usually 500 mb, is considered. (For this model, the vorticity equation reduces to the statement that a fluid element conserves its absolute vorticity or spin throughout its history). Alternatively, BAROCLINIC models may be used which involve consideration of several surfaces, allowing vertical variations of winds and temperatures to be represented more realistically; some primitive-equation models use ten or more surfaces. The equations are solved by FINITE-DIFFERENCE METHODS, careful selection of grid size and time interval being necessary to avoid computational instability; initial data are based on observations made both at ground level and in the upper air using various methods of OBJECTIVE ANALYSIS.

Most numerical methods in routine operational use do not take proper account of the sources and sinks of atmospheric energy; they are therefore applicable only to forecasts for periods of not more than about three days. The normal end products are PROGNOSTIC CHARTS for times some 24 to 48 hours ahead. For the simple vorticity models these charts are usually only of the heights of isobaric surfaces and require separate subjective interpretation for forecasts of weather elements such as cloudiness and rainfall, but the more elaborate primitive-equation models (such as the Bushby–Timpson 10-level model* developed in the Meteorological Office) provide in addition direct numerical forecasts of rainfall, humidity, cloudiness, winds and vertical motion.

Nusselt number: A non-dimensional number (Nu) which occurs in respect of the transfer of heat by free CONVECTION from a heated surface immersed in a fluid. It is given by

$$(Nu) = lq/k \triangle T$$

where l is a length characteristic of the system, q is the rate of heat flow per unit area of the surface, k is the thermal conductivity, and $\triangle T$ the characteristic temperature difference between the heated surface and the fluid.

---

* BENWELL, G. R. A., GADD, A. J., KEERS, J. F., TIMPSON, MARGARET S. and WHITE, P. W.; The Bushby–Timpson 10-level model on a fine mesh. *Scient Pap Met Off*, London, No. 32, 1970.

O

objective analysis: In synoptic meteorology, an ANALYSIS of initial observational data by a predetermined numerical (or graphical) method, such that the results obtained from a given set of data are independent of the analyst.

Methods of objective analysis so far developed have been mainly in connection with NUMERICAL WEATHER FORECASTS made by machine computation, the objects being to eliminate entirely the subjective element and to effect an economy in time, the analysis itself being carried out by machine. The aim of the analysis in such cases is to obtain appropriate values of initial data (mainly contour heights) at fixed grid points.

One method consists essentially of achieving a best possible fit (e.g. by the method of LEAST SQUARES) between the observational data and a polynomial function representing the data. Tests show that the accuracy attainable is much the same as that obtained by hand analysis. A requirement of the objective method, as with the hand method, is to eliminate, preferably also by an objective method, any gross errors in the observational data received.

objective forecast: A weather forecast which is entirely based on the application of a single rule or equation, or combination of rules or equations, to selected observed meteorological elements; personal judgement on the part of the forecaster is thus completely eliminated.

A NUMERICAL WEATHER FORECAST is an example of a forecast which is (almost entirely) objective. Simpler examples include forecasts of night minimum temperature or radiation fog by the application of (largely) empirical rules based on observed values of such elements as temperature, dew-point, wind speed, and cloud amount.

oblique visibility: Oblique visibility, or 'slant visibility', is the greatest distance at which a given object can be seen and identified with the unaided eye along a line of sight inclined to the horizontal.

Oblique visibility in a downward direction, an important element in aircraft operation, is generally different from the VISIBILITY measured at the earth's surface owing (i) to height variations of atmospheric EXTINCTION COEFFICIENT in the layer concerned, and (ii) to the fact that objects are then viewed against a terrestrial background.

occlusion: A FRONT which develops during the later stages of the life-cycle of a frontal DEPRESSION. The term arises from the associated occluding (shutting-off) of the warm air from the earth's surface.

As convergence takes place at the fronts and in the WARM SECTOR of the depression, the COLD FRONT closes in on the WARM FRONT. The warm sector is thus reduced to a TROUGH line called the line of occlusion and is subsequently lifted from the surface of the earth. The trough line is marked by a belt of cloud and precipitation and by a wind shift. In those cases where there is a substantial temperature difference between the cold air mass in advance of the warm front and that behind the cold front a 'warm occlusion' (less cold air behind) or 'cold occlusion' (less cold air in advance) forms; the effect is to extend the cloud and precipitation well in advance of the surface occlusion in the former case, and behind the occlusion in the latter case. Occlusions are common in north-western Europe, the warm type being the more common in winter, the cold type in summer.

ocean current: See CURRENT, OCEAN.

oceanity (or **oceanicity**): In meteorology, a measure of the extent to which the climate at any place is subject to maritime, as opposed to land, influences. See CONTINENTALITY.

oceanography: The study of the seas and oceans, including their physical, chemical, and dynamical properties (CURRENTS, TIDES, etc.).

The seas and oceans are by far the main source of atmospheric water vapour and are also a major reservoir of heat. Their interaction with the atmosphere is of great importance in controlling the distribution of climate over the earth as a whole, and also in affecting the day-to-day weather elements in neighbouring land areas.

ocean waves: Away from coasts a wind-generated ocean-wave system normally covers a wide area of sea and changes its characteristics only slowly with distance. A general division of wave systems is made into 'sea' and SWELL. The system of waves raised by the local wind blowing at the time of observation is usually referred to as 'sea'. Those waves not raised by the local wind blowing at the time of observation, but due either to winds blowing at a distance or to winds that have ceased to blow, are known collectively as 'swell'. Sea and swell are separately reported only in clearly distinguished cases—see Plate 26.

For each distinguishable system the reported characteristics are the direction from which the waves come (scale 01–36 as for wind direction), the period (seconds) and the height (metres). The reported height refers to 'characteristic' or 'significant' waves, being the mean height of the highest one-third of waves; the reported period also refers to significant waves.

ocean weather stations: By international agreement the existing designations and locations of the nine North Atlantic ocean weather stations are:

A	(Alfa)	62° 00'N 33° 00'W
B	(Bravo)	56° 30'N 51° 00'W
C	(Charlie)	52° 45'N 35° 30'W
D	(Delta)	44° 00'N 41° 00'W
E	(Echo)	35° 00'N 48° 00'W
I	(India)	59° 00'N 19° 00'W
J	(Juliett)	52° 30'N 20° 00'W
K	(Kilo)	45° 00'N 16° 00'W
M	(Mike)	66° 00'N 02° 00'E

The four most westerly stations (B, C, D and E) are manned by ships of the U.S.A., while the remainder are manned by ships of some of the European countries on a rotational basis.

ogive: A graph of cumulative frequency (or percentage frequency) versus a selected element, from which the frequency of occurrence above or below any specified value of the element may be read. The term, referring to the inflected shape of the curve, is borrowed from architecture.

okta: Unit, equal to area of one eighth of the sky, used in specifying cloud amount.

omega (ω) equation: A differential equation for the variation with hydrostatic pressure of dp/dt—conventionally denoted by ω—which plays the part of vertical velocity when the equations of motion are written in PRESSURE CO-ORDINATES. If we assume the hydrostatic relationship and geostrophic motion, the vorticity and thermodynamic equations yield the ω equation in the form

$$\nabla^2(\sigma\omega) - \frac{f(\zeta+f)}{g}\frac{\partial^2\omega}{\partial p^2} + \omega\nabla^2\frac{\partial^2 z}{\partial p^2} + \nabla\omega\cdot\nabla\frac{\partial^2 z}{\partial p^2} + \nabla\frac{\partial\omega}{\partial p}\cdot\nabla\frac{\partial z}{\partial p}$$

$$= \frac{g}{f}J\left(z,\frac{\partial z}{\partial p}\right) - \frac{\partial}{\partial p}J(z,\zeta+f).$$

In this equation, σ is a measure of atmospheric stability given by $(g\rho\theta)^{-1}\partial\theta/\partial p$ where θ is the potential temperature and ρ the density; f is the Coriolis parameter; ζ is the relative vorticity; z is the geopotential height; g is the gravitational acceleration and J indicates a Jacobian operator such that

$$J(u,v) \equiv \frac{\partial u}{\partial x}\frac{\partial v}{\partial y} - \frac{\partial u}{\partial y}\frac{\partial v}{\partial x}.$$

The ω equation is purely diagnostic and relates vertical velocities to existing pressure or contour patterns: it says nothing about rates of change. The last three terms on the left-hand side are small and are usually omitted in obtaining numerical solutions.

opacus (op): One of the CLOUD VARIETIES (Latin, *opacus* shady).

'An extensive cloud patch, sheet or layer, the greater part of which is sufficiently opaque to mask completely the sun or moon.

This term applies to ALTOCUMULUS, ALTOSTRATUS, STRATOCUMULUS and STRATUS.'* See also CLOUD CLASSIFICATION.

open system: An open (thermodynamic) system is one in which there is exchange of matter between the system and its environment, e.g. a precipitating cloud.

opposition, astronomical: A planet or other heavenly body is said to be in opposition when it is in line with earth and sun, and in the direction opposite to that of the sun, as viewed from the earth.

optical air mass: The length of the path of the sun's rays through the earth's atmosphere, measured in terms of the path length when the sun is in the ZENITH. For ZENITH DISTANCES Z up to about 80° the optical air mass (m) is approximately given by sec Z. Corrections are required, especially at zenith angles greater than 80°, for atmospheric refraction and for the earth's curvature.

Accurate values are given in the *Smithsonian meteorological tables*.†

optical mass: A term used, in calculations of emission or absorption of radiation, to signify the total mass of a given emitting or absorbing substance which lies in a vertical column of unit cross-sectional area between two specified levels (frequently, the earth's surface and the top of the atmosphere). It is also termed the 'optical thickness' or 'optical depth'.

optical phenomena: See ATMOSPHERIC OPTICS.

---

* Geneva, World Meteorological Organization. International cloud atlas, Vol. 1. Geneva, WMO, 1956, p. 16.
† LIST, R. J.; Smithsonian meteorological tables. *Smithson Misc Collns, Washington*, **114**, 1968, p. 422.

orientation: From the Latin, *oriens* the rising of the sun—the east. The direction of an object referred to the points of the compass.

orographic cloud: Cloud which is formed by forced uplift of air over high ground. The reduction of pressure within the rising air mass produces ADIABATIC cooling and, if the air is sufficiently moist, CONDENSATION. Lenticular wave clouds, including those formed well to leeward of the high ground, are common orographic clouds; stratus, cumulus and cirrus clouds are also sometimes orographic in origin.

orographic depression: A non-frontal DEPRESSION (or TROUGH of low pressure) formed by purely dynamical processes to leeward of a range of mountains which presents a barrier to the airflow. Well-broken cloud usually characterizes the central region of such a depression because of the action of the FÖHN effect. It is also termed a 'lee depression'.

orographic rain: Rain which is caused, entirely or in major part, by the forced uplift of moist air over high ground. The formation of OROGRAPHIC CLOUD is followed, in the event of continued uplift of the air, by PRECIPITATION which in the British Isles generally reaches the ground as rain. The warm sector of a vigorous depression is the synoptic situation in which the orographic influence on rainfall is generally seen to best effect.

Even where rainfall is predominantly cyclonic or convectional in nature, the orographic influence is always present to some extent. Its dominant influence in mean RAINFALL distribution is readily seen on mean rainfall charts where, to a first approximation, high ground corresponds to high rainfall amount. In those (normal) cases in which a definite prevailing wind direction may be defined, larger rainfall amounts to windward than to leeward of high ground (over regions of average upward and downward air motion, respectively) are apparent. Empirical relationships, differing with locality, may be derived between mean rainfall amount and height of ground. Such relationships may, in the absence of detailed rainfall information, be usefully employed to obtain an estimate of spot or areal rainfall, provided either that the period is long enough to ensure, or it is otherwise confirmed, that departures of wind velocity from average values for the region were small during the period concerned.

orthogonal curvature: The curvature of ORTHOGONAL LINES. It is a measure of the extent to which STREAMLINES or CONTOURS are confluent or diffluent, and is taken as being positive for confluence. See NATURAL CO-ORDINATES.

orthogonal functions: Functions are orthogonal over a field of points in one or more dimensions when the products of corresponding values at these points add up to zero when summed over the whole field. Examples are simple harmonic functions in one dimension and spherical harmonics in two.

The representation of fields of data in terms of the coefficients of a chosen set of orthogonal functions has great advantages provided that the functions chosen agree with the natural patterns present in the data; in this case the method enables the significant information to be separated from the NOISE. The most efficient separation is achieved by the use of 'empirical orthogonal functions' calculated from the data themselves by means of PRINCIPAL COMPONENT ANALYSIS.

orthogonal lines: Lines drawn so as to intersect the STREAMLINES or CONTOURS at right angles. See also ORTHOGONAL CURVATURE.

oscillations: Alternating departures in opposite senses from a mean value. A simple dynamical illustration is the motion of a pendulum.

In meteorology, the term is usually applied to processes which are not strictly periodic, e.g. oscillations of tropical stratospheric winds. Compare CYCLE.

oscillations, atmospheric: This term generally signifies the tidal movements undergone by the atmosphere. See ATMOSPHERIC TIDES.

overlapping means: An alternative for RUNNING MEANS.

overseeding: In CLOUD SEEDING, the hypothetical artificial production of an excessive number of ice crystals. The consequent multiple sharing out of available water among the crystals prevents any of them from growing big enough to fall through the cloud updraught and so inhibits precipitation.

overturning: A rapid exchange of air between different levels effected by BUOYANCY forces. Such an exchange occurs by vigorous CONVECTION in an atmosphere in which there is a superadiabatic lapse rate.

oxygen: A chemically active gas which in the molecular form (O_2), of molecular weight 31·999, comprises 20·95 and 23·14 parts per 100 parts of dry air by volume and weight, respectively.

Oxygen also occurs in the atmosphere as atomic oxygen (O) and as OZONE (O_3). The dissociation of O_2 by ultra-violet radiation results in a rapidly increasing proportion of unattached atoms relative to molecules upwards of about 80 km, the atoms predominating above about 100 km. The dissociation process operates to a decreasing extent down to about 20 km, but in the denser air at these levels the atoms are quickly lost by attachment to oxygen molecules.

That part of the absorption spectrum of oxygen of chief meteorological interest is the strong Schumann–Runge region from about 0·13 to 0·17 μm, with a peak at 0·146 μm. Strong absorption by O_2 and O of wavelengths below 0·1 μm is important in the formation of the ionosphere.

ozone: The triatomic form (O_3) of OXYGEN, of molecular weight 47·998, which is present in the atmosphere in very small amounts ranging from about 0·2 to 0·6 cm equivalent thickness at normal temperature and pressure.

The presence of ozone is due mainly to photochemical processes in the high atmosphere. Downward diffusion brings the gas in very small concentration (generally less than 5×10^{-2} part per million) to the lower atmosphere where it is reduced to oxygen by contact with various organic substances. Minor and local sources of ozone exist close to the earth's surface, produced by the oxidation by ultra-violet light of exhaust gases of motor vehicles. Local low-level formation in lightning discharges and in connection with radioactivity has also been suggested.

Ozone is formed and destroyed in the high atmosphere by the absorption of ultra-violet radiation by oxygen and ozone, respectively, and by subsequent particle collision processes. The main absorption processes concerned are absorption by molecular oxygen of radiation in the Schumann–Runge region (wavelength about 0·13 to 0·17 μm) to form atomic oxygen, and by ozone in the Hartley region (about 0·20 to 0·30 μm) to form molecular and atomic oxygen. The main collision processes are (i) a triple collision between molecular and atomic oxygen and a third molecule to form ozone, and (ii) collision between ozone molecules and oxygen atoms to form oxygen molecules.

The absorption processes are so intense that the associated temperature rise is largely concentrated near the top of the OZONE LAYER at about 50 km. The ozone which is formed at, or transferred to, levels below about 35 km is, in large measure, protected from destruction. The result is that some 90 per cent of atmospheric ozone is at levels below 35 km, with maximum concentration at about 25 km; and that total ozone or, more precisely, the ozone mixing ratio at various levels, is a useful tracer of horizontal and vertical air motion in the stratosphere.

The standard instrument for surface measurement of total ozone amount is the DOBSON SPECTROPHOTOMETER. The use of this instrument, in conjunction with the UMKEHR EFFECT method, yields a smoothed picture of the vertical distribution of ozone. Optical and chemical types of instrument, carried aloft by balloon, rocket or aircraft, have been used to obtain the ozone profile in finer detail and have shown, for example, a jump to higher ozone concentration on passage upwards through the tropopause.

The systematic space and time variations of total ozone also do not accord with conditions of photochemical equilibrium but largely reflect the corresponding large-scale vertical and horizontal transport mechanisms which are at work in the atmosphere. The main features are large amounts of ozone in high relative to low latitudes, especially in spring, and, in middle and high latitudes, a spring maximum and autumn minimum. Day-to-day changes which are correlated with surface and upper air synoptic features also occur.

Ozone is an important gas in the radiation balance of the stratosphere. The main features of its absorption spectrum are: the intense Hartley region (0·20 to 0·30 μm, with a maximum at 0·25 μm) and the weak Huggins bands (0·30 to 0·35 μm) in the ultra-violet; the weak Chappuis bands (0·45 to 0·65 μm) in the visible; and bands centred at 4·7, 9·6 and 13·0 μm in the infra-red.

ozone layer: That layer of the atmosphere, also termed the 'ozonosphere', in which the concentration of OZONE is greatest. The term is used in two ways: (i) to signify the layer from about 10 to 50 km in which the ozone concentration is appreciable; (ii) to signify the much narrower layer from about 20 to 25 km in which the concentration generally reaches a maximum.

ozonosphere: An alternative for OZONE LAYER.

P

pack ice: Term used in a wide sense to include any area of sea ice other than fast ice no matter what form it takes or how disposed.
    Close pack ice: composed of floes mostly in contact.
    Open pack ice: floes seldom in contact, with many leads and pools.

paleoclimatology: The study of the nature of and reasons for the types of climate that are inferred, from a variety of evidence, to have obtained over the earth in the course of geological time. See CLIMATIC CHANGES.

paleomagnetism: The study of the nature of and reasons for the changes (more especially directional changes) in the earth's magnetic field that are inferred, from studies of remanent rock magnetism, to have occurred in the course of geological time. Deductions of possible significance in the theory of CLIMATIC CHANGES have been made from such studies. See GEOMAGNETISM.

pallium: An obsolete term for NIMBOSTRATUS.

pampero: A name given in the Argentine and Uruguay to a severe storm of wind, sometimes accompanied by rain, thunder and lighting. It is a LINE-SQUALL, with the typical arched cloud along its front. It heralds a cool south-westerly wind in the rear of a depression; there is a great drop of temperature as the storm passes.

pannus (pan): An accessory cloud (Latin, *pannus* shred).
    'Ragged shreds sometimes constituting a continuous layer, situated below another cloud and sometimes attached to it.
    This accessory cloud occurs mostly with ALTOSTRATUS, NIMBOSTRATUS, CUMULUS and CUMULONIMBUS.'* See also CLOUD CLASSIFICATION.

parallax: An apparent change in the position of an object caused by a change in the position of the observer. In connection with the reading of meteorological instruments, an error of parallax may arise whenever the indicator of the instrument (e.g. end of column of mercury or water, pointer, etc.) and the scale against which the indicator is to be read are at a distance from one another which is comparable with the length of the smallest readable scale division; for in such a case a movement of the observer's head may cause his line of vision to the indicator to intersect the scale at different points and so give rise to different readings. The error is eliminated by ensuring that the line of vision to the indicator is at right angles to the scale when the reading is made.

parameter: A quantity related to one or more variables in such a way that it remains constant for any specified set of values of the variable or variables.

---

* Geneva, World Meteorological Organization. International cloud atlas, Vol. 1. Geneva, WMO, 1956, p. 18.

paranthelion: A PARHELION ('mock sun') at the same elevation as the sun and in azimuth greater than 90° from the sun may be called a paranthelion. White paranthelia at 120° from the sun are fairly common. Paranthelia at about 140° from the sun have been recorded on rare occasions.

paraselene (plural, **paraselenae**): An image of the moon, also termed 'mock moon', produced in a way analogous to the PARHELION; because of the weak intensity of the moon's light relative to that of the sun the paraselene is more weakly coloured and less frequently observed than the parhelion.

parcel method: The estimation of vertical STABILITY in the atmosphere by a method based on the assumption that individual 'parcels' of air move upwards without disturbing their environment. See also ADIABATIC, SLICE METHOD.

parhelic circle: A bright but colourless circle passing through the sun parallel to the horizon. The phenomenon is explained by the REFLECTION of sunlight from hexagonal ice crystals whose axes are vertical. REFRACTION of light through such crystals produces PARHELIA with which, therefore, the parhelic circle is often associated.

parhelion (plural, **parhelia**): An image of the sun, coloured or white; it is also termed 'mock sun'.

The parhelia seen most frequently are at the same elevation as and on both sides of the sun and are coloured with red nearest the sun. When the sun is near the horizon the angular distance of the sun from the parhelia is equal to the radius of the ordinary HALO, i.e. 22°. When the sun is higher the distance is greater so that if halo and parhelion are both seen the parhelion is outside the halo; for a solar elevation of 55° the angular difference is about 14°. Parhelia are occasionally seen at points on the PARHELIC CIRCLE other than near 22°, notably at 120° (PARANTHELION), less frequently at 46°, 90° and 140°.

Parhelia are caused by the REFRACTION of sunlight within hexagonal ice crystals whose axes are vertical. Oblique rays (sun above the horizon) do not lie in a plane perpendicular to the axes of such crystals and emerge at an angle greater than that corresponding to MINIMUM DEVIATION.

Parry arcs: Rare small arcs observed above and below the small HALO at angular distances varying with solar elevation. They are ascribed to REFRACTION of light through hexagonal ice crystals floating with principal axis and a pair of opposite sides horizontal.

partial pressure: In a mixture of gases, that part of the total gas pressure which is exerted by a specified constituent gas; it is the pressure that each would exert if it alone were present and occupied the same volume as the whole mixture.

According to Dalton's law, the total pressure of a mixture of gases is the sum of the partial pressures, as defined above.

pascal: The name given to the unit of pressure in the International System (SI) of units. The pascal (Pa) is thus 1 N/m^2 and is equivalent to 10^{-2} mb. See SI UNITS.

pastagram: An AEROLOGICAL DIAGRAM, designed by J. C. Bellamy, in which the ordinate is a combined linear scale of height (Z_p) and corresponding pressure (p) in the STANDARD ATMOSPHERE, and the abscissa is the temperature anomaly $(T - T_p)/T_p$ where T is the actual temperature at pressure p and T_p the temperature at pressure p in the standard atmosphere.

path method: A term applied in synoptic meteorology to signify the method of extrapolation, for forecasting purposes, of the path of a pressure system or other set of isopleths (e.g. ISALLOBARS) drawn on a synoptic chart.

pearl-necklace lightning: A rare form of LIGHTNING, also termed 'chain lightning' or 'beaded lightning', in which variations of brightness along the discharge path give rise to a momentary appearance similar to pearls on a string.

pentad: A period of five consecutive days. Five-day means are often used in meteorological work, as five days form an exact subdivision (1/73) of the ordinary year, an advantage not possessed by the week.

percentile: A convenient term for denoting thresholds or boundary values in FREQUENCY DISTRIBUTIONS. Thus the 5-percentile is that value which marks off the lowest 5 per cent of the observations from the rest, the 50 percentile is the same as the MEDIAN and the 95 percentile exceeds all but 5 per cent of the values. When percentiles are estimated by ranking the items of a finite sample, the percentile generally falls between two of the observed values, and the midway value is often taken.

The terms TERCILE, QUARTILE, QUINTILE and DECILE should refer to the percentiles which divide the distribution into 3, 4, 5, or 10 equal parts, respectively. These terms are, however, often used to signify a specified fraction of the total number of ranked values. Thus the value of a particular element may be predicted to be in a specified tercile or quintile of the appropriate distribution. See also QUANTILE.

percolation: The downward passage of surface water through the soil. Part of the rain which falls on the land surface re-evaporates, part runs off into streams and rivers to the sea, while part percolates through the soil. Measurements of the amount of rain-water which percolates through certain depths of soil have been published in the annual volumes of *British Rainfall*. Usually the gauge consists of a cubic yard of natural earth inserted in a metal container and sunk in the hole formed by removing this earth. The rain-water which percolates through is drained off and measured daily at 09h, access to the receiver being obtained by means of a trap door at the side of the gauge. Such or very similar types of gauge are sometimes referred to as a 'drainage gauge' or 'seepage gauge'. The results are usually published as a depth in hundredths or thousandths of an inch (0·01 in = 0·254 mm). See EVAPORATION.

perfect gas: A hypothetical gas which obeys the gas laws of Boyle and Charles perfectly. For practical purposes the gases which compose unsaturated air may be considered perfect gases.

perigee: That point of the orbit of a satellite, natural or artificial, which is nearest the earth.

perihelion: That point of the orbit of a planet or comet which is nearest the sun. Perihelion for the earth occurs on about 1 January; the sun–earth distance is then 1·5 per cent less than the yearly mean distance.

period: A function which varies with time and which repeats itself exactly after a constant time interval (say, T) is said to be 'periodic', and T, the time of a complete oscillation, is termed the 'period' of the function; it is the reciprocal of the FREQUENCY.

periodicity: A time variation of a function comprising a single fixed PERIOD, or combination of fixed periods.

The standard methods of identifying periodicity in a variable quantity are HARMONIC ANALYSIS, PERIODOGRAM analysis, CORRELOGRAM analysis and FILTERING. A (simple) periodicity requires for its complete determination the length of the period, the amplitude (i.e. half the total range) of the variation, and the time of occurrence of the maximum ('phase').

While periodicity uncomplicated by other factors is not found in any meteorological element, the process of averaging over a large number of periods tends to remove non-periodic factors and to allow certain periodic phenomena to emerge. Examples are the average diurnal variation of surface pressure and the average annual variation of surface temperature, both of which average variations are almost entirely explained by a combination of the first two harmonic components of periods 24 and 12 hours, and 12 and 6 months, respectively.

An exhaustive search for periodicity, other than diurnal or annual, in meteorological elements has been almost entirely unsuccessful. More probable than true periodicity in these phenomena is 'quasi-periodicity' of the type shown by annual SUNSPOT NUMBERS (i.e. rather variable period and amplitude but apparently little or no change in phase), and by a RECURRENCE TENDENCY as shown, for example, by certain types of ionospheric and magnetic storms (little change in period but abrupt changes in phase). While some effects, more especially of the latter type, have been found, none is yet so striking as to be of much significance in the problem of long-range weather forecasting.

periodogram: A diagram used in a method devised by Schuster for the investigation of possible hidden PERIODICITIES in a series of observations. The amplitudes (R) corresponding to different trial periods (T) are first obtained by the standard methods employed in HARMONIC ANALYSIS. The periodogram consists of a plot of the various values of R^2 (or R) as ordinate against the corresponding values of T as abscissa. The values of T corresponding to any conspicuous peaks in the graph (high values of R) are the most likely periods. If the original n observations formed a random distribution, with standard deviation σ, the expectation (or mean value) of R^2 would be $4\sigma^2/(n-1)$. Schuster showed that the probability that any particular value of R^2 should exceed k times $4\sigma^2/(n-1)$ is e^{-k}. This expression may be used to test the reality of any period suggested by the periodogram. Where, as in most geophysical data, the observations compose a TIME SERIES, the value of n used in the test should be the 'effective number of independent observations' which, because of PERSISTENCE, may be several times smaller than the total number of observations in the series.

perlucidus (pe): One of the CLOUD VARIETIES (Latin, *perlucidus* allowing light to pass through).

'An extensive cloud patch, sheet or layer, with distinct but sometimes very small spaces between the elements. The spaces allow the sun, the moon, the blue of the sky or over-lying clouds to be seen.

This term applies to ALTOCUMULUS and STRATOCUMULUS.'* See also CLOUD CLASSIFICATION.

permafrost: Soil which remains permanently frozen, summer heating being insufficient to raise above freezing-point the lower part of a frozen layer formed during the winter. The limit of permafrost is considered to accord very approximately with an annual mean air temperature of $-5°C$.

---

* Geneva, World Meteorological Organization. International cloud atlas, Vol. 1. Geneva, WMO, 1956, p. 16.

permanent aurora: An obsolete alternative for NIGHTGLOW.

permanent gas: A gas which is at a temperature above its 'critical temperature', i.e. at a temperature at which it cannot be liquefied by pressure alone. The gases in the air other than water vapour, sulphur dioxide, and carbon dioxide are permanent gases.

persistence: In meteorology, a term used of a synoptic feature or meteorological condition that is unusually long-lasting.

In meteorological and other geophysical TIME SERIES, persistence (sometimes termed 'coherence' or 'conservation') signifies a greater-than-random tendency for relatively high (or low) values to occur in succession. The degree of persistence varies with meteorological element and generally decreases with increase of time interval between successive members of the series.

The persistence which is inherent in most time series is fundamentally important in questions of statistical SIGNIFICANCE. Thus, for example, the estimate of average seasonal pressure at a specified place which is provided by a mean of, say, 20 successive daily pressure values is much less reliable than that provided by a mean of 20 such values selected at random in the season; in the former case the average may be biased by the dominance of a particular synoptic pattern during the period concerned.

In statistical investigations of time series, unnecessary labour may be saved by confining attention to statistically independent data, i.e. to data spaced at intervals greater than the 'persistence interval', which may be defined as that interval beyond which AUTOCORRELATION becomes negligibly small. When this is not done allowance must be made, in assessing significance, for the fact that the total number of observations employed may be far in excess of the number of statistically independent data. The 'equivalent number of repetitions' ($\varepsilon(n)$) in a series of n values may be obtained from the formula

$$\varepsilon(n) = 1 + 2/[n\{(n-1)r_1 + (n-2)r_2 + \ldots + r_{n-1}\}]$$

where r_1, r_2, r_3, etc. are the correlation coefficients between successive terms, terms two apart, terms three apart, etc. An approximate expression for $\varepsilon(n)$ is

$$\varepsilon(n) \approx 1 + 2/[n\{(n-1)r_1 + (n-2)r_1^2 + (n-3)r_1^3 + \text{etc.}\}].$$

The 'effective number of independent observations' in the series of n values is given by $n/\varepsilon(n)$.

persistence forecast: A type of FORECAST, often used as a basis of comparison in the assessment of the success attained in forecasts made by conventional methods, in which the assumption is made that meteorological conditions during the forecast period will remain unchanged from those obtaining at the beginning of the forecast period.

personal equation: An expression used to denote the error of an observer's readings of an instrument which is due to an unconscious tendency on his part to read too high or too low. The tendency is usually nearly constant for any given observer reading a given instrument. PARALLAX is a common source of personal equation.

perturbation method: A method, widely applied in meteorology, by which a formal solution to the fundamental, non-linear equations of motion is obtained by the superposition of small disturbances on basic steady fluid flow. Such solutions have the form of waves, the stability and speed of which are found to depend on the wavelength and on the characteristics of the undisturbed flow. Though strictly applicable only to waves of very small amplitude, the solutions are often found to apply, with minor modifications, to disturbances of appreciable amplitude.

phase: The phase of a periodic function is its arrangement of maximum and minimum point or points with regard to a specified initial or starting point. It is measured by a 'phase angle' in which a complete revolution (360°) is equated to a complete PERIOD. Two periodic functions are said to display 'phase reversal' with respect to each other if a maximum value of one function corresponds to a minimum value of the other.

The term 'phase' is also used synonymously with 'state' of matter—solid, liquid, or gaseous.

phenology: The study of the sequence of seasonal changes in nature. All natural phenomena are included, seed-time, harvest, flowering, ripening, migration, and so on, but the observations are often limited to the time at which certain trees and flowering plants come into leaf and flower each year, and to the dates of the first and last appearances of birds and insects.

The phenology of plant flowering in the British Isles is contained in *The Phenological Reports* (1877–1948) of the Royal Meteorological Society. Long-term means are contained in a paper by E. P. Jeffree.*

phenomenon: Word used in meteorology to denote either (i) an unusual intensity of some occurrence, e.g. 'ugly' sky, high rainfall, low temperature, high pressure, gale, or (ii) occurrences which are only occasionally noted, e.g. thunder, halo, fog, glaze.

photochemistry: The study of chemical reactions involved in the absorption or emission of RADIATION. In meteorology, such reactions are mainly confined to high atmospheric levels as, for example, in the absorption of radiation by oxygen to form ozone. Reactions also occur locally near the earth's surface in connection with certain products of atmospheric pollution.

photodissociation: The splitting of a molecule into atoms or into smaller molecules by the absorption of RADIATION.

photoelectron: An ELECTRON ejected from an atom or molecule which is exposed to RADIATION of a frequency higher than a critical value. The process of 'photoionization', important for example in the formation of the IONOSPHERE, is involved in such release of electrons.

photoionization: See PHOTOELECTRON.

photometeor: A little used generic term for the luminous phenomena produced by the reflection, refraction, diffraction and interference of light from the sun or the moon. Photometeors are observed in clear air (mirage, green flash, etc.), on or inside clouds (corona, halo, etc.), and on or inside certain HYDROMETEORS and LITHOMETEORS (rainbow, glory, etc.).

photosphere: The bright disc of the SUN from which continuous emission of solar radiation takes place.

physical meteorology: Because of overlapping at many points with other 'branches' of meteorology this term cannot be defined precisely but is often used to signify all those directly physical aspects of meteorology which are not normally dealt with in DYNAMICAL METEOROLOGY.

---

* JEFFREE, E. P.; Some long-term means from The Phenological Reports (1891–1948) of The Royal Meteorological Society. *Q J R Met Soc*, London, **86**, 1960, pp. 95–103.

piezotropy: In a change of the state of the pressure and density of an individual element of the atmosphere the condition that there exists a relationship

$$B = \frac{d\rho}{dp}$$

where $d\rho/dp$ is the change of density with pressure of the individual element and B is the 'coefficient of piezotropy'.

B is a function of the thermodynamic variables and so depends on the initial state of a selected particle and varies for different particles. The condition of piezotropy is implicitly assumed in drawing an adiabatic 'path curve' of a selected particle on an AEROLOGICAL DIAGRAM. In this important special case of piezotropy, the coefficient B has the value $1/\gamma RT$ where γ is the ratio (c_p/c_v) of the specific heats of air at constant pressure and constant volume, R is the specific gas constant, and T is the absolute temperature.

pileus (pil): (Latin, *pileus* cap).

'An accessory cloud of small horizontal extent, in the form of a cap or hood above the top or attached to the upper part of a cumuliform cloud which often penetrates it. Several pileus may fairly often be observed in superposition.

Pileus occurs principally with CUMULUS and CUMULONIMBUS.'* See also CLOUD CLASSIFICATION, CAP.

pilot balloon: This term is generally applied to the smaller meteorological balloons. Their main use is the determination of upper winds, although the smallest balloon, one weighing about 10 g, is most commonly used for the measurement of cloud-base height by timing the duration of flight below cloud base and assuming a constant rate of ascent. For this reason, 10-g balloons are sometimes known as 'ceiling balloons'. The theoretical rate of ascent V of a balloon filled with hydrogen (as are most balloons) is approximately

$$V = qL^{\frac{1}{2}}/(L + W)^{\frac{1}{3}}$$

where L is the FREE LIFT in grammes, and W is the total weight (grammes) of the balloon and any attachments; q is a constant and has a value of 275 if V is expressed in feet/minute (1·40 if V is in metres/second)—see *Handbook of meteorological instruments*.†

Larger balloons, such as the 30-g balloon, are used for wind finding. In the 'single theodolite' method, the height of the balloon is estimated by timing as for cloud-base determination. The 'tail' method employs one (or two) piece(s) of paper attached to the balloon by a known length of string. The height of the balloon at any instant can then be deduced from the angle subtended by the tail (assumed vertical) at the theodolite; this angle is measured on a divided scale (graticule) fitted to the eyepiece of the theodolite. The 'double theodolite' method employs simultaneous observations of the balloon's position by two theodolites at known relative positions; greater accuracy is possible by this method since not only wind direction and speed but also height are determined.

The calculation of wind velocity from the theodolite readings is normally performed by a special slide-rule, although graphical methods can also be used.

When absence of cloud permits optical wind finding at greater heights, balloons as large as 100 g may be used up to about 16 km.

---

* Geneva, World Meteorological Organization. International cloud atlas, Vol. 1. Geneva, WMO, 1956, p. 18.
† London, Meteorological Office. Handbook of meteorological instruments. Part II. London, HMSO, 1961, p. 139.

pitot tube: An instrument for determining the velocity of a stream of fluid by measuring the increase of pressure, above the 'static' or undisturbed pressure, in an open tube facing the stream. The velocity is computed from the relationship $p = \frac{1}{2}\rho v^2$ (where p is pressure, ρ density and v velocity). Suitably mounted, a pitot tube may be used as an ANEMOMETER.

Planck's law: See RADIATION.

planetary albedo: See ALBEDO.

planetary temperature: See RADIATION.

platykurtic: See KURTOSIS.

pluvial period: A geological period of large amounts of rainfall relative to earlier and later periods. Evidence of such periods, which are generally thought to have coincided in time with the glacial periods of ICE AGES, is found in those land regions which lay equatorwards of the ice-covered regions during the glacial periods.

poise: The unit of VISCOSITY in the C.G.S. SYSTEM of units, being the tangential stress in dynes/centimetre^2 when the velocity gradient is 1 cm/s cm. The dimensions are $ML^{-1}T^{-1}$.

$$1 \text{ poise} = 0.1 \text{ N s/m}^2.$$

Poisson distribution: If in a large number of trials an event occurs rarely so that the probability of its occurring during any one trial is small, then the frequency of occurrence of the event during a number of sets of trials follows a Poisson distribution. If over a large number of sets of trials the mean number of occurrences per set is m, then the fraction, $\Pi(i, m)$, of the number of sets in which the number of occurrences will be chosen whole number i, is given by

$$\Pi(i, m) = \frac{e^{-m} m^i}{i!}$$

where $m = np$ is the expected number of events in the period. It may be shown that m is both MEAN and VARIANCE of this FREQUENCY DISTRIBUTION.

The probabilities of many rare meteorological events vary with time and are often higher once an event has actually occurred. This feature limits or complicates the application of the Poisson distribution to meteorological events. See also BINOMIAL DISTRIBUTION, FREQUENCY DISTRIBUTION, NORMAL FREQUENCY DISTRIBUTION.

polar air: Air originating in high latitudes, normally subdivided in synoptic meteorology into 'maritime polar' (*mP*) air and 'continental polar' (*cP*) air, according to the nature of the surface over which the AIR MASS originates. Air which moves almost directly equatorwards from the ice-bound areas of the Arctic is now usually termed ARCTIC AIR.

During the motion of *mP* air away from its source region, heat and moisture may be convected into the air mass from a warmer underlying sea surface. In the British Isles *mP* is associated generally with a west or north-west wind, steep lapse rate, low freezing-level, good visibility, instability showers (sometimes with thunder), and surface temperature below seasonal normal.

Maritime polar air which travels far to south and then returns to the British Isles from the south-west is termed 'returning maritime polar air' (*rmP*); typical conditions associated with this air mass are much cloud but fewer showers than in *mP* air and temperatures close to the seasonal average.

Easterly winds are often associated in the British Isles with the less common *cP* air mass. In winter, typical conditions are very cold but dry apart from light snow showers near the east coast; in summer, conditions are dry and rather warm except on the east coast where coastal fog ('haar') is often widespread. (The *cP* air is warmed in winter, cooled in summer, by its passage over the North Sea.)

polar-air depression: A SECONDARY DEPRESSION of a non-frontal character which forms, more especially in winter, within an apparently homogeneous polar air mass; near the British Isles the development is usually within a northerly or north-westerly airstream. The chief characteristics of this type of depression, which seldom becomes intense, are a movement in accordance with the direction of the general current in which the depression forms and the development of a belt of precipitation near the depression centre and along a trough line which often forms on the side farther from the parent depression where also the pressure gradient (and surface winds) is increased.

polar climate: A type of CLIMATE which obtains, in general, within the polar regions (polewards of 66° 33' N and S). In the Köppen classification, the polar climate is subdivided into TUNDRA climate (mean temperature of warmest month between 0°C and 10°C), and 'perpetual frost' climate (mean temperature of warmest month below 0°C).

polar continental air: See POLAR AIR.

polar co-ordinates: A system of co-ordinates in which the position of a point is specified by its distance (r) from the origin ('pole') and by the angle (θ) made by the line joining point and origin with a reference line ('polar axis').

polar distance: An alternative for COLATITUDE.

polar front: A FRONT which divides 'polar' and 'tropical' air masses and on which most of the travelling disturbances of middle latitudes form. In the North Atlantic, for example, this front, which can often be traced as a continuous line over thousands of kilometres, extends in winter, on average, in a north-easterly direction from a position off the east coast of the United States of America (at about 30°N). In summer the front is less well marked and has little tendency for a preferred position.

polarization: A state of ELECTROMAGNETIC RADIATION in which the transverse vibrations which compose the wave motion occur wholly or in part ('partial polarization') in a specified manner—for example in a plane, circle or ellipse—that is, they do not occur in all the possible planes which contain the direction of propagation of the radiation. The plane of polarization is defined as that in which the wave motion is a minimum, or, in terms of electromagnetic theory, that in which the electric vector is a minimum.

Polarization of emitted radiation may be effected by a suitable aerial array. Since SCATTERING causes polarization of initially unpolarized radiation, the phenomenon is observed naturally in the atmosphere in the DIFFUSE RADIATION which reaches the earth's surface, such radiation being polarized in the plane which contains the sun, the observer and the observed point of the sky. In accordance with the theory of RAYLEIGH SCATTERING the polarization is strongest in the solar zenith (i.e. in the light scattered from a point 90° from the sun) and in the antisolar point. Because of multiple and non-molecular scattering, the polarization disappears at the 'neutral points' discovered by Arago, Babinet and Brewster—see ARAGO'S POINT.

polarization, electric: The separation of positive and negative charges within a particle in response to an electric field acting on the particle.

The occurrence of this effect within falling water and ice particles, due to the presence of the atmospheric electric field, was advanced by C. T. R. Wilson as leading to selective ion capture and hence the separation of charge in a thunderstorm.

polar low: An alternative for POLAR-AIR DEPRESSION.

polar maritime air: See POLAR AIR.

polar-night jet: See JET STREAM.

polar vortex: The circumpolar westerly cyclonic circulation which affects middle and high latitudes of both hemispheres.

In summer the circulation affects the middle and upper troposphere; it decreases in the lower stratosphere and reverses at higher stratospheric levels, mainly because of maximum OZONE heating at high levels over the summer pole. In winter the westerly circulation in the higher troposphere is much stronger than in summer. Since, also, the upper atmosphere over the winter pole is a heat sink, the westerly circulation increases further with increasing height in the stratosphere in this season and comprises the 'polar stratospheric vortex' or 'polar-night jet'. Westerly winds of about 150 knots are attained, on average, in higher middle latitudes near the top of the stratosphere near mid-winter; individual speeds in excess of 200 knots are not uncommon and values greater than 350 knots have been observed. (On the other hand, easterly winds of over 100 knots may occur in association with major stratospheric disturbances.) See also THERMAL WIND, JET STREAM.

polar wandering: Hypothetical movement of the earth's axis of rotation relative to the earth's surface, in the course of geological time. Such movement has been advanced as a possible cause of climatic changes on this time-scale. Among the evidence advanced in support of the hypothesis is that of remanent rock magnetism (see GEOMAGNETISM), on the assumption that the earth's magnetic axis has always been close to the axis of rotation. (Further evidence of the same kind is to the effect that CONTINENTAL DRIFT has also been involved.)

polar year: In the First International Polar Year (FPY), from 1 August 1882 to 1 September 1883, and the Second (SPY), from 1 August 1932 to 31 August 1933, near times of sunspot maximum and minimum, respectively, stations were manned by co-operating nations in the Arctic, and to a much smaller extent in the Antarctic regions. Observational programmes covered mainly meteorology, geomagnetism and aurora.

pole: The earth's geographical poles are the points of intersection of the earth's axis of rotation (polar axis) with the earth's surface.

For an observer situated at the centre of the CELESTIAL SPHERE, the 'celestial poles' are points on the celestial sphere in the direction parallel to the earth's axis of rotation. In the northern hemisphere the pole star is about 1° away from the north celestial pole. The altitude of an observer's celestial pole is equal to his latitude.

See GEOMAGNETISM for an explanation of magnetic and geomagnetic poles.

pollution: See ATMOSPHERIC POLLUTION.

ponente: A westerly wind which blows in the Mediterranean.

population: A statistical term used to denote the (generally hypothetical) stock from which a sample of data available for analysis may be assumed to be drawn. Statistical arguments often seek to infer the properties of the population from those of the sample.

potato blight warning: A short period of mild humid weather, such as that defined by a BEAUMONT PERIOD, is a factor which favours the spread of the potato blight disease.

From a network of about 50 full 24-hour stations distributed over the United Kingdom reports of the occurrence of Beaumont periods are passed to various agricultural authorities. Plant pathologists use this information, together with other facts in their possession, to advise potato growers in appropriate areas to spray their crops in order to guard against the spread of the disease.

potential energy: The ENERGY possessed by a body by virtue of its position. It is measured by the amount of work required to bring the body from a standard position, where its potential energy is zero, to its present position. A common example is that of 'gravitational potential energy', mean sea level being then the normal selected standard level.

The term 'total potential energy' is employed (first by M. Margules) to signify the sum of gravitational potential energy, defined above, and INTERNAL ENERGY. See also AVAILABLE POTENTIAL ENERGY, KINETIC ENERGY.

potential evapotranspiration: See EVAPOTRANSPIRATION.

potential gradient: Atmospheric potential gradient is defined as the difference of electric potential between two points vertically disposed with respect to each other and separated by unit distance; it is expressed in volts/metre and is, by convention, reckoned positive if directed downwards.

Surface potential gradient—the potential difference between a conductor at a height of one metre and the level ground—is the most regularly measured element of ATMOSPHERIC ELECTRICITY.

potential instability: See STABILITY.

potential temperature: That temperature (θ), readily obtained from an AEROLOGICAL DIAGRAM, which a given sample of air would attain if transferred at the dry ADIABATIC lapse rate to the standard pressure, 1000 mb. If the pressure (mb) and absolute temperature of the air are p and T, respectively, and γ is the ratio of the specific heat of air at constant pressure to that at constant volume, then

$$\theta = T \left(\frac{1000}{p}\right)^{(\gamma - 1)/\gamma} = T \left(\frac{1000}{p}\right)^{0.286}.$$

θ is related to ENTROPY (S) by the equation

$$S = c_p \log \theta + \text{constant}$$

where c_p is the specific heat of air at constant pressure.

The 'partial potential temperature' (θ_d) of an air sample, as defined by C. G. Rossby and used in the ROSSBY DIAGRAM, is the potential temperature appropriate to the temperature and partial pressure (p_d) exerted by the dry air of the sample ($p_d = p - e'$, where $e' =$ vapour pressure);

$$\theta_d = T \left(\frac{1000}{p_d}\right)^{0.286}.$$

potential transpiration: See TRANSPIRATION.

potential vorticity: In ADIABATIC motion of a column of air the quotient of the absolute VORTICITY of the air column $(\zeta + f)$ to the pressure difference between the top and bottom of the column $(\triangle p)$ is constant (potential vorticity theorem), i.e.

$$\frac{\zeta + f}{\triangle p} = \text{constant}$$

where ζ is the relative vorticity of the column and f is the CORIOLIS PARAMETER.

The value of the absolute vorticity of a column which corresponds to a standard value of $\triangle p$ (say, 50 mb) is termed the potential (absolute) vorticity. This is a value which, on analogy with the definition of the POTENTIAL TEMPERATURE of an air element, is conserved in adiabatic motion of an air column.

power spectrum: See EDDY SPECTRUM.

PPI: Abbreviation for 'plan position indicator'. See RADAR METEOROLOGY.

praecipitatio (pra): (Latin, *praecipitatio* fall).
'A supplementary cloud feature, appearing as an extension of certain clouds, comprising precipitation (rain, drizzle, snow, ice pellets, hail, etc.) falling from a cloud and reaching the earth's surface.

This supplementary feature is mostly encountered with ALTOSTRATUS, NIMBO-STRATUS, STRATOCUMULUS, STRATUS, CUMULUS and CUMULONIMBUS.'* See also CLOUD CLASSIFICATION.

Prandtl number: The non-dimensional ratio (σ) defined by the relationship

$$\sigma = v/a$$

where v is the kinematic VISCOSITY and a the thermometric CONDUCTIVITY of a fluid; σ has the approximate value 0·7 in air near the earth's surface.

precipitable water: The precipitable water of a column of air is the depth of water (alternatively expressed as the total mass of water) that would be obtained if all the water vapour in the column, of unit area cross-section, were condensed on to a horizontal plane of unit area. 'Precipitable water' is a useful measure of the water vapour content of an air column. The term is not, however, to be regarded as implying that the amount of water may, in fact, be precipitated by an actual physical process.

The depth of precipitable water of an atmospheric column, of pressure p_1 (mb) at the bottom and p_2 (mb) at the top and of mean mixing ratio $\bar{r}$ (g/kg), is given, in tenths of millimetres, by the approximate formula

$$\text{Precipitable water} = \frac{\bar{r}(p_1 - p_2)}{g}$$

where g is the gravitational acceleration (9·8 m/s²).

The precipitable water for an entire atmospheric column is found by applying this formula to selected successive layers; a typical value for temperate-latitude summer conditions is 20–30 mm.

precipitation: Used in meteorology to denote any aqueous deposit, in liquid or solid form, derived from the atmosphere.

---

* Geneva, World Meteorological Organization. International cloud atlas, Vol. 1. Geneva, WMO, 1956, p. 17.

The main problem in the 'precipitation process' is the explanation of the manner of growth of drops from the size commonly associated with non-precipitating clouds (diameter about 15 micrometres) to that found in most forms of precipitation; growth by direct CONDENSATION on to a NUCLEUS is too slow to account for the transformation. Two mechanisms are considered to operate.

(i) Ice-crystal (Bergeron) process. In a cloud of predominantly supercooled water droplets ice crystals form, in increasing numbers with decrease of temperature below 0°C, because of the presence of natural ICE NUCLEI. The formation of the ice crystals occurs at saturation with respect to water. Since the saturation VAPOUR PRESSURE over ice is less than that over water at the same temperature (maximum deficit of 0·27 mb at $-12°C$), the ice crystals increase rapidly in size because of diffusion of water vapour from neighbouring water droplets to the ice crystals. (Growth is subsequently accelerated by 'coalescence' of the relatively heavy ice crystals with other ice crystals or with water drops.)

(ii) Coalescence (accretion) process. If, within a cloud, there are some liquid drops appreciably larger than the great majority of drops, the slower rate of rise of such large drops in a cloud updraught leads to collisions and, in some cases, coalescence with the smaller liquid drops. Factors which promote this process are appreciable cloud depth and updraught speed which permit of collision growth to a size sufficient to ensure that the drop will not evaporate at the top of the cloud but will fall back through the cloud, growing further by collision and reaching the ground as rain. The cause of the initial differences in drop size which are essential to the coalescence mechanism—whether, for example, the differences are due to the presence of rare giant nuclei or are associated with variations of updraught velocity and nuclei concentration near the condensation level—is yet uncertain.

For many years it was considered that the ice-crystal process was much the more important of the two mechanisms. It is now clear that both play an important part. Attempts have been made, by appropriate CLOUD SEEDING, to promote precipitation by either mechanism.

precipitation static: See ST ELMO'S FIRE.

pressure: Pressure is the force per unit area exerted on a surface by the liquid or gas in contact with it. Pressure at any point in a fluid is exerted equally in all directions. The dimensions are $ML^{-1}T^{-2}$.

The pressure of the atmosphere at any point is the weight of the air which lies vertically above unit area centred at the point; on average, the weight above each square metre of the earth's surface is about 10^4 kg (10 tons). Pressure may be expressed in 'inches' or 'millimetres', being the equivalent height of a column of mercury of STANDARD DENSITY (13·5951 g/cm^3) under conditions of STANDARD GRAVITY (980·665 cm/s^2) required to balance atmospheric pressure. In meteorology the MILLIBAR (mb), equivalent to 10^3 c.g.s. units of pressure, has almost entirely supplanted the inch or millimetre as the unit of pressure. In SI units the unit is the newton/metre^2.

The pressure exerted by the wind is relatively small; a wind of force 6 on the BEAUFORT SCALE, for example, exerts only about one thousandth part of the pressure of the atmosphere.

Conversion formulae for pressure units are:

millibars = inches × 2·54 × 13·5951 × 980·665 × 10^{-3}
millibars = millimetres × 13·5951 × 980·665 × 10^{-4}
millibars = 10^2 newtons/metre^2.

pressure altitude: The height of a given level in the STANDARD ATMOSPHERE above the level corresponding to a pressure of 1013·2 mb.

pressure co-ordinates: A system of co-ordinates in which the independent variables are x, y and p. x and y are horizontal rectangular co-ordinates while position in the vertical is defined by the hydrostatic pressure (p). This system has advantages over the closely similar CARTESIAN CO-ORDINATE system where, as is normal in dynamical meteorology, isobaric analysis is preferred to constant-level analysis.

pressure gradient force: That force which acts on air by virtue of the variation of pressure in space. It is a three-dimensional vector, denoted $-\nabla p$ or $-\text{grad } p$, and equals $-\left(\dfrac{\partial p}{\partial x}\mathbf{i} + \dfrac{\partial p}{\partial y}\mathbf{j} + \dfrac{\partial p}{\partial z}\mathbf{k}\right)$, where $\mathbf{i}$, $\mathbf{j}$, $\mathbf{k}$, are unit vectors in the x, y, z directions. The force is normal to surfaces of constant pressure; the sign is negative because the force acts from high to low values of pressure.

The horizontal pressure gradient force is a horizontal vector which is perpendicular to horizontal isobars. It is denoted, for example, $-\nabla_H p$, and equals

$$-\left(\dfrac{\partial p}{\partial x}\mathbf{i} + \dfrac{\partial p}{\partial y}\mathbf{j}\right).$$

In meteorological dynamics, the pressure gradient force acting per unit mass of air is the significant force and is generally referred to simply as the 'pressure gradient force'. This force is $-\dfrac{1}{\rho}\nabla p$ or $-\dfrac{1}{\rho}\nabla_H p$ in three or two dimensions, respectively, where ρ is air density.

pressure jump: A sudden and short-lived rise of pressure, of the order millibars per minute which often accompanies the arrival of a LINE-SQUALL or similar phenomenon.

pressure-pattern flying: The planning of a flight route in such a way as to complete the flight, for a given aircraft and load, in the shortest possible time. The normal synoptic aids in such planning are isobaric charts on which contours are drawn.

pressure-plate anemometer: See ANEMOMETER, ANEMOGRAPH.

pressure-tube anemometer: See ANEMOMETER, ANEMOGRAPH.

prevailing wind: That direction of wind which, at a given place, occurs more frequently than any other during a specified period.

Over all parts of the British Isles statistics show the prevailing wind to be, on an eight-point compass, south, south-west or west. There is, however, an appreciable annual variation, e.g. in some places the prevailing wind in spring and early summer has an easterly component.

primitive equations: The fundamental EQUATIONS OF MOTION of a fluid modified by the assumption of HYDROSTATIC EQUILIBRIUM. The term is used particularly in the context of numerical weather forecasting where the assumption of hydrostatic equilibrium is necessary to eliminate high-speed oscillations in the vertical from the solutions of the equations. See also QUASI-GEOSTROPHIC MOTION, NUMERICAL WEATHER FORECAST.

principal component analysis: A powerful method for the statistical analysis of fields of data; it has the effect of representing each field as the linear sum of multiples of standard uncorrelated components, with the feature that the main patterns of variation are shown by the least possible number of components. The determination of the principal components, or 'empirical orthogonal functions', involves finding the eigenvectors of the covariance matrix derived from the data, and is only practicable by electronic computer. See also ORTHOGONAL FUNCTIONS.

probability: A numerical measure of the expectation that a particular event will occur. Its value ranges from 0 (impossibility) up to 1 (certainty).

This fundamental statistical concept developed from the probabilities which can be associated with apparatus such as dice and roulette wheels (with which a trial must have one of a number of results) which are produced by processes which do not favour one result rather than another. For this limited class of probabilities, the laws for the addition and multiplication of probabilities can be built up, arguing from considerations of symmetry and the enumeration of cases. The calculus of probabilities defined in this way is found to be applicable to the much wider class of probabilities which cannot be estimated by the enumeration of cases but only in terms of relative frequencies. Thus if in a large number N of trials an event has occurred n times, and has failed to occur $(N-n)$ times, the probability of occurrence is n/N. Probabilities estimated in this way form the basis of most of STATISTICAL INFERENCE.

The process of judging the SIGNIFICANCE of the result of an experiment consists in estimating the CHANCE EXPECTATION or 'probability' of the result, assuming the starting values of the quantities determining the result to have been chosen 'at random', that is in a way which does not favour one possible result rather than another. If the chance expectation is very small the result is regarded as 'significant'.

Estimation of the chance expectation is usually a matter of mathematical inference from a certain FREQUENCY DISTRIBUTION, and is greatly simplified by the use of statistics which fall into the frequency distributions which have been studied and tabulated in advance. These include the CHI-SQUARE DISTRIBUTION, the F-distribution (see F-TEST), the NORMAL DISTRIBUTION and the t-distribution (see STUDENT'S t-TEST). The experimenter, when deciding which statistic to use, is well advised to keep to statistics which conform to one of these distributions or to the BINOMIAL DISTRIBUTION or the POISSON DISTRIBUTION for which the chance expectation can be derived from first principles.

probability distribution: A somewhat more accurate term for what is often termed a FREQUENCY DISTRIBUTION, in which the probability density is plotted against each possible value of a variable, so that the PROBABILITY of the variable taking a value in a certain range is given by the integral of the probability distribution through the range.

probability integral: See NORMAL (FREQUENCY) DISTRIBUTION.

probable error: The range on either side of the mean which contains half the items in a NORMAL DISTRIBUTION. It is found by multiplying the STANDARD DEVIATION by 0·6745. The probable error is now little used as a measure of dispersion since it has no advantage over the STANDARD DEVIATION from which it is derived.

probable maximum precipitation: An estimate of the upper limit of precipitation, of specified duration, over a particular region. Such an estimate, used for example in dam design, may be based on physical reasoning or on past statistics; a very long RETURN PERIOD applies in either case.

prognostic chart: A SYNOPTIC CHART depicting, for a specified future hour, the expected distribution of isobars or contours and usually also of fronts. The term is often used of a forecaster-produced chart as distinct from a computer-produced forecast chart.

projection: This term is used in connection with maps in a sense wider than that of geometrical perspective. It denotes any relationship establishing a correspondence between a domain of the earth's surface and a domain of a plane surface, the map, such that to each point of one corresponds one and only one point of the other. The projection is completely represented by constructing, on the plane surface, a graticule formed by two intersecting systems of lines, corresponding respectively to parallels of latitude and meridians of longitude on the earth. The position on the map of any features on the earth's surface is then determined by reference to this graticule.

The scale of the map is the ratio of the distance between two neighbouring points on the map to the corresponding distance on the earth. A perfect map in which the scale is uniform throughout is not possible. A class of projections termed 'orthomorphic' or 'conformal' has the property that, at any point, the scale in all directions is the same, though varying from point to point. This is equivalent to the property that the angle of intersection of any two lines on the earth (such as an isobar and a meridian) is preserved unchanged on the map, or the shape of any small area is preserved. Orthomorphic projections are not much used generally, but owing to the above properties, they enter into meteorological practice as base maps for the representation of meteorological elements.

The following projections should be used for weather charts:

(i) The stereographic projection for the polar areas on a plane cutting the sphere at the standard parallel of latitude $60°$.
(ii) Lambert's conformal conic projection for middle latitudes, the cone cutting the sphere at the standard parallels of latitude $10°$ and $40°$ or $30°$ and $60°$.
(iii) Mercator's projection for the equatorial areas, with true-scale standard parallel of latitude $22\frac{1}{2}°$.

The projection recommended in (i) is a special case of conical orthomorphic projection in which the meridians converge at their true angle and in which the scale is correct on any one chosen parallel. It is particularly suitable for the polar regions but can be extended to cover a hemisphere. It is used in most working charts which are exchanged internationally and is suitable for use with computer-printed charts.

The projection recommended in (ii) is very suitable, especially for middle latitudes, and was at one time used for the majority of the working charts of the Meteorological Office. The meridians are straight lines converging to the pole and the parallels of latitude are circles centred at the pole. The scale is correct at all points along two chosen parallels and any two meridians converge at an angle which is a fraction of the angle between them on the earth, the fraction depending solely on the choice of standard parallels. The spacing of the other parallels is then uniquely determinate to secure the orthomorphic property. The scale is somewhat too low between the standard parallels and increases rather rapidly outside them.

The projection recommended in (iii) is another special type of orthomorphic projection in which the angle between the meridians is zero. The meridians are equally spaced parallel straight lines and the parallels of latitude are straight lines at right angles to the meridians and spaced so as to secure the orthomorphic property. This projection is most suitable for the equatorial zone.

Conical and zenithal projections can be made with other than orthomorphic properties by suitably altering the spacing of the parallels. Preservation of areas, or a compromise between the equal area and orthomorphic properties, may be secured. The latter is the case in Clarke's zenithal projection which was at one time used in the *Daily Weather Report* (British Section) of the Meteorological Office.

There are, in addition, many other projections, each to serve a special purpose. Of these Mollweide's 'equal-area projection' is often useful when a world map is required. The whole globe is represented within an ellipse whose major axis is twice the minor axis.

A chart required to give bearings for thunderstorm location uses the oblique gnomonic projection which has the valuable property that all great circles on the globe are reproduced as straight lines.

For further information see *An introduction to the study of map projections** and *Principles of meteorological analysis*.† The mathematical properties of orthomorphic projections are to be found under 'conformal representation' in books on the theory of functions of a complex variable.

proton: A constituent particle of all atomic nuclei, itself comprising the hydrogen nucleus, which carries unit positive charge (equal but opposite to that of the ELECTRON) and has a mass about 1850 times that of the electron.

pseudo-adiabatic (or **pseudo-adiabat**): Line on an AEROLOGICAL DIAGRAM representing the pseudo-adiabatic lapse rate.

See ADIABATIC.

pseudo-equivalent temperature: The pseudo-equivalent temperature (T_{se}) of a sample of air at any level is found from an AEROLOGICAL DIAGRAM by dry adiabatic expansion to the lifting CONDENSATION LEVEL of the sample, followed by ascent along the saturated adiabatic till all the water vapour is condensed, and finally dry adiabatic descent to the initial pressure level.

The 'pseudo-equivalent potential temperature' (θ_{se}) is found by progressing along the dry adiabatic line from T_{se} to the 1000-mb level.

T_{se} for an air sample exceeds the (isobaric) EQUIVALENT TEMPERATURE (T_e), and θ_{se} exceeds the equivalent potential temperature (θ_e), by an amount which is not negligible.

See also ADIABATIC.

pseudo wet-bulb temperature: A temperature (T_{sw}) obtained most readily from an AEROLOGICAL DIAGRAM by ascent of the sample at the dry adiabatic lapse rate until saturation is reached (at the 'lifting condensation level'), followed by descent at the saturated adiabatic lapse rate till the original pressure is reached. The temperature so attained is T_{sw}.

The 'pseudo wet-bulb potential temperature' (θ_{sw}) is found by progressing along the saturated adiabatic from T_{sw} to the 1000-mb level.

T_{sw} for an air sample is less than the WET-BULB TEMPERATURE T_w and θ_{sw} is less than the WET-BULB POTENTIAL TEMPERATURE (θ_w), to a degree which is in practice negligible (usually less than 0·5 degC).

See also ADIABATIC.

psychrograph: A recording PSYCHROMETER.

psychrometer: A type of HYGROMETER (also termed the 'dry-and-wet-bulb hygrometer' or 'Mason's hygrometer') in which two similar thermometers are used; one, the 'dry-bulb', gives the air temperature (T), while the other, the 'wet-bulb', whose

---

* STEERS, J. A.; An introduction to the study of map projections. 14th edn. London, University of London Press, 1965.

† SAUCIER, W. J.; Principles of meteorological analysis. Chicago, University Press, 1955.

bulb is covered with muslin wetted with pure water, gives a reading T_w. In unsaturated air T_w is lower than T by an amount (the 'wet-bulb' depression) which, at a specified temperature, depends mainly on the relative humidity of the air but also, to a small extent, on the degree of ventilation of the wet bulb. The lower temperature is explained by the fact that the latent heat required to evaporate water from the muslin is supplied by the air which is in contact with the wet bulb.

At air pressure p, the vapour pressure (e') of an air sample is related to the saturation vapour pressure at the wet-bulb temperature (e'_w) and to the wet-bulb depression ($T - T_w$) by the semi-empirical 'psychrometric formula'

$$e' = e'_w - Ap(T - T_w)$$

where A is a 'constant'.

The *Hygrometric tables** and humidity slide-rule issued by the Meteorological Office for the purpose of obtaining values of vapour pressure, dew-point and relative humidity from readings of T and T_w observed in a THERMOMETER SCREEN ('light air' conditions of ventilation) are based on a value of the product Ap of 0·799 for $T \geqslant 0°C$ and of 0·720 for $T < 0°C$ for readings in degrees Celsius (0·444 for $T \geqslant 32°F$ and 0·400 for $T < 32°F$ for readings in degrees Fahrenheit); in this product p is assigned the value 1000 mb. The saturation vapour pressure with respect to water is used as the standard at all temperatures.

In the ASSMANN PSYCHROMETER and WHIRLING PSYCHROMETER (or 'sling psychrometer') the rate of ventilation of the bulbs is controlled and values of A are used in the reduction tables and slide-rule other than those appropriate to thermometer screen readings.

psychrometric formula: See PSYCHROMETER.

pumping: Unsteadiness of the mercury in the barometer caused by fluctuations of the air pressure produced by a gusty wind, or due to the oscillation of a ship.

purga: See BURAN.

purple light: Shortly after the sun has set below the western horizon a brighter patch appears on the darkening sky about 25° directly above the position where the sun has disappeared. This patch appears brighter as the sky darkens and takes on a purple tone. The patch expands into a disc and when the sun is about 4° below the horizon it reaches its maximum brilliancy, when it may be so bright that white buildings in the east which are lit up by it glow with a purple colour which corresponds to the AFTERGLOW seen on the peaks of snow-covered mountains. The disc of purple light sinks downwards at twice the rate at which the sun sinks while at the same time its radius expands and its light becomes less intense. It finally sets behind the bright segment of the TWILIGHT ARCH. Occasionally when the first purple light has passed below the horizon, the phenomenon repeats itself with less intensity. The second patch of light appears at a slightly lesser altitude than the first but otherwise follows the same course.

pyranometer: A term applied both to the type of instrument used in measuring the DIFFUSE RADIATION (direct sun excluded) on a horizontal surface and to that which measures the total radiation ('sun plus sky') received on a horizontal surface; the latter type of apparatus is also termed 'solarimeter'.

The instrument generally consists of a THERMOPILE which is under a protective hemispherical glass cover and is connected to a recorder.

---

* London, Meteorological Office. Hygrometric tables. Parts I and II. London, HMSO, 1964.

pyrgeometer: An instrument for measuring the NOCTURNAL RADIATION. That designed by Ångström uses the fact that the radiation from a gilded strip of manganin is less than that from a blackened strip.

pyrheliometer: An instrument for measuring the direct solar RADIATION (DIFFUSE RADIATION excluded) at normal incidence.

Three main types are used. In Ångström's form, the rate of absorption of heat by a thin strip of blackened platinum, normal to the sun's rays, is found by measuring the electric current necessary to heat a similar strip to the same temperature. In Abbot's silver-disc pyrheliometer, the rise of temperature in a silvered disc exposed normal to the sun's rays is measured directly and the intensity of radiation determined by reference to calibration data of the instrument. In a third type, a THERMOPILE, covered by a flat glass plate and fitted in a HELIOSTAT, is used; direct and continuous recording is employed with this instrument.

Q

Q-code: A letter code used by aircraft in requests for information: it is also used in the supply of information to aircraft. Certain items in the code relate to meteorological information, e.g. QFE refers to station-level pressure, QFF to mean-sea-level pressure, QNH to ALTIMETER SETTING. See *Handbook of weather messages.**

quantile: A generic term used to denote one of several equal sections into which a FREQUENCY DISTRIBUTION may be divided. Examples are TERCILE, QUARTILE and QUINTILE. See also PERCENTILE.

quartile: One of four equal sections into which a ranked FREQUENCY DISTRIBUTION may be divided—see PERCENTILE, QUANTILE.

In an alternative usage, the 'lower quartile' is that value below which one quarter of the ranked elements lie, the 'upper quartile' is that value above which one quarter of the elements lie, and the 'interquartile range' is the difference between these values.

quasi-biennial oscillation: A well-defined OSCILLATION of the zonal wind component in the equatorial stratosphere, the 'period' being about 27 months. It is also termed 'stratospheric oscillation'.

The range of the fluctuations, greatest at a level of about 25 km, decreases with distance from the equator. There is a phase lag from higher to lower levels. At a given level the phase does not appear to vary with longitude.

Similar oscillations exist in stratospheric temperature and in total ozone amount.

quasi-geostrophic assumption: Air motion which is known, or more frequently assumed, to be closely approximate to the GEOSTROPHIC WIND. Such an assumption is generally valid in those cases where large-scale atmospheric motion is considered, but it becomes increasingly inaccurate as the scale of the motion decreases and cannot be made, for example, in considering FRONTS where the ageostrophic motions are very important. See also NUMERICAL WEATHER FORECAST.

In describing the motion as 'quasi-geostrophic' rather than 'geostrophic', it is usually implied that derived quantities such as the horizontal DIVERGENCE which depend on small horizontal gradients of wind speed or direction cannot be taken from the gradients of the geostrophic wind, but must be determined in other ways.

quasi-hydrostatic approximation: In dynamical meteorology, the assumption that the vertical component of the EQUATIONS OF MOTION is given by the HYDROSTATIC EQUATION. It has the effect of filtering out those solutions of the equations which correspond to sound waves, these being of no meteorological interest.

quasi-stationary front: A FRONT whose position is (almost) unchanged on successive synoptic charts. There is a strong tendency in such cases for wave-like disturbances of the front to form.

quintile: A term used for a section of a ranked FREQUENCY DISTRIBUTION divided into five equal parts. See also PERCENTILE, QUANTILE, QUARTILE.

---

* London, Meteorological Office. Handbook of weather messages. Part II. London, HMSO.

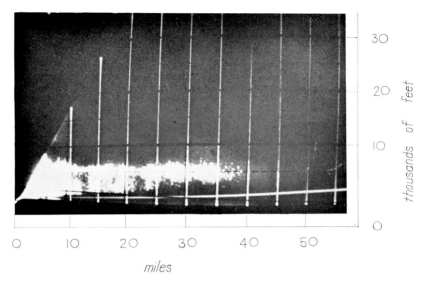

PLATE 22 RHI pattern in warm-front rain at 1750 GMT on 5 February 1957 (East Hill near Dunstable). Range-markers are at 5-mile (8 km) intervals and height lines at 5000-foot (1500 m) intervals.

In this photograph, taken with reduced receiver sensitivity, the strong echo from the melting level at 5000 ft (see MELTING BAND) is clearly revealed. (In a photograph taken immediately prior to this, with the receiver on full sensitivity, echoes from snow were obtained up to a height of 20 000 ft (6000 m) at short ranges; the melting band was then correspondingly reduced in relative intensity.)

PLATE 23 PPI pattern of widespread showers and thunderstorms at 1345 GMT on 12 August 1958 (East Hill, near Dunstable). Range circles are at 5-mile (8 km) intervals (maximum 80 miles (128 km)) and azimuth lines at 20° intervals with magnetic north accentuated. The display has been slightly off-centred.

The well-defined edges of the shower echoes are a notable feature of the pattern.

R

radar: A system of detection and location of 'targets' which are capable of reflecting high-frequency radio waves (microwaves), generally in the wavelength range from a fraction of a centimetre to some tens of centimetres.

The system consists of sending from a transmitter a narrow beam of radio waves and obtaining any reflected signal ('echo') in an adjacent receiver which generally uses the same aerial system as the transmitter. The distance of the target is obtained from the time interval between transmission and reception, and its direction from that of the aerial system employed. The information is presented visually on a cathode-ray tube. Detection of the echo requires a very sensitive receiver and portrayal requires great amplification since much of the emitted radio energy is lost by atmospheric absorption and scattering. See also RADAR METEOROLOGY.

radar meteorology: The main applications of RADAR in meteorology are in the measurement of RADAR WINDS and in the detection of cloud and precipitation elements. The latter is used as a direct forecasting aid and in detailed studies of the structure of precipitation regions.

Various types of radar display are used in the surface investigation of precipitating clouds. They include: a 'plan position indicator' (PPI), which employs a horizontally scanning aerial system and gives a picture of the distribution of precipitation regions in all directions round the observing point to a distance dependent on the characteristics of the set and the intensity of precipitation—for moderate precipitation this distance may be about 150 km; a 'range-height indicator' (RHI), in which the scanning is in the vertical plane; and an 'A-scope indicator', which shows the variation of amplitude of reflected signal with range of target. See Plates 22 and 23.

Since the amount of back-scattered energy and the degree of atmospheric attenuation of the waves which travel to and from the target both increase with decrease of wavelength, a compromise choice of wavelength is required. For detection of precipitation a choice is made in the approximate range 3 to 20 cm, while detection of the larger cloud particles even at very short range requires a wavelength in the millimetre range.

For a given wavelength, the strength of the echo increases with the concentration, and very rapidly with the size (proportional to the sixth power of the diameter), of cloud or precipitation particles. The echo strength is, therefore, almost entirely due to the energy back-scattered by any precipitation particles, as opposed to the much smaller cloud particles, which may be present. Empirical relations have been obtained between precipitation rate and corresponding echo intensity.

The vertical and horizontal extent, development and movement of a precipitation region may be indicated by radar. Various complicating factors greatly reduce, however, the amount of quantitative work possible and make very difficult even qualitative assessment as to the precise nature of the reflecting particles. These factors include: particle size distribution relative to radio wavelength in use (determining whether RAYLEIGH SCATTERING or MIE SCATTERING obtains); particle shape (affecting 'radar cross-sections'); and particle phase (whether, for example, ice or water, or dry or wet hailstones). See also DOPPLER RADAR, BRIGHT BANDS, MELTING BAND, RAYLEIGH SCATTERING.

radar storm-detection: An obsolescent term for the detection of 'storms'—in practice, regions of precipitation—in the atmosphere, by means of ground-based or airborne RADAR. See RADAR METEOROLOGY.

radar wind: A wind in the upper atmosphere determined by means of radar reflections from a 'radar target' carried aloft by a free balloon, the observed elements being the range, elevation, and azimuth of the target.

radian: A unit of angular measure, being the angle subtended at the centre of a circle by an arc equal in length to the radius of the circle. Thus, π radians = 180°.

radiation: The transmission of energy by electromagnetic waves. See ELECTROMAGNETIC RADIATION. (The term is also used to signify the emission of particles by a source, as in 'cosmic radiation' and 'solar particle radiation'.)

Radiation flux across a surface is the energy which crosses unit area of the surface in unit time. The following relations hold for the various units of flux which have been used in meteorology (the SI UNIT is the watt/metre^2 (W/m^2) but WMO recommended practice is the mW/cm^2):

$$1 \text{ IT cal/cm}^2\text{min} = 4\cdot1868 \text{ J/cm}^2\text{min}$$
$$= 69\cdot8 \text{ mW/cm}^2.$$

A valuable concept in radiation is that of the 'black body'. A perfect black body is one which absorbs all the radiation falling on it and which emits, at any temperature, the maximum amount of radiant energy. The term arises from a correlation between darkness of colour and proportion of visible light absorbed. A body which appears white because it scatters the visible light falling on it may, however, act nearly as a black body to radiation of different wavelength. Snow is an example, being effectively a black body for wavelengths greater than 1·5 micrometres.

The properties of a black-body radiator are expressed in a number of laws:

(i) Planck's law. The distribution of energy with temperature (T) and wavelength (λ) for a perfect radiator was represented by Planck as

$$E_\lambda = c_1 \lambda^{-5}/(e^{c_2/\lambda T} - 1)$$

where E_λ is the energy emitted in unit time from unit area within unit range of wavelength centred on λ, and c_1 and c_2 are constants. The corresponding energy distribution curve is illustrated in Figure 26 for four radiation temperatures; it shows, for example, that the solar radiation spectrum barely overlaps that of ground radiation. The area bounded by the curve, the wavelength axis and any pair of selected wavelengths gives a relative measure of the energy contained in the corresponding part of the spectrum.

(ii) Wien's (displacement) law. The variation of wavelength of maximum energy emittance (λ_{max}) with temperature of radiator is given by

$$\lambda_{max} T = 2\cdot898 \times 10^{-3} \text{m degK}.$$

(iii) Stefan–Boltzmann law. The amount of energy emitted in unit time from unit area of a black body is proportional to the fourth power of absolute temperature, i.e.

$E = \sigma T^4$ where σ (Stefan's constant) = $5\cdot670 \times 10^{-8}$ mW/cm^2 (degK)^4.

(iv) Kirchoff's law. The ratio of emissive to absorptive power, for a particular wavelength and temperature, is the same for all types of body and is numerically equal to the emissive power of a black body (whose absorptive power is, by definition, unity). Thus, if a body at a given temperature strongly absorbs radiation of a certain wavelength, it also radiates strongly this wavelength, provided it is present in the radiation spectrum for the temperature. The wavelength dependence is here crucial, e.g. fresh snow absorbs very little direct (short-wave) solar radiation but acts very nearly as a black body to long-wave radiation from the atmosphere.

The spectrum of solar radiation outside the earth's atmosphere, inferred from ground observations or measured directly by rockets and satellites, extends from the X-ray region through the ultra-violet, visible (0·4 to 0·7 μm) and infra-red to the radio-wave region. About half the total solar energy is in the form of visible light. The spectrum observed at the ground is sharply cut off in the near ultra-violet at about 0·29 μm, owing to the complete ABSORPTION of shorter wavelengths, containing about 5 per cent of the total energy, by gases in the high atmosphere; there is also selective absorption of higher wavelengths by atmospheric constituents.

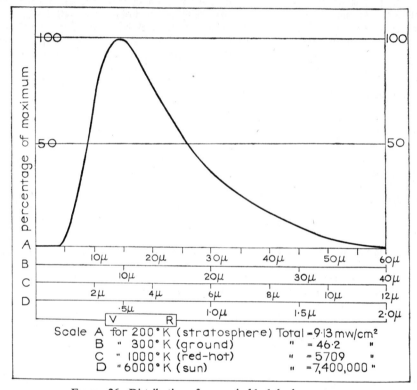

FIGURE 26—Distribution of energy in black-body spectrum.

Application of the inverse square and Stefan–Boltzmann laws, with the assumption of a SOLAR CONSTANT of 139·6 mW/cm² (2·00 cal/cm² min), shows that the sun has a black-body radiation temperature of about 5800 K; while a sun 'colour' temperature of about 6100 K is obtained by insertion of the value $\lambda_{max} = 0.474\mu m$ into Wien's formula. Detailed examination of the extraterrestrial solar spectrum in terms of Planck's law reveals, however, that there is no close fit at all points of the spectrum with a perfect radiator of any given temperature, probably because the radiation originates at different levels in the sun's atmosphere and therefore at different temperatures.

The radiant energy delivered by the sun to the fringe of the earth's atmosphere, about 7 per cent greater at PERIHELION (early January) than at APHELION (early July) is 139·6 mW/cm² at the earth's mean distance ('SOLAR CONSTANT'). The mean flux perpendicular to the earth's surface is about 34·9 mW/cm². About 40 per cent of this incident radiation is diffusely reflected to space without change of wavelength (ALBEDO of earth–atmosphere system), about 15 per cent heats the atmosphere by direct absorption by constituents, and the remaining 45 per cent is absorbed at the earth's surface as both direct ('sun') and diffuse ('sky') radiation. The ratio of sky to sun radiation increases with cloudiness and latitude and is greater in winter than in summer; in middle latitudes it averages 30–40 per cent.

Most of the solar energy absorbed at the earth's surface is transferred as heat to the atmosphere initially by conduction, then by turbulence and convection and by evaporation; radiation in the wave band appropriate to the temperature of the surface is smaller but also important. The total amount of long-wave radiation emitted to space by the atmospheric gases and direct from the earth's surface (terrestrial radiation) equals, on balance, the 60 per cent of solar radiation which is effective in heating the earth and atmosphere; the black-body radiation 'planetary' temperature appropriate to this radiation balance is about 250 K. Study of the conditions of balance within the earth–atmosphere system shows that, except in higher latitudes in winter, the earth's surface radiates less heat than it absorbs (heat source) and the troposphere everywhere radiates more heat than it absorbs (heat sink); in the stratosphere local radiative equilibrium more nearly obtains.

Direct absorption of solar radiation has been found by estimation and measurement to cause a daily heating of about 0·1 degC in the low stratosphere and of 0·1 to 0·6 degC (depending on season and water vapour content) in the lower troposphere in middle latitudes. In the high stratosphere, solar heating effects are much greater—notably at about 50 km, associated with absorption by oxygen and ozone. The daily net cooling rates, associated with the divergence of terrestrial radiation, is about 1 to 2 degC throughout a cloudless troposphere and low stratosphere; cooling is much reduced below a cloud layer but is increased at the top of such a layer.

See also TERRESTRIAL RADIATION.

radiation balance: The resultant flux of the solar and terrestrial RADIATION through a horizontal surface. It is considered positive if the flux downwards exceeds that upwards. It is also termed 'net radiation'.

In general, the radiation balance at the earth's surface is positive by day and negative by night. The average annual surface balance is everywhere positive. It is highest in low latitudes and, in a given latitude, is greater over an ocean than over a continent. The positive balance for the surface of the globe is estimated to average about 3×10^9 J/m² per year (0·951 mW/cm².) Over-all radiative equilibrium for the earth and atmosphere system is achieved by a negative balance of the same amount in the atmosphere. See also ENERGY BALANCE.

radiation chart: A chart (for example, that of W. M. Elsasser) which, by providing a graphical method of numerical integration of the equations of radiative transfer in the atmosphere, permits of the calculation of the upward and the downward fluxes of radiation at any level, the vertical distribution of temperature and humidity being known.

radiation fog: A common type of FOG which forms overland on nights characterized by light wind, clear sky, and moist air in the lower levels of the atmosphere. The first two conditions lead to the formation of a RADIATION INVERSION. Since, however, loss of water from air in contact with cold ground proceeds rather more quickly than loss of heat, some turbulent interchange of air, with associated adiabatic cooling, and downward transfer of water vapour by EDDY DIFFUSION, are required to produce condensation in an appreciable layer. The presence of hygroscopic nuclei, as in industrial areas, facilitates fog formation by allowing condensation to occur in unsaturated air. In the British Isles fog is often caused in part by radiation and in part by advection processes—see ADVECTION FOG.

radiation inversion: An INVERSION of temperature through an atmospheric layer extending upwards from the earth's surface, such a condition developing in the course of a RADIATION NIGHT over a land surface as a result of radiational cooling of the surface. The depth of the inversion layer increases in the course of the night owing to downward conduction of heat from the atmosphere.

radiation night: A night, characterized by absence of cloud and wind, on which there is marked radiational cooling of the ground and, by conduction of heat from air to ground, of the surface layers of air. Absence of cloud ensures that there is relatively little compensating radiation directed downward to ground (see ATMOSPHERIC WINDOW). Absence of wind confines cooling to a shallow layer near the ground and produces a low minimum surface temperature. The occurrence on some occasions of minimum air temperature at a few centimetres above the ground rather than at the ground itself is well substantiated but has as yet no accepted explanation.

Other factors which favour a low temperature minimum on a radiation night are relatively dry air, low heat conductivity of the ground, and long hours of darkness.

radiation pressure: The pressure exerted on a body by the electromagnetic RADIATION incident on it. That exerted by solar radiation which is incident on the earth's surface is minute relative to atmospheric pressure. In solar physics, radiation pressure is considered to play an essential part in the emission of atoms from the sun.

radiative equilibrium: State of balance between the absorption and emission of RADIATION.

Turbulent transfer, evaporation and condensation are the processes which chiefly inhibit the occurrence of the state of radiative equilibrium. The state is therefore of little or no significance at the earth's surface and in the troposphere. At higher levels, evaporation and condensation are unimportant and the effect of turbulent transfer on the temperature distribution is, in general, secondary to that of radiation processes. Thus, for example, approximate radiative equilibrium obtains in the upper part of the ozone layer where ultra-violet radiant energy is strongly absorbed by oxygen and ozone in certain lines and bands and is shared by collision with other atmospheric gases. The air temperature is thus raised to a level at which the terrestrial radiation emitted by the gas mixture (mainly by the constituents ozone, carbon dioxide and water vapour) is in balance with the incoming radiation. See also RADIATION BALANCE.

radiatus (ra): One of the CLOUD VARIETIES (Latin, *radiatus* having rays).

'Clouds showing broad parallel bands or arranged in parallel bands, which, owing to the effect of perspective, seem to converge towards a point on the horizon or, when the bands cross the whole sky, towards two opposite points on the horizon, called radiation point(s).

This term applies mainly to CIRRUS, ALTOCUMULUS, ALTOSTRATUS, STRATOCUMULUS and CUMULUS'.* See also CLOUD CLASSIFICATION.

---

* Geneva, World Meteorological Organization. International cloud atlas, Vol. 1. Geneva, WMO, 1956, p. 15.

radioactive carbon (radiocarbon): Carbon–14 (^{14}C) is a radioactive isotope which is important in geophysics. Bombardment of the atmosphere by cosmic rays produces neutrons which react with nitrogen (^{14}N) to form ^{14}C at a rate which is greatest at about 10 km. Measurements indicate that about 2·4 ^{14}C atoms are produced in this way per second per square centimetre of the earth's surface. ^{14}C has a radioactive HALF-LIFE of about 5500 years and reverts to ^{14}N by emission of a beta particle.

Radiocarbon is distributed, like the vastly more plentiful non-radioactive carbon, throughout the earth–atmosphere system. Only 1 or 2 per cent of ^{14}C is stored (as radioactive carbon dioxide) in the atmosphere. About 90 per cent is contained in the carbonaceous materials of the oceans which are in exchange equilibrium with carbon dioxide. The remaining 8 or 9 per cent which is retained by plant and animal life gives rise to the technique of CARBON DATING.

Local artificial injections of ^{14}C into the atmosphere have been caused by nuclear explosions. The ^{14}C introduced in this way has been used as a tracer element of atmospheric motion, by the measurement of the ^{14}C content of air samples collected at various levels and of vegetation.

radioactive fallout: The descent to the earth's surface of radioactive material produced in a nuclear explosion.

A distinction is made between two classes of fallout: (i) 'close-in fallout' (or 'local fallout'), i.e. material that descends close to 'ground zero' (point on earth's surface at or above which the device is exploded); (ii) 'delayed fallout' (or 'world-wide fallout'), i.e. material that reaches the earth's surface after a long delay, most of it far from 'ground zero'. Meteorological processes are important in both types of fallout.

(i) *Close-in fallout*

If a nuclear explosion occurs at a level sufficiently low for the associated FIREBALL to intersect the earth's surface (technically, a 'ground-burst'), fused ground material is sucked into the rising cloud and made radioactive. The average size of cloud particle is, in these circumstances, relatively large and the average TERMINAL VELOCITY correspondingly great. The largest (fastest falling) particles reach the earth's surface about 20 minutes after the explosion, the smaller particles proportionately later and farther downwind. Since the particles start their descent from a range of levels within the cloud, they are subject to horizontal movement by mean vector winds extending through various layers (all terminating at the surface) as well as to horizontal diffusion and so give rise to a fallout pattern on the ground, the edges of which are likely to be irregular if there are strong vertical currents or precipitation in the region. The pattern is long and narrow if the upper winds are strong and have little vertical shear and much more nearly circular if the upper winds are light and have pronounced shear. For a large thermonuclear ground-burst device (5–10 megatons), the height reached by the top of the cloud is of the order 25–30 km.

The radioactive dust which settles on the ground emits GAMMA RADIATION which is relatively intense downwind from ground zero along that line which corresponds to the direction of the mean vector wind in the layer from the surface to the most active part of the cloud (the so-called 'hot line') and which is relatively weak towards either edge of the plume. After fallout is complete at any place within the plume, the intensity of emitted radiation decreases at an approximately exponential rate, reckoned from the time of burst. The limits of close-in fallout are arbitrarily selected on the basis of a minimum measured dose-rate of radiation at a fixed time interval after the explosion; the selected limits are such that it is only particles of diameter greater than about 50 micrometres, falling in the period up to some 36 hours after the explosion, which are involved. Close-in fallout comprises more than half the total radiochemical energy of a ground-burst weapon.

(ii) *Delayed fallout*

If a nuclear explosion occurs at such a level that the fireball does not intersect the surface of the earth (an 'air-burst'), the size of the particles is very small and terminal velocities are low. There is then negligible close-in fallout (except, perhaps, in the event of rain at the time of the explosion), but a delayed fallout, partly by deposition of radioactive dust and mainly by WASHOUT with rain. If the original cloud is confined to the troposphere, most of the delayed fallout occurs within a period of weeks. If, however, the material reaches the stratosphere, most of the fallout is delayed for months or years (depending on the height, season and latitude of injection). The term 'residence half-time' or 'storage half-time', signifying the time required for one half of the material to be deposited, is used with respect to delayed fallout.

During the period when the material is stored in the atmosphere, it acts as a 'tracer element' in respect of air movement. Other meteorological factors, however, —notably the distribution and intensity of rainfall—play an important part in determining the eventual pattern of deposition of the material on the earth's surface.

Radioactive fallout has been found to cause long-period changes in atmospheric electrical conductivity—see ATMOSPHERIC ELECTRICITY.

radioactivity: The property of spontaneous disintegration of unstable, into more stable, elements, accompanied by the emission of ALPHA PARTICLES, BETA PARTICLES, or GAMMA RAYS. It is a property possessed by some of the naturally occurring elements; those in the earth's crust play a significant part in the IONIZATION of air at low levels over land areas.

radio direction-finder: An instrument, also termed 'radiogoniometer', for determining the azimuth from which radio waves are received. The term is sometimes used when elevation also is determined, being then a RADIO-THEODOLITE.

radio direction-finding (RDF): Measurement of the direction of arrival of radio waves. It is also termed 'radiogoniometry'. See RADIO DIRECTION-FINDER.

radio duct: See ANOMALOUS RADIO PROPAGATION.

radiogoniometer: An alternative for RADIO DIRECTION-FINDER.

radiogoniometry: An alternative for RADIO DIRECTION-FINDING.

radiosonde: A small radio transmitter by means of which observations, usually of pressure, temperature and humidity, may be obtained from the upper atmosphere.

In the Meteorological Office Mark 2B radiosonde there are sensitive elements consisting of an aneroid capsule for pressure, bimetallic strip for temperature and gold-beater's skin for relative humidity, and these control the armatures of variable inductors. As the balloon ascends, a small windmill-driven switch causes each inductor to be connected in turn to a valve oscillator. The completed circuit generates a note whose frequency (pitch), determined by the variable gap between the inductance and armature, modulates the carrier frequency generated in the output circuit. The transmitted note is picked up by a radio receiver on the ground; the frequency or period is measured, and these readings are converted into values of pressure, temperature or humidity by means of calibration curves or tables.

A new Meteorological Office radiosonde, the Mark 3, is due to replace the present sonde, the Mark 2B, during 1972. The new design uses a resistance wire of diameter $13 \cdot 5$ μm to measure temperature, an improved aneroid capsule for pressure and continues to use gold-beater's skin for relative humidity. The change in resistance is used to control an audio-oscillator directly, while the deflexion of the capsule and gold-beater's skin are used to vary the penetration of a ferrite rod within a ferrite pot. Each of these three devices is connected to the oscillator, one at a time, to generate a note in the frequency range 3500–6800 hertz. The frequency generated by the audio-oscillator is used, via an isolating amplifier to amplitude modulate a small crystal-controlled radio transmitter operating between 27·5 and 28·0 megahertz. The instrument uses discrete components together with six transistors. The effect of environmental changes upon this circuit is monitored by three reference signals which together with the three meteorological data signals are selected by an electric motor-driven switch. An interlaced signal pattern is transmitted which provides a temperature signal every 4 seconds, a relative humidity signal every 8 seconds and a pressure signal every 16 seconds. With the normal rate of ascent of 6·25 metres per second, the corresponding height intervals are 25, 50 and 100 metres respectively. Each reference signal occurs at intervals of 48 seconds. Additional brief signal pulses of two fixed frequencies are sent prior to the temperature and pressure signals to identify these signals for automatic ground-equipment processing. The transmitted notes are picked up by a radio receiver and the periodicity of each is measured. The readings may be converted to values of air pressure, temperature and relative humidity by human computation, using calibration curves and tables. However, it is planned to use an automatic ground-recording system together with a small digital computer to calculate the results.

In practice, wind velocities may also be obtained in the course of a radiosonde ascent, either by tracking the balloon by means of a RADIO-THEODOLITE or by radar reflections from a 'radar target' carried by the balloon; the latter method is used in the Meteorological Office.

The height attained by the sonde is about 21 km; a large balloon used in favourable circumstances will give observations to 30 km or above.

radio-theodolite: An apparatus for determining the direction (angles of elevation and azimuth) from which radio waves reach a receiver. It consists essentially of a receiver coupled to an aerial which can be rotated about horizontal and vertical axes. Used with a RADIOSONDE, it gives upper-level winds with a lower degree of accuracy than is obtained by the radar method ('radar wind') in which slant range is measured.

radio waves: ELECTROMAGNETIC RADIATION of wavelengths greater than about 1 mm, i.e. beyond the rather indefinite upper limit of INFRA-RED RADIATION.

The radio band of waves is subdivided according to frequency, and ranges from 'extremey high' to 'very low' frequency bands. Those of shortest wavelength (highest frequency) are termed 'microwaves' and are used, for example, in RADAR. For the locating of lightning sources (see SFERICS FIX) special radio receivers, tuned to receive long waves of about 30 km, are used.

radio wind: A wind in the upper atmosphere determined by radio means (see RADIO-THEODOLITE).

radon: One of the INERT GASES, radon is emitted by radioactive materials in the earth's crust and comprises a minute constituent of air at low levels. Its atomic mass is 222 and atomic number 86. The ALPHA PARTICLES which are emitted by the decay of radon (together with those emitted by its isotopes THORON and ACTINON) are responsible for part of the IONIZATION of the air at low levels over land.

PLATE 24 Rainbow: Inverness-shire.

Primary and secondary bows, with opposite sequences of colour, are seen. A faint and rare 'reflection rainbow', produced by light which has been reflected from Loch Morlich (behind the observer) is visible in the relatively dark segment of sky between the primary and secondary bows.

rain: Liquid PRECIPITATION in the form of DROPS of appreciable size (by convention, of diameter greater than about 500 µm, that is 0·5 mm, which is the limiting size of DRIZZLE drops).

For synoptic purposes, rain (other than in showers) is classified as 'slight', 'moderate', or 'heavy', for rates of accumulation less than 0·5 mm/h, 0·5 to 4 mm/h, and greater than 4 mm/h, respectively. See also RAINDROPS.

rainbow: A rainbow is seen when the sun shines upon raindrops. The drops may be at any distance from the observer from a few metres to several kilometres. When sunlight falls upon a drop of water the light is refracted on entering the drop, is reflected from the far side and emerges with further refraction, from the near side. The light which is reflected in this way does not come out in all directions but only in directions lying within about 42° from the direction of the sun. The reflected light is most intense near the limit. Accordingly an observer looking towards the raindrops receives a certain amount of light from all directions within 42° from the shadow of his head but most light along rays which make about 42° with the central line. The limiting angle depends on the colour of the light and insomuch as white light is compounded of light of different spectral colours the observer sees a number of concentric arches of different colours, generally with violet to the inside and red to the outside. See Plate 24.

Some of the light falling on a drop does not emerge until after it has been reflected twice. None of the twice reflected light which reaches an observer makes an angle of less than 50° with the line to the centre of his shadow. The colours of the outer bow formed in this way are in the reverse order to those of the inner bow. The space between the inner and outer ('primary' and 'secondary') bows appears darker than the space inside the primary bow or beyond the secondary bow. Several 'supernumerary' bows may also appear within the primary bow.

The coloration of a rainbow depends on the size of the drops. Drops larger than 1 mm in diameter yield brilliant bows about 2° in width, in which the limiting colour is distinctly red. With drops about 0·3 mm in diameter the limiting colour is orange and inside the violet there are bands in which pink predominates. With smaller drops 'supernumerary' bows appear to be separate from the primary bow. With still smaller drops about 0·05 mm in diameter the rainbow degenerates into a white fogbow with faint traces of colour at the edges. The variations of colour with drop size, and also the appearance of supernumerary bows, are due to DIFFRACTION.

Rainbows are not infrequently observed by moonlight but as the human eye cannot distinguish colour with faint lights the lunar rainbow appears to be white.

rain day: Defined for statistical purposes as a period of 24 hours, commencing normally at 09 GMT, on which 0·2 mm (0·01 in) or more of RAINFALL is recorded.

See also WET DAY.

raindrops: Liquid-water drops of diameter greater than about 0·5 mm. The term is sometimes used to include also DRIZZLE drops, extending then to a lower limit of about 0·2 mm.

Surface tension forces act on small raindrops to minimize the surface-to-volume ratio and so make them spherical. The lower surface of larger drops is, however, appreciably flattened by aerodynamic forces. Raindrops of greater (equivalent) diameter than about 6 or 7 mm break up into smaller drops.

Both 'median volume diameter' (defined as the drop diameter such that half the total water is contained in larger drops) and drop concentration (drops/m^3) tend to increase with rate of rainfall. There is, however, considerable variability and dependence, in particular, on type of rainfall. The following are typical values: at a rate of rainfall of 0·5 mm/h there are about 250 drops/m^3 and drops of about 1 mm diameter contribute most of the water (the average drop size being smaller); while at a rainfall rate of 25 mm/h there are about 1200 drops/m³ and drops of about 2 mm diameter contribute most of the water.

rainfall: The total liquid product of PRECIPITATION or CONDENSATION from the atmosphere, as received and measured in a RAIN-GAUGE. SNOW, SLEET* and HAIL, in addition to RAIN, make up much the greater part of the total 'rainfall', as defined above. There are also small additions due to the deposition of DEW, HOAR-FROST and RIME on to the collecting surface of the rain-gauge. One inch of rainfall is equivalent to about 100 tons of water per acre (1 mm is equivalent to 1 kg/m²).

Rainfall is classified into three general types: OROGRAPHIC, CYCLONIC and CONVECTIONAL types. These types, discussed under their individual headings, are by no means mutually exclusive. Other terms, such as 'frontal rainfall', are sometimes also employed.

The average monthly and annual rainfall during the current standard period, 1916–50, for the larger divisions of the United Kingdom are given in Table X. These values replace the provisional figures published in the fourth edition of this glossary.

TABLE X—*Average rainfall* 1916–50

	England	Wales	England and Wales	Scotland	Northern Ireland	Great Britain
			millimetres			
January	83	149	92	154	109	112
February	60	105	66	106	76	80
March	52	85	57	89	66	68
April	57	82	60	88	67	70
May	59	86	63	87	72	72
June	52	79	55	87	71	66
July	75	106	79	114	96	92
August	75	116	81	122	102	95
September	69	114	76	128	96	93
October	83	148	92	158	111	114
November	88	145	95	143	104	113
December	79	142	88	143	111	106
Year	832	1357	904	1419	1081	1081

Table XI shows the rainfall over England and Wales, Scotland and Northern Ireland for the 66 years 1901–66 together with the 1916–50 average for easy reference.

In England and Wales the wettest year of the period 1901–60 was 1960, with 1171 mm; in the longer series from 1727 there were two wetter years, 1872 (1288 mm) and 1852 (1265 mm). The driest years were 1731 (582 mm), 1788 (584 mm), 1741 (607 mm) and 1921 (627 mm). In Scotland the wettest years (in a series of observations from 1869) were 1872 (1715 mm), 1877 (1669 mm) and 1903 (1654 mm); the driest year was 1933 (1024 mm). Data for Northern Ireland are available only from 1900: the wettest year was 1928 (1300 mm) and the driest 1933 (770 mm). Over Great Britain the wettest and driest years were 1872 and 1887 respectively. Extreme annual values at individual stations are a maximum of about 6528 mm (at Sprinkling Tarn, Cumberland, in 1954) and a minimum of 236 mm (at one station in Margate in 1921).

---

* The term 'sleet' is commonly used in this country to describe precipitation of snow and rain (or drizzle) together, or of snow melting as it falls, but it has no agreed international meaning.

TABLE XI—*Annual rainfall for each year 1901–69 and the 1916–50 average*

	England and Wales	Scotland	Northern Ireland		England and Wales	Scotland	Northern Ireland
	millimetres				*millimetres*		
Average 1916–50	904	1419	1081	Average 1916–50	904	1419	1081
Year				Year			
1901	790	1207	1026	1941	859	1113	958
1902	752	1105	970	1942	841	1278	1138
1903	1146	1654	1212	1943	833	1354	1046
1904	798	1194	983	1944	897	1346	1113
1905	770	1234	894	1945	833	1295	993
1906	904	1412	1011	1946	1057	1270	1097
1907	886	1323	1031	1947	823	1240	1062
1908	813	1267	1036	1948	953	1575	1135
1909	940	1285	975	1949	785	1382	1013
1910	1011	1341	1062	1950	1021	1455	1214
1911	841	1260	886	1951	1110	1392	1026
1912	1118	1382	1158	1952	902	1196	914
1913	876	1184	1019	1953	757	1237	851
1914	968	1311	1049	1954	1085	1626	1260
1915	986	1232	955	1955	785	1041	935
1916	1019	1501	1092	1956	869	1306	988
1917	876	1240	1095	1957	899	1356	1085
1918	958	1359	1107	1958	1029	1361	1123
1919	940	1194	963	1959	805	1290	930
1920	975	1346	1173	1960	1171	1374	1153
1921	627	1265	940	1961	881	1461	1062
1922	942	1204	975	1962	790	1438	960
1923	1011	1534	1224	1963	854	1276	1017
1924	1074	1341	1196	1964	709	1279	966
1925	947	1278	1046	1965	999	1405	1130
1926	912	1417	1064	1966*	1024	1491	1217
1927	1100	1458	1021	1967	982	1506	1148
1928	1026	1575	1300	1968	981	1300	1037
1929	894	1290	1049	1969	909	1240	960
1930	1052	1387	1080				
1931	975	1328	1092	* Values for 1966 onwards are provisional.			
1932	922	1379	978				
1933	726	1024	770				
1934	851	1407	1059				
1935	1011	1379	1003				
1936	975	1201	1102				
1937	986	1163	1016				
1938	886	1572	1179				
1939	1013	1212	1029				
1940	904	1252	1140				

The wettest months were: Great Britain, October 1903 (221 mm); England and Wales, October 1903 (211 mm); Scotland, December 1867 (264 mm); Northern Ireland, September 1950 (196 mm). The driest months were: England and Wales, March 1742, February 1891, June 1925 (each 2·5 mm); Scotland, August 1947 (5 mm); Northern Ireland, February 1932 (5 mm). The largest individual monthly value was at Snowdon (Llyn Llydaw) in Caernarvonshire in October 1909 (1436 mm). Since 1901 no rain was recorded at a number of stations in 18 different months: they include since 1940, June 1942, August 1946, February and August 1947, March 1953, July 1955, April 1957 and February and September 1959. The largest area over which there was no measurable rain in a calendar month was 16 595 square kilometres in June 1925. The longest periods on record, with no rainfall, were in 1893, when some 20 stations in south-east England recorded no rain for a period of 50 days or more and locally there was a two-months drought from 17 March to 16 May.

The following list of maximum recorded falls in specified short periods refers to the period 1860 to 1966:

(i) 31·8 mm in 5 minutes at Preston, Lancashire, on 10 August 1893.
(ii) 80·0 mm in 30 minutes at Eskdalemuir, Dumfriesshire, on 26 July 1953.
(iii) 92·2 mm in an hour at Maidenhead, Berkshire, on 2 July 1913.
(iv) 154·7 mm in 2 hours at Hewenden Reservoir, Bradford, Yorkshire, on 11 July 1956.

The annual publication *British Rainfall** contained, until 1960, detailed information of rainfall amounts and duration and of the incidence of RAIN SPELLS, WET SPELLS, DRY SPELLS, DROUGHTS, etc. during each year. From 1961, with the 101st volume of *British Rainfall*, there have been great changes in the contents of the book and in the presentation of the material. In the *General Table* stations are arranged in a new order within hydrological areas, monthly and annual totals have been included, and for most stations, the wettest day in the year. Heavy falls continue to be classified, in ascending order of intensity, as 'noteworthy', 'remarkable', or 'very rare'. The classification concerned is described in *British Rainfall*, 1935. Checks with more recent data have not suggested that there is a need for a revision of the criteria. More detailed analysis has been made possible by use of a computer, and a seven-point frequency distribution of daily falls replaces the old simple grouping into RAIN DAYS and WET DAYS.

Information about authorities and daily, monthly and recording gauges, the commencing date of records, mean annual rainfall and the altitude of each station, is being published in 5-year supplements to *British Rainfall*, the first of which is for the period 1961–65.

Metric units for station altitude and rainfall will be used in *British Rainfall*, 1964 and later years, and in the 5-year supplement for the period 1966–70, instead of the British units used hitherto.

Averages of rainfall for Great Britain and Northern Ireland, 1916–1950† contains monthly and annual averages for 719 stations. Local and seasonal distributions of rainfall are illustrated on the Ministry of Housing and Local Government map of average annual rainfall 1916–50‡ for Great Britain which has an accompanying explanatory text.

rainfall station: A STATION at which the only regular measurements made are those of rainfall. The large majority of the total of about 6400 stations in Great Britain and Northern Ireland which make returns of measured rainfall to the Meteorological Office are of this category.

rain-gauge: An instrument for measuring rainfall.

In its ordinary form a funnel is used to collect the rain and a tube, made of fairly narrow bore in order to minimize evaporation, leads the collected water to a receiving vessel. The rim of the funnel should have a sharp edge, bevelled outside and

---

* London, Meteorological Office. *British Rainfall*. London.
† London, Meteorological Office. Averages of rainfall for Great Britain and Northern Ireland, 1916-1950. London, HMSO, 1958 (reprinted 1967).
‡ London, Ministry of Housing and Local Government, Scottish Development Department and Meteorological Office. Rainfall: Annual average 1916–1950. London, Ordnance Survey, 1967.

falling away vertically inside. The vertical walls should be sufficiently deep and the shape of the funnel sufficiently steep to prevent rain splashing in and out. In the United Kingdom the funnel has usually been made of copper with a brass rim, generally five inches, but occasionally eight inches, in diameter, and mounted in an open situation with the rim horizontal and 12 inches above ground level.

The recently developed Meteorological Office 'standard' rain-gauge has a number of interchangeable units used with a choice of aperture area of 150 or 750 cm^2. The material is glass-fibre reinforced polyester resin with stainless steel for outlet pipes. The height of the rim above the ground is standardized at 30 cm. The associated units are collecting bottles, measuring flasks, tipping-bucket switches, totalizer counters, incremental recorders and magnetic recorders.

In self-recording gauges, such as the HYETOGRAPH and the 'tilting-siphon rain-gauge', the collected rainfall is usually made to raise a float to which is attached a pen which records on a chart wound on a clock-driven drum. Alternatively, as for example with the tipping-bucket switch, the record shows the intervals of time which have elapsed between the falling of successive small amounts, generally 0·1 mm, of rain. See also EXPOSURE, TIPPING-BUCKET SWITCH.

rain shadow: An area with a relatively small average rainfall due to sheltering by a range of hills from the prevailing rain-bearing winds. The phenomenon is noticeable in rainfall maps for months in which unusually strong westerly winds have predominated, e.g. to the east of Wales.

rain spell: A period of at least 15 consecutive days to each of which is credited 0·2 mm (0·01 in) or more of RAINFALL. Commencing in 1905, the occurrences of rain spells at selected stations were tabulated in the annual volumes of *British Rainfall*.* From 1919 onwards details of 'wet spells' at the same stations were included, a 'wet spell' being a period of at least 15 consecutive days to each of which is credited 1 mm (0·04 in) or more of rainfall. The tabulation of both rain spells and wet spells ceased after 1960 when a new system of presenting rainfall deficiencies and excesses was used in *British Rainfall*.

rainy season: A period of a month or more, recurring every year, which is characterized in a given region (generally tropical or subtropical) by relatively large amounts of precipitation for the region. Thus, for example, the period of the south-west MONSOON is the rainy season (or 'wet season') in most parts of south-east Asia, while winter is the rainy season in regions with a MEDITERRANEAN-TYPE CLIMATE.

random forecast: A type of FORECAST, sometimes used as a basis of comparison in the assessment of the success attained in forecasts made by more conventional methods, which is based on chance selection of values of the meteorological elements concerned.

randomization: The arrangement of material in a statistical experiment with the general object of ensuring that the initial conditions do not favour one outcome rather than another; in such a case the actual PROBABILITY of occurrence of any combination is equal to the CHANCE EXPECTATION. This object is generally achieved by numbering the items in any convenient way, and then placing them in the order of items taken from a table of RANDOM NUMBERS.

randomness: A basic statistical concept implying an absence of plan or pattern, or of any tendency to favour one consequence rather than another. See, for example, PROBABILITY.

---

* London, Meteorological Office. *British Rainfall*. London.

random numbers: Sets of numbers, usually produced by mechanical or electronic means or by mathematical congruential methods, which have been tested and found to have the statistical property of RANDOMNESS. Complete and exhaustive testing for randomness is impossible, and the important thing in any application is that the numbers should behave as random numbers in the respects necessary for that application. Tables of ostensibly random numbers which have been tested extensively without giving evidence of non-random behaviour have been published by Fisher and Yates,* for example.

Random numbers are used in RANDOM SAMPLING and other processes of RANDOMIZATION, and in MONTE CARLO METHODS.

random sampling: The chance selection of one or more items from a much larger group or 'population'. The object of such sampling is generally to determine, within defined limits, the average characteristics of the entire population.

range: The difference between the extreme values in a group of measurements. As a measure of variability, range has the disadvantages of systematic dependence on size of sample and of large differences between different samples of the same size.

The range may be used to obtain a quick estimate of the STANDARD DEVIATION of a large sample. Thus, for example, an estimate is obtained by dividing the items into groups of 9, finding the range in each group, averaging the results and dividing by 3. The dividing factor changes with group size.

rare gases: An alternative for INERT GASES.

ravine wind: A wind which blows through a ravine or narrow valley penetrating a mountain barrier because of the existence of a pressure gradient directed from one side of the barrier to the other. Such winds may attain great strength because of FUNNELLING. An example is the ravine wind at Genoa, caused by a pressure difference between the Po valley and the Gulf of Genoa.

Rayleigh number: A non-dimensional parameter (Ra) which is critical in the static stability of fluids. It is defined by the equation

$$(Ra) = g \triangle T a h^3 / va$$

where g is the gravitational acceleration, $\triangle T$ the initial temperature difference between the bottom of the fluid and the fluid at height h, α the coefficient of expansion, v the kinematic VISCOSITY, and a the thermometric CONDUCTIVITY of the fluid. If the fluid state is such that (Ra) is less than a critical value, which depends to some extent on the nature of the fluid boundary conditions (free or rigid); any tendency for convection is damped out by viscosity and conductivity.

Rayleigh scattering: SCATTERING of ELECTROMAGNETIC RADIATION effected by spherical particles of radius less than about one-tenth the wavelength of the incident radiation. Two important cases arise in meteorology: (i) scattering of incident solar radiation by air molecules in a manner which explains the BLUE OF THE SKY; (ii) scattering of radar waves by raindrops in the atmosphere. Solid particles of appropriately limited radius also conform to this type of scattering.

According to Lord Rayleigh's theory of molecular scattering, the scattering coefficient (β) is given by

$$\beta = \frac{32\pi^3(n-1)^2}{3N\lambda^4}$$

---

* FISHER, R. A. and YATES, F.; Statistical terms for biological, agricultural and medical research, 6th edn. London, Oliver and Boyd, 1963.

where n is the refractive index of the gas, N its number of molecules per unit volume, and λ the wavelength. The wavelength dependence of n is so slight that to a close approximation molecular scattering varies inversely as the fourth power of λ. Thus, molecular scattering of white sunlight is such that the scattering of blue light is about five times greater than that of the red light contained in the incident beam.

Rayleigh's theory of molecular scattering includes also the angular distribution of intensity of scattered light (symmetrical about a plane normal to the incident beam with maxima in the 'forward' and 'backward' directions) and of its state of POLARIZATION.

The radar reflectivity of raindrops of a size which satisfies the condition for Rayleigh scattering is given by

$$\frac{\pi^5}{\lambda^4} \left(\frac{n^2 - 1}{n^2 + 2}\right)^2 \Sigma N d^6$$

where λ is the radar wavelength, n the refractive index of the particle, N the number of particles per unit volume and d their diameter.

RDF: Abbreviation for RADIO DIRECTION-FINDING.

Réaumur scale: A scale of temperature, now almost obsolete, introduced in 1731 by the French physicist Réaumur. On it the freezing-point of water is 0°, and the boiling-point 80°.

recombination: The various processes by which positive and negative IONS, or positive ions and ELECTRONS, recombine to form neutral particles. The rate of recombination is expressed by a 'recombination coefficient' with dimensions L^3T^{-1}.

recovery factor: A thermometer exposed in a rapidly moving airstream will usually record a temperature different from the free-stream stagnation temperature (T_s). If the indicated temperature is T_r and the true air temperature is T_a, then the recovery factor, r, is defined by the equation

$$r = \frac{T_r - T_a}{T_s - T_a}.$$

More detailed discussion is given in WMO *Technical Note* No. 39.*

recurrence tendency: A recurrence tendency in a TIME SERIES is a feature which, though not strictly periodic, implies a greater-than-random frequency of separation of relatively high (or low) values in the series by a specific 'recurrence interval' of time.

A well-known example in geophysics is the 27-day recurrence tendency in geomagnetic disturbance. This feature would not be revealed by PERIODOGRAM analysis of a geomagnetic disturbance time series, since it is associated with a solar cause which, after persisting for a few solar rotation periods of about 27 days, dies out before reappearing, after a variable time interval, very probably out of phase with the solar cause previously in operation. A SINGULARITY in meteorology is an example of a type of recurrence tendency which, though yet only quasi-periodic in nature, is not subject to such changes of phase.

recurvature (of tropical storm): See TROPICAL CYCLONE.

red flash: See GREEN FLASH.

---

* JONES, R. F.; Ice formation on aircraft. *Tech Notes Wld Met Org, Geneva*, No. 39, 1961.

reduction: In meteorology, the substitution of computed values for those directly observed, the purpose being to eliminate the effect of some particular factor or factors.

The reduction process is most commonly used to eliminate the effects of varying height on observed surface values of air temperature and pressure and is termed 'reduction to sea level'. Isobars drawn on surface synoptic charts, often also isotherms drawn on climatological charts, refer to values reduced in this way.

Mean temperature values are sometimes reduced to sea level by addition to observed values at the rate of 1 degC per 165 metres or 1 degF per 300 feet of station elevation. Observed pressures are reduced to sea level by application of the ALTIMETER equation; the normal assumption in the latter case is that the mean temperature of the 'missing' air column is the same as the screen temperature, and involves negligible errors except for stations at heights greater than about 300 metres above sea level.

reflection: The return to the original medium of the RADIATION incident on a boundary between two media; 'total reflection' is said to occur when all the incident radiation is returned. Reflection is said to be 'specular' (or 'regular') if the reflecting boundary has irregularities which are small relative to the wavelength of the radiation, 'diffuse' if they are large relative to the wavelength. The two laws of reflection—(i) incident ray, reflected ray and normal to the reflecting surface at the point of incidence lie in the same plane, and (ii) angle of incidence (that is, angle between incident ray and normal) equals angle of reflection (angle between reflected ray and normal)—apply to specular reflection. Radiation which is diffusely reflected, on the other hand, emerges in many different directions which are unrelated to the angle of incidence.

In meteorology, the reflecting power (ALBEDO) of surfaces is of fundamental importance in the heat balance that is achieved, either locally or in the earth–atmosphere system as a whole. The albedo has some important dependence on wavelength and on the angle of incidence of radiation.

Reflection of light waves plays an important (in some cases the sole) part in some atmospheric optical phenomena; the reflection of radio waves (radar) has a number of meteorological applications.

refraction: The change of direction to which energy waves (light, sound or radio waves) are subject on passing through a medium of varying density (gradual bending) or through a boundary separating media of different densities (sudden bending).

The two laws of refraction state that (i) the incident and refracted rays and the normal to the surface of separation of two media at the point of incidence lie in the same plane and (ii) (Snell's law) the ratio of the sine of the angle of incidence (angle between incident ray and normal) to the sine of the angle of refraction (angle between refracted ray and normal) is a constant for any two media. See also REFRACTIVE INDEX.

Among the many atmospheric optical phenomena in which refraction plays at least the major part are the delay of apparent SUNSET relative to the sunset time indicated by geometry, the apparent flattening of the sun or moon close to the horizon, the HALO, RAINBOW, and MIRAGE. The associated colouring is due to the fact that the amount of bending suffered by light waves is wavelength dependent.

ANOMALOUS AUDIBILITY and ANOMALOUS RADIO PROPAGATION are examples of the refraction of sound and radio waves, respectively.

refractive index: The refractive index (n) of a medium is a non-dimensional measure of the degree of the REFRACTION of energy waves passing through the medium. It is given by the ratio of the velocity (c) of an electromagnetic wave in a vacuum to its velocity (v) in the medium, that is

$$n = c/v.$$

Changes (continuous and discontinuous) of refractive index of the air with height are associated with lapse rates of temperature and humidity and cause various atmospheric optical effects and anomalous propagation of radar waves. Wavelength dependence of n causes dispersion of visible light and coloration of various of the optical phenomena.

The variation of n for dry air with air density (ρ) is represented by

$$(n^2 - 1)/(n^2 + 2)\,\rho = \text{constant}.$$

Since $n \approx 1$ the simple relation $(n - 1)/\rho = \text{constant}$ is normally assumed. The equation imples that n generally decreases with height and that rays which pass through the atmosphere, in a direction other than a normal, acquire curvature towards the denser part of the medium. For a ray travelling nearly horizontally in a horizontally stratified atmosphere the ray curvature is proportional to dn/dh (height gradient of refractive index).

For dry air n is for practical purposes given by the same formula for light and radio waves, i.e.

$$n - 1 = 79 \left(\frac{p}{T}\right) \times 10^{-6}$$

where p is in millibars and T in kelvins.

For moist air at optical wavelengths n is given by

$$(n - 1)\,(Tp_0/T_0 p) = 0{\cdot}0002918 - 0{\cdot}000035r/(1 + r)$$

where T and p are air temperature and total pressure, T_0 and p_0 are the 'standard' values 273 K and 1013 mb, respectively, and r is humidity mixing ratio.

At radio wavelengths, n for moist air is given by

$$n - 1 = \frac{79}{T}\left(p - \frac{e}{7} + \frac{4800e}{T}\right) \times 10^{-6}$$

where p and e are total pressure and vapour pressure (mb) and T is absolute temperature.

In the atmosphere ($n - 1$) is of the order 300×10^{-6}; for convenience this is normally expressed as 300 M units, signifying 300 millionths.

For the standard conditions $T = 273$ K and $p = 1013$ mb and for wavelength $= 0{\cdot}5893$ μm (sodium D line) the following are the values of n: dry air, $1{\cdot}0002918$; water vapour $1{\cdot}000257$. At 15°C and for the sodium D line relative to air, n for liquid water $= 1{\cdot}333$ and n for ice $= 1{\cdot}31$.

See also DIELECTRIC CONSTANT, MODIFIED REFRACTIVE INDEX.

refractometer: In meteorology, an instrument which employs a microwave radio technique for measuring the REFRACTIVE INDEX of the atmosphere. The instrument, carried in an aircraft, gives continuous recording of the resonant frequency of a cavity exposed to the ambient air, corresponding to changes of the quantity N, defined by $(n - 1) \times 10^6$ where n is the refractive index of the air.

regelation: Ice at a temperature near its MELTING-POINT may be melted by the application of excess pressure, owing to the reduction of the melting-point effected by such pressure. The re-solidification of the ice which accompanies the removal of the excess pressure (as in the making of snowballs) is known as 'regelation'.

Regional Association: The Regional Associations of the WORLD METEOROLOGICAL ORGANIZATION each comprise those members of the organization the networks of which lie in or extend into one of the six meteorological Regions of the world. The Regions comprise Africa, Asia, South America, North and Central America, south-west Pacific, and Europe. Each Association normally meets at intervals not exceeding four years.

regression equation: An approximate relation, generally linear, connecting two or more quantities, derived from the CORRELATION coefficient.

relative contour: An alternative for THICKNESS line.

relative humidity: The relative humidity U (per cent) of moist air is defined by

$$U = 100 \frac{e'}{e'_w}$$

where e' is the VAPOUR PRESSURE of the air and e'_w the saturation vapour pressure with respect to water at the same temperature. To a close approximation the corresponding ratios of the MIXING RATIO or of SPECIFIC HUMIDITY may be used. The actual relationship in terms of mixing ratio, for example, is

$$U = 100 \frac{r}{r_w} \frac{0 \cdot 62197 + r_w}{0 \cdot 62197 + r}$$

where r and r_w are mixing ratio and saturation mixing ratio, respectively.

Relative humidity may be measured indirectly from wet- and dry-bulb temperature readings, with the aid of humidity tables, or directly, as with a hair hygrometer. At temperatures below 0°C, relative humidity is evaluated with respect to super-cooled water and not with respect to ice.

Relative humidity has a marked systematic diurnal variation opposite in phase to that of temperature, that is it has a daily maximum around dawn and a minimum in the afternoon. It has a less well-marked annual variation, more especially in afternoon hours, also of opposite phase to that of temperature.

relative isohypse: An alternative for THICKNESS line.

relative vorticity: See VORTICITY.

relaxation of a trough: A decrease in the amplitude of a trough. When the term is used in a westerly synoptic situation in middle latitudes it also implies a poleward movement of the isopleths which define the trough (e.g. contours or thickness lines). It is the opposite of MERIDIONAL EXTENSION of a trough.

report, meteorological: See WEATHER REPORT.

representativeness: A representative air-mass property is one that is typical of the air mass as a whole and so may be useful in AIR-MASS ANALYSIS. Some of the surface meteorological elements, in contrast to the same elements measured in the upper atmosphere, are readily changed by purely local influences and so are not representative; surface temperature is an example of such an element.

Réseau Mondial: An annual publication, now discontinued, of the Meteorological Office. It contained climatological data for all parts of the world on the basis of two stations per 10-degree square of latitude and longitude. The last volume published (in 1957) was that for 1934.

reshabar or **rrashaba**: A name, meaning 'black wind', given to a strong, very gusty, north-easterly wind which blows down certain mountain ranges in southern Kurdistan. It is dry, comparatively hot in summer and cold in winter.

residence half-time: See RADIOACTIVE FALLOUT.

residual: The difference between an individual observation and the mean of a series, or the difference between an individual observation and the value derived from the adopted values of the constants which have been obtained by a discussion of the observations.

Thus an observed quantity may be known to be a function of variables x, y, z, and constants a, b, c, of the form of $ax + by + cz = l$. If a number n of observed values of l are given for known values of x, y and z, there will be n equations to determine the three constants a, b and c. The equation will not in general be accurately satisfied for any one observation, and the value of $l - (ax + by + cz)$ is the residual.

resonance: If a periodic force is applied to a system, resonance is said to occur when the period of the force comes into close accord with a FREE PERIOD of the system, resulting in an increase in the amplitude of vibration of the system. Resonance was previously advanced as an explanation of the large mgnitude of the semi-diurnal atmospheric pressure wave but is not now favoured. See ATMOSPHERIC TIDES.

resultant: The sum of a number of directed quantities or vectors. See VECTOR.

retrograde system: In synoptic meteorology, a pressure system which reverses its direction of movement.

An atmospheric wave is said to be retrograde if it moves in a direction opposite to that of the flow in which it lies.

return period: The number of years within which a particular extreme value of an element, for example gust speed, is likely to be exceeded only once. The best available estimate is usually based on past records.

return stroke: See LIGHTNING.

returning polar maritime air: See POLAR AIR.

reversal (spring, autumn): The stratospheric circulation of middle and high latitudes is dominated in winter by strong westerly winds while the summer circulation is characterized by weaker easterlies. The summer régime is of considerably shorter duration than the winter régime. The changes of circulation type which usually occur in late April and early September in the northern hemisphere are called the spring and autumn reversals.

reversing layer: The lower part of the atmosphere of the SUN, comprising a layer of relatively cool gas extending about 1000 km outwards from the PHOTOSPHERE.

revolving storm: A term synonymous with an intense TROPICAL CYCLONE.

Reynolds number: An important parameter in the flow pattern of fluids, designated (Re) and defined by the non-dimensional quantity $\bar{u}d/v$, where $\bar{u}$ is a characteristic fluid velocity, d a characteristic length, and v the kinematic viscosity of the fluid.

Reynolds (1883) showed experimentally that turbulent, as opposed to laminar, flow is not sustained for a value of (Re) less than about 2000. A critical value of (Re)

similarly exists for the onset of TURBULENCE as the speed of fluid flow past a smooth body increases.

Reynolds stresses: Fundamental stresses (τ), also termed 'eddy shearing stresses', which operate within a turbulent fluid to transport momentum. τ has the dimensions $ML^{-1}T^{-2}$.

If u', v', w' are turbulent components of velocity (instantaneous departures from average) in rectangular co-ordinate directions x, y, z, respectively, and ρ is the fluid density at a point, then the additional shearing stresses in the x, y plane are $-\overline{\rho u'w'}$, $-\overline{\rho v'w'}$ in the x and y directions, respectively; corresponding shearing stresses act in the other two planes. (Bars represent time averages.)

Non-zero values of τ depend on the existence of correlation between the component eddy velocities in the above products as, for example, a general association of gusts (u' positive) with descending air (w' negative). τ is generally assumed constant within about the lowest 25 metres of the atmosphere.

RHI: Abbreviation for range-height indicator. See RADAR METEOROLOGY.

ribbon lightning: A name applied to a markedly tortuous LIGHTNING discharge from cloud to ground.

Richardson number: An important parameter in atmospheric TURBULENCE, designated (Ri) and defined as the non-dimensional quantity

$$(Ri) = \frac{g}{T} \frac{(\partial T/\partial z) + \Gamma}{(\partial u/\partial z)^2} = \frac{g}{\theta} \frac{\partial \theta/\partial z}{(u\partial/\partial z)^2}$$

where g is the gravitational acceleration, Γ the dry adiabatic lapse rate, and $\partial T/\partial z$, $\partial \theta/\partial z$, $\partial u/\partial z$ the vertical height gradients of air temperature, potential temperature, wind velocity, respectively.

An equivalent 'flux form' of the parameter is

$$(Ri) = -\frac{gH}{c_p T \, \tau(u\partial/\partial z)}$$

where H is the flux of sensible heat, c_p is the specific heat of air at constant pressure, and τ is the REYNOLDS STRESS.

Richardson (1925)* investigated the rate of consumption of energy which is implied in vertical motion of fluid elements in relation to the rate of energy production associated with vertical wind shear. On certain assumptions he derived the criterion that turbulence will persist in a fluid if (Ri) < 1 and will subside if (Ri) > 1. While subsequent measurements have not yet verified as critical the value (Ri) = 1, or any other single value, (Ri) is employed as a fundamental stability parameter in the study of atmospheric turbulence and dynamical meteorology.

ridge: A ridge (of high pressure), also termed a 'wedge', is an extension of an ANTICYCLONE or high-pressure area shown on a weather chart, corresponding with a ridge running out from the side of a mountain (see Figure 27). It is the converse of a TROUGH of low pressure and is generally associated with fine anticyclonic-type weather.

Maximum curvature of isobars occurs along the 'axis' of a ridge. One in which such maximum curvature is relatively small is termed a 'flat' ridge and tends to be a faster-moving isobaric feature than one in which the curvature is great.

---

* RICHARDSON, L. F.; Turbulence and vertical temperature difference near trees. *Phil Mag*, London, **49**, 1925, p. 81.

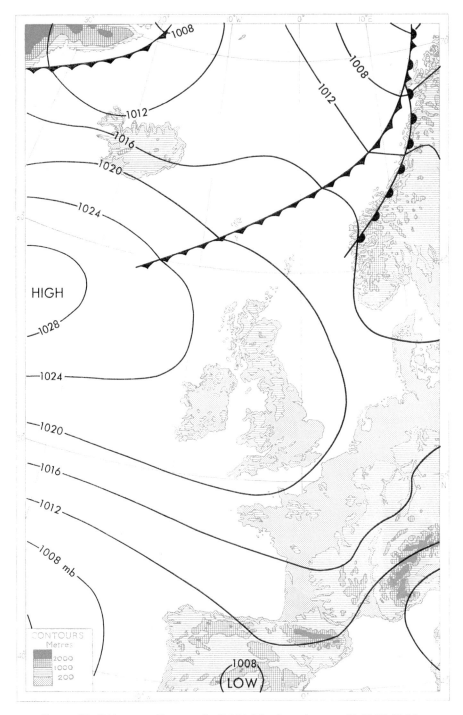

FIGURE 27—Ridge extending over the British Isles from the west, 06 GMT, **22 May 1961.**

rime: Deposit of white, rough ice crystals which forms when supercooled water droplets of fog come into contact with a solid object at a temperature below 0°C. The deposit grows out on the windward side of the object. The phenomenon seldom occurs at low levels in the British Isles because supercooled fogs are uncommon at these levels. It occurs, however, much more frequently on mountain tops. It is often popularly confused with HOAR-FROST.

Ringelmann shades: A scale of shades, varying in degree of blackness, which is used by an observer to form a subjective comparison with the blackness of SMOKE emitted by a chimney and so afford an estimate of the concentration of solid material which is being emitted.

A numerical measure of the average smoke content of air over a period of time is obtained by measuring photoelectrically the reflectance of a stain made on white filter-paper by the passage of a measured quantity of air and comparing with the measured reflectance of the various Ringelmann shades.

roaring forties: A nautical expression used to denote the prevailing westerly winds of temperate latitudes (below 40°S) in the oceans of the southern hemisphere.

rocket lightning: A very rare and unexplained form of LIGHTNING in which the speed of propagation of the lightning stroke is slow enough to be perceptible to the eye.

rocketsonde: The instrumented payload which is ejected from a METEOROLOGICAL ROCKET. It is equipped to measure one or more meteorological variables. Temperature sensors are most frequently carried but pressure and density gauges and ozone sensors are also flown.

The sonde descends attached to a parachute which is tracked by radar for wind measurement. A typical temperature sensor is a very fine tungsten wire, a change of resistance of which causes a change of frequency of the note generated by an audio-oscillator. The carrier frequency of the sonde's transmitter is frequency-modulated with this note. The signal transmitted is received at a ground station equipped with the standard RADIOSONDE reception equipment. Thus wind and temperature measurements from about 65 km to 15 km are obtained.

rocket sounding: Exploration of the earth's atmosphere up to heights of several hundred kilometres by means of instruments carried by a rocket. Among the geophysical measurements obtained from such soundings are those of atmospheric pressure, air temperature, density and composition, winds, solar radiation, electrical properties, and the earth's magnetic field. The data are obtained mainly by radio telemetering, or by recovery of photographic record after descent by parachute.

rockoon sounding: Exploration of the earth's atmosphere by a small instrumented rocket which is carried aloft by a large balloon to a pre-determined height where it floats till fired by means of a radio relay operated from the ground at the moment when some particular event, e.g. a SOLAR FLARE or AURORA, is seen to occur.

roll cumulus: An obsolete term for a form of STRATOCUMULUS in which long, parallel rolls of cloud alternate with clear spaces.

röntgen: A unit of intensity of GAMMA RADIATION (or X-RAYS), defined as that quantity of gamma radiation which will form $1 \cdot 61 \times 10^{12}$ ION pairs when absorbed in 1 g of air. The biological hazard of gamma radiation is closely related to this unit of dosage since harmful chemical effects follow the ionization of molecules present in animal cells.

Rossby diagram: An AEROLOGICAL DIAGRAM, devised by C. G. Rossby for air-mass identification, in which the abscissa is MIXING RATIO (r) and the ordinate is partial POTENTIAL TEMPERATURE (θ_d).

Rossby number: A dimensionless parameter (Ro) found in studies of the DISHPAN EXPERIMENT type to be important in the form of relative fluid motion (U) generated in a rotating pan of radius r and of angular velocity Ω; (Ro) is defined by the equation
$$(Ro) = U/r\Omega.$$
The analogous ratio for large-scale relative air motion on the rotating earth is found to be about 0·1 to 0·2.

Rossby parameter: The northward variation of the CORIOLIS PARAMETER, arising from the spherical shape of the earth. It is given by
$$\beta = \frac{d}{dy}(2\Omega \sin \varphi) = \frac{2\Omega \cos \varphi}{a}$$
where a is the earth's radius. It is sometimes assumed to have the constant value appropriate to $\varphi = 45°$, i.e. $\beta = 1·619 \times 10^{-11}$ m^{-1} s^{-1}.

Rossby wave: An alternative for LONG WAVE.

rot: An alternative for CURL. See VORTICITY.

rotor: A large, closed EDDY with a horizontal axis which is produced in the lee of a range of mountains or hills crossed by a stable airstream. Such closed eddies are sometimes formed under LEE WAVES of large amplitude, the surface wind under the wave crest being reversed. Horizontal dimensions are 3 to 10 km and vertical dimension 1 to 3 km. A very turbulent 'rotor cloud' forms in the upper part of the closed eddy when the air is sufficiently moist, with an apparent motion round its centre; the axis of the eddy is usually below the base of such a cloud.
  Examples of rotors occur in connection with the HELM WIND ('helm bar' cloud) and in the lee of the Sierra Nevada in California.

roughness length: A quantity (z_0), also called the 'roughness coefficient' or 'roughness parameter', which enters as a constant of integration into the form of the LOGARITHMIC VELOCITY PROFILE appropriate to 'fully rough' flow near a surface. z_0 is proportional to the average height of the roughness elements of the surface; wind-tunnel measurements of flow over grains of sand suggest the value $z_0 = \varepsilon/30$, where ε is the average height of the obstacles.
  The equation which defines z_0 is
$$\frac{u}{u_*} = \frac{1}{k} \log_e \frac{z}{z_0}$$
where u is the mean velocity at distance z from the boundary, u_* is the FRICTION VELOCITY, and k is Kármán's constant (about 0·4). See also AERODYNAMIC ROUGHNESS.

running means: An alternative for MOVING AVERAGES.

run-off: That portion of the rainfall over a DRAINAGE AREA which is discharged from the area in the form of a stream or streams.

run-of-wind anemometer: See ANEMOMETER.

runway visual range: The maximum distance in the direction of take-off or landing at which the runway or the specified lights or markers delineating it can be seen from a position above a specified point on its centre line at a height corresponding to the average eye-level of pilots at touchdown.

A height of approximately 5 metres is regarded as corresponding to the average eye-level of pilots at touchdown.

In practice, runway visual range cannot be measured directly from the position specified in the definition but is an assessment of what a pilot would see from that position.

S

St Elmo's fire: 'A more or less continuous, luminous electrical discharge of weak or moderate intensity in the atmosphere, emanating from elevated objects at the earth's surface (lightning conductors, wind vanes, masts of ships) or from aircraft in flight (wing tips, propellors, etc.).'*

This discharge can also occur on an aircraft where the static charge has been induced by the impact of ice or snow crystals, rain, dust, or sand. The build-up of static electricity on an aircraft in flight, whether by friction or by induction, is loosely termed precipitation static.

The phenomenon is usually bluish or greenish in colour, sometimes white or violet. It is accompanied by a crackling sound and occurs when the electrical field in the neighbourhood of the object becomes very strong, as when a CUMULONIMBUS cloud is overhead. The phenomenon is also termed 'corposant' (holy body).

St Luke's summer: A period of fine weather which is popularly supposed to occur about the time of St Luke's day, 18 October.

St Martin's summer: A period of fine weather which is popularly supposed to occur about the time of St Martin's day, 11 November.

St Swithin's day: A well-known example of British weather lore is to the effect that if rain falls on St Swithin's day (15 July) then, in the same locality, each of the next 40 days will also have some rain. Rainfall records lend no support to this tradition.

salinity: A measure of the salts dissolved in a given solution. The salinity of a natural water surface, such as sea water, is usually expressed in parts per thousand by weight. Thus a salinity of 35 per mille (written $35°/_{oo}$) indicates that there are 35 g of salts in 1000 g of sea water. Since the total dissolved solids are difficult to determine directly with accuracy, salinity is derived in practice by applying factors to the specific gravity or to the halide content, which can be exactly measured (dissolved solids are present in constant ratios).

The salinity value of 35 per mille is a rough average for surface ocean water. Salinity varies systematically by a few per cent with latitude and is subject also to small casual and systematic time variations.

Depression of the freezing-point temperature and of the temperature of maximum density increase with increase of salinity: for a salinity of 35 per mille the freezing-point temperature is $-1.9°C$ and the temperature of maximum density is $-3.5°C$ (compared with 0°C and 4°C, respectively, for pure water).

sand pillar: A rarely used alternative for DUST WHIRL.

sandstorm, duststorm: 'An ensemble of particles of dust or sand energetically lifted to great heights by a strong and turbulent wind.'†

---

* Geneva, World Meteorological Organization. International cloud atlas, Vol. 1. Geneva, WMO, 1956, p. 76.

† Geneva, World Meteorological Organization. International cloud atlas, Vol. 1. Geneva, WMO, 1956, p. 71.

Surface visibility is reduced to low limits; the qualification for a synoptic report is visibility below 1000 m.

sastruga (also **zastruga**): (From the Russian.) An irregularity or wave formation caused by persistent winds on a snow surface. The size varies according to the force and duration of the wind, and the state of the snow surface in which it is formed.

satellite sounding: Satellites launched from the earth have, since October 1957, yielded much important geophysical information both directly (by means of automatic instruments) and indirectly (by inferences drawn from the precise path of the satellite).

A satellite of mass m, in a circular orbit at distance R from the earth's centre, has a critical velocity (v) such that the CENTRIFUGAL FORCE is exactly balanced by the force of GRAVITY acting on it. Where r is the earth's radius and g the gravity at the earth's surface, the balance equation is $mv^2/R = mg(r/R)^2$, whence $v^2 = gr^2/R$. For $(R - r) = 480$ km, v is about 27 000 km/h and the corresponding period of revolution ($2\pi R/v$) is about 103 minutes. Since the achievement of a circular orbit requires great precision of the final velocity imparted to the satellite, the orbit is, in practice, more or less elliptical; from the geophysical viewpoint this has the advantage that a range of heights is covered by the satellite. The frictional drag exerted on the satellite by the earth's very high atmosphere causes its orbital radius gradually to decrease till it is eventually burnt up in the denser lower atmosphere.

Among the direct meteorological measurements made by satellites are those of solar and terrestrial radiation, earth albedo, and cloud distribution. Air density at great heights and the shape and size of the earth have been calculated from accurate tracking of the orbit. Other geophysical measurements have included those of the earth's magnetic field, cosmic rays, meteor impact, and properties of the ionosphere.

saturated soil: A saturated soil, or waterlogged soil, is one in which all the soil pores, including those which in a healthy soil contain air, are filled with water.

saturation: A moist air sample is said to be saturated, with respect to water or to ice, if its composition is such that it can coexist in neutral equilibrium with a plane surface of pure condensed phase, water or ice, at the same temperature and pressure as the sample.

To a high degree of approximation the capacity of air to hold water in the form of vapour depends only on temperature; at temperatures below 0°C, however, an important factor in the equilibrium conditions is the phase (liquid or solid) of the water substance (see VAPOUR PRESSURE). A sample of air is said to be 'supersaturated' if it contains more than enough water vapour to saturate it at its existing temperature. Owing to the presence of condensation nuclei, an appreciable degree of supersaturation with respect to water is rarely observed in the atmosphere.

saturation (or saturated) adiabatic (or adiabat): Line on an AEROLOGICAL DIAGRAM representing the saturation adiabatic lapse rate. See ADIABATIC.

saturation deficit: The difference between the actual VAPOUR PRESSURE of a moist air sample at a given temperature and the saturation vapour pressure corresponding to that temperature.

savanna: The term applied to a type of tropical CLIMATE, with a wet and a dry season, in which the most common form of vegetation is the tall tropical grass 'savanna'.

SCA: An abbreviation for 'sudden cosmic (noise) anomaly'. When a SOLAR FLARE occurs, ionospheric absorption of the cosmic radio waves entering the earth's atmosphere suddenly increases in the sunlit hemisphere and there is an associated decrease of the radio noise level (SCA) recorded at the earth's surface. See also SID.

scalar: A scalar quantity is one that is completely specified by its magnitude, expressed in a given system of units (as opposed to a directed or VECTOR quantity such as wind velocity). Meteorological examples of a scalar quantity are pressure, temperature, and divergence of wind velocity.

scale height: The equivalent height (H) of a hypothetical atmosphere in which air density is constant (see HOMOGENEOUS ATMOSPHERE). It is defined by the equation

$$\frac{dp}{p} = -\frac{1}{H} dh$$

where $H = kT/mg$ (k = Boltzmann's constant = 1.3805×10^{-23} J/degK, m is the mean molecular mass, T the absolute temperature at base of atmosphere and g the gravitational acceleration).

The height variations of atmospheric scale height (termed then 'local scale height') provide a measure of the proportional variations of pressure with height, and mainly reflect corresponding height variations of temperature and composition (to a minor extent of gravity). Local scale height is a concept mainly used in studies of the high atmosphere and is a quantity which may be inferred from the vertical sounding of the ionosphere.

scatter diagram: A graphical point plot of corresponding pairs of associated values of two variables (ordinate and abscissa). A diagram on which, for example, the points are closely grouped round a line inclined to both axes indicates a near-linear relationship between the two variables. A diagram on which the points are randomly distributed indicates no relationship between the variables.

scattering: The process by which some of the electromagnetic RADIATION incident on particles, of molecular size upwards, which are suspended in a medium, is dispersed in all directions. The scattering process is one which gives rise to a diminution of the intensity of an incident beam of radiation; the measure of this effect—the 'scattering coefficient' (β)—is defined by the equation (analogous to the case of ABSORPTION)

$$I = I_0 e^{-\beta x}$$

where I_0 is the intensity of incident radiation and I the intensity after a path of length x through the scattering (non-absorbing) medium. The 'scattering cross-section' of a scattering particle is the area normal to a beam of radiation which would intercept the same amount of radiation as that actually scattered by the particle.

In meteorology, radiation which has been subject to scattering is generally termed DIFFUSE RADIATION. Such radiation may have been scattered once or more than once—so-called single or multiple scattering, respectively; the terms primary, secondary, tertiary scattering are also used, as appropriate, while in some usages secondary scattering signifies multiple scattering.

Scattering is a complex phenomenon which depends mainly on the ratio of the size of scattering particle to the wavelength of the incident radiation but depends also on the refractive index, shape and composition of the scattering particle. Atmospheric scattering is usually classified as either molecular scattering or that effected by haze particles or water droplets. See RAYLEIGH SCATTERING, MIE SCATTERING, POLARIZATION.

scintillation (of stars): Rapid variations of apparent brightness ('twinkling') of stars, much more marked in stars near the horizon than in those near the zenith. Variations of colour may also occur at altitudes less than about 50°. The phenomenon is caused by small variations of REFRACTIVE INDEX of air associated with atmospheric inhomogeneities, mainly in the low atmosphere. A similar effect is visible at times in terrestrial objects, for example the shimmering of objects near the earth's surface on a hot day.

scirocco: A warm, southerly wind in the Mediterranean region. Near the north coast of Africa the wind is hot and dry and often carries much dust. After crossing the Mediterranean, the scirocco reaches the European coast as a moist wind and is often associated with low stratus.

Scotch mist: A combination of thick mist and drizzle, so called because it is most commonly experienced in the hillier districts of much of Scotland; it also effects at times low-lying districts of the west and north. In its most typical form it is associated with a moist stream of maritime tropical air.

In the uplands of the Devon–Cornwall peninsula the same phenomenon, which is there very frequent, is known as 'mizzle'.

scud: A mainly nautical term for ragged fragments of low cloud, often moving rapidly in a strong wind below rain clouds. The meteorological term is stratus fractus.

SEA: An abbreviation for 'sudden enhancement of atmospherics'. When a SOLAR FLARE occurs, the extra ionization of the D-layer in the sunlit hemisphere makes it a more efficient reflector of the radio waves emitted at times of lightning flashes; there is, therefore, an associated sudden increase in the recorded level of distant atmospherics. See also SID.

sea-breeze: See LAND- AND SEA-BREEZES.

sea disturbance: The degree of sea disturbance is reported in a 'state of sea' code in which the scale number increases from 0 to 9 according to the average wave height. The specifications are: 0, glassy; 1, rippled; 2, smooth; 3, slight; 4, moderate; 5, rough; 6, very rough; 7, high; 8, very high; 9, phenomenal. Scale number 5, for example, corresponds to waves of average height 2·5 to 4 metres, and scale number 9 to an average height of over 14 metres.

sea fret: A local name in parts of north-east England for a sea fog in coastal districts. This is especially a spring and summer feature. See ADVECTION FOG, HAAR.

sea level: Owing to waves, swell, tides and varying atmospheric static pressure, the actual LEVEL of the sea is constantly changing. A 'mean sea level' at any place may be determined, such that short-period fluctuations of level are eliminated, by averaging coastal observations of tide level over a period of years. The length of period required to obtain a suitable mean value varies considerably from place to place because of local variation of the amplitude of fluctuation about the mean position.

The present datum mean sea level— often referred to simply as 'sea level'—used in Great Britain, with reference to which contour levels on Ordnance Survey maps of Britain are shown, is based on observations at Newlyn in Cornwall, on the edge of the Atlantic. This datum is 0·13 feet (40 mm) below the Liverpool datum previously used. The permanent land survey datum is not the mean-sea-level datum itself but is referred to permanent bench marks in the neighbourhood of the tidal gauge. Land survey datums of other countries do not all depend on mean sea level. That used in Ireland, for example, refers to a particular low-water datum in Dublin bay and is estimated to be about 8 feet (2·4 m) lower than that of Newlyn.

World sea level, as given by the average of observations in many places, varies in response to (i) changes of average temperature of the oceans in depth, with accompanying expansion or contraction, and (ii) melting or accretion of ice-caps and glaciers. There is evidence that a world-wide ('eustatic') rise of almost 100 metres occurred between 15 000 and 5000 years ago because of the melting of the ice sheets of the last glaciation, and that subsequent changes have been relatively slight, the present level being lower than that 5000 years ago by a few metres. On a shorter time-scale, it is considered that a eustatic fall between about 1680 and 1850 was followed (up to at least 1930) by a more rapid rise to about the 1680 level; in the early years of the present century the rise was at the rate of about 10 mm/decade.

Mean world changes of sea level may be locally masked because of local ('isostatic') changes in response to movements of the land and in the ocean floor and by modification of coastline. Thus, for example, sea level in the Thames Estuary is apparently some 3 metres higher than in Roman times. This is a local effect produced mainly by down-warping of the land—possibly a form of compensatory movement as Scotland and Scandinavia continue to rise by isostatic recovery from the depression caused by the former load of ice.

sea smoke: An alternative for ARCTIC SEA SMOKE.

seasons: In meteorology, the manner of the division of the year into seasons for climatological purposes varies with latitude. In middle latitudes the normal division corresponds to that of the 'farmer's year'; in the northern hemisphere the divisions made are: autumn—September, October, November; winter—December, January, February; spring—March, April, May; summer—June, July, August.

In the tropics, the terms 'winter' and 'summer' lose their higher-latitude significance and a division into seasons is usually made in terms of rainfall amount or, in places, the associated wind direction—thus, 'dry season' and 'rainy season' or 'north-east monsoon' and 'south-west monsoon' in India. In the continental subtropical regions the natural seasons are usually defined in terms of temperature (cold and hot), or rainfall (dry and rainy), or both.

In a country of temperate climate such as the British Isles, the seasonal temperature changes progress much more gradually than in continental regions of the same latitude. In polar regions, the transition from winter to summer and vice versa is so sudden that spring and autumn largely disappear.

sea temperature: The normal methods of measuring sea temperature are: (i) to draw water in a specially designed bucket from the ship's side, forward of all ejection pipes, and to read the temperature of the sample with a specially designed thermometer ('bucket method'); or (ii) to read the temperature of the engine-room intake water ('condenser intake method'). The water temperatures so measured are, respectively, a mean value in the surface layer of depth about one foot (1 ft $\approx$ 0·3 m), and a value at a depth of several feet.

A third method recently introduced is to measure the temperature of the ship's hull, in a forward unheated compartment, with a thermometer which is thermally insulated from the air in the compartment. Each of these measurements approximates to a 'sea surface temperature' except when the sea is calm.

Radiation thermometry and infra-red photography, employed from low-flying aircraft or from satellites, are able to give a clear indication of the boundaries of water masses of different temperatures.

The mean annual sea surface temperature exceeds 27°C (80°F) over a broad

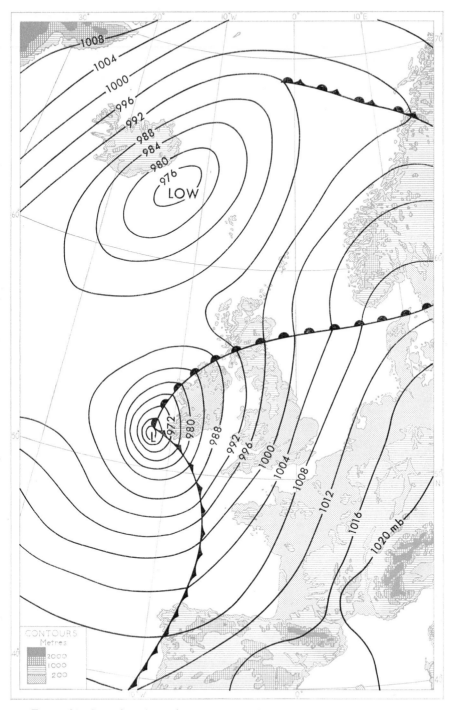

FIGURE 28—Secondary depression centred off Irish coast, 06 GMT, 16 September 1961.

belt of the equatorial region, and is somewhat less than $-1\,°C$ (30°F) in the polar regions. The run of the isotherms varies in the two hemispheres and in the different oceans. The seasonal range of temperature is about 6 degC (10 degF) in both polar and equatorial regions, and is greater in middle latitudes, where for the most part it lies between 6 degC and 16 degC (10 degF and 30 degF). The greatest range, some 30 degC (50 degF or more), is found in small areas, extending to the coast of the western North Atlantic and western North Pacific Oceans. The diurnal variation of sea surface temperature is very small, 0·5 degC (1 deg)F or less. See also N.S.R.T.

secondary cold front: The development of a TROUGH or troughs of low pressure within the cold AIR MASS lying in the rear of a deep depression is relatively common. On those occasions on which a trough appears to mark the line of advance of colder air (owing to rather different recent histories of the air masses on either side of the trough) the trough line is termed a 'secondary cold front'.

secondary depression: A secondary depression is one which, forming within the region of circulation of another depression, is, at least on initial formation, of higher central pressure than the other ('primary') depression. The formation of the closed circulation which defines the 'secondary' is preceded by a widening of the isobars in the region concerned.

Most secondaries of middle latitudes form at fronts—see COLD-FRONT WAVE, WARM-FRONT WAVE, COLD-OCCLUSION DEPRESSION, and WARM-OCCLUSION DEPRESSION —but non-frontal secondaries also form, more especially in an unstable AIR MASS, as, for example, the POLAR-AIR DEPRESSION. In general, a secondary deepens at the expense of the primary depression; on some occasions, mainly with the cold-front wave, the secondary deepens to such an extent as to absorb the original primary depression. (See Figure 28.)

secular trend: In statistics, a persistent tendency for a variate to increase or decrease with the passage of time, apart from irregular variations of shorter period. A secular trend is generally revealed more clearly if the data are smoothed.

seepage gauge: See PERCOLATION.

seiche: A tidal oscillation of the waters of inland lakes, very variable in period and amplitude. Among probable causes of the phenomenon are winds, earth tremors, and atmospheric oscillations of the type revealed by a microbarograph. Temperature seiches—abrupt changes of temperature below a lake surface, with associated wave motion in the layer of transition—have also been observed.

seismogram: The record of a SEISMOGRAPH.

seismograph: An instrument which records ground tremors, in particular, EARTHQUAKES. It consists essentially of a damped pendulum, the axis of which is rigidly fixed to the ground and which oscillates in response to a specific component of ground movement. At a seismological observatory three instruments are used to measure the independent components in the north–south, east–west, and vertical directions, respectively. Recording is made on a very open time-scale.

seismology: The science concerned with the measurement and analysis of earth tremors and, indirectly, with the nature of the earth's interior revealed by such analysis.

Observational data on earth tremors are provided by the SEISMOGRAPH recordings operated at some 600 observatories distributed over the world. Such data are found to accord with the mathematical theory which relates to the propagation of disturbances in a deformable, perfectly elastic medium. This theory postulates that the shock waves or 'seismic waves', which transport energy from the focal regions of an earthquake, are of two main types: (i) waves which travel over the earth's surface, and (ii) 'body' waves which travel downwards and which subdivde into the primary or 'push' (P) longitudinal waves, and the secondary or 'shake' (S) transverse waves. P waves move faster than S waves by some 50 per cent, while S waves move faster than surface waves. Thus, when the epicentre of an earthquake is at a distance of one quarter of the earth's circumference, the times of passage of the P, S and surface waves are, respectively, about 13 min 16 s, 24 min 14 s, and 43 min 30 s. The P and S waves usually constitute 'preliminary tremors' to the surface waves which are of much larger amplitude; in the more unusual case of a deep-focus earthquake, the surface waves are relatively small.

The P and S waves are subject to both gradual and rapid changes of direction and speed (refraction and reflection), corresponding to associated changes in the physical properties of the interior of the earth. A complex situation results in which waves frequently take a large number of paths from a single earthquake to a given observatory, with associated time differences of arrival. The study of such data from many observatories over a large number of earthquakes, together with the use of the theoretical inference that S waves do not pass through a fluid, has led to a coherent picture of the physical nature and properties of the interior of the EARTH. Analysis of artificially produced earth tremors, as by large explosions, has helped in these studies.

A direct link with both oceanography and meteorology is provided by the study of the quasi-regular small oscillations ('microseisms') which appear on seismograms. See also EARTHQUAKE, MICROSEISMS.

seisms: Oscillations of the crust of the earth. See EARTHQUAKES, MICROSEISMS.

seistan wind: A strong northerly wind which blows in summer in the province of Seistan, in eastern Iran. It continues for about four months and is, therefore, known as the 'wind of 120 days'. It sometimes reaches hurricane force.

sensible heat: See HEAT, ENTHALPY.

serein: Fine rain falling from an apparently clear sky. In this rare phenomenon, the cloud droplets are presumably evaporated when the larger precipitation drops are formed.

Much more commonly, in conditions of strong vertical wind shear between cloud and ground, precipitation arrives when the sky overhead is clear and the shower cloud is visible at a lower angle of elevation.

serial correlation: An alternative for AUTOCORRELATION.

sferics fix: The estimated location (foyer) of a LIGHTNING flash, deduced by combining the observations of its direction (azimuth), made at several stations. The Meteorological Office network uses four stations in the British Isles and three in the Mediterranean. Two directional aerials are used at each station and the signal received on arrival of the electro-magnetic wave radiation from the flash is displayed on a cathode-ray tube. A frequency of about 9 kHz (33 000-m wavelength) is used. The range of the system is about 2000 km. See also ATMOSPHERICS.

shade temperature: The temperature of the air indicated by a thermometer sheltered from precipitation, from the direct rays of the sun and from heat radiation from the ground and neighbouring objects, and around which air circulates freely. A standard shelter such as the THERMOMETER SCREEN is intended to satisfy these conditions.

shadow of the earth: A steely-blue segment darker than the rest of the sky rises from the eastern horizon just after sunset, encroaching on and soon obliterating the COUNTERGLOW. This is the shadow thrown by the solid earth on the atmosphere; all light received by an observer from that part of the atmosphere within the earth's shadow has been scattered more than once. The edge of the shadow weakens as TWILIGHT progresses and, except in a very clean atmosphere, is indistinguishable well before its passage through the zenith. A similar shadow, descending to the western horizon, occurs just before sunrise.

shamal: A hot, dry north-westerly wind which blows with special persistence in summer over Iraq and the Persian Gulf. It is often strong during the day-time but decreases at night.

Shaw week: A unit of time used by the Meteorological Office since January 1935 for agrometeorological purposes.
The unit is based on a division of the year into four quarters each centred on a solstice or equinox. The 'Shaw year' begins on 6 November; the seven-day periods therefore start on a different day of the week in successive years. Account is taken of the one or two days additional to 52 weeks in the normal or leap years, respectively.

shear: See WIND SHEAR.

shear-gravity wave: A wave disturbance which forms at the boundary between two atmospheric layers of different densities and moving with different speeds. Theory based on an incompressible atmosphere, specifies a critical wavelength (of about 10 km for typical atmospheric values of density and wind discontinuities) below which such waves are unstable and above which they are stable. Wind shear is therefore inadequate by itself to account for the development of frontal wave depressions.

shear hodograph: See HODOGRAPH.

shearing instability: That type of dynamical instability which arises at the boundary between two atmospheric layers moving with different speeds or in a layer containing WIND SHEAR.

shearing stress: An alternative for REYNOLDS STRESS.

shear wave: An unstable type of wave which forms at the boundary between two atmospheric layers moving with different speeds.

sheet clouds: See LAYER CLOUDS.

sheet lightning: The popular name applied to a 'cloud discharge' form of LIGHTNING in which the emitted light appears diffuse and there is an apparent absence of a main channel because of the obscuring effect of the cloud.

shelter-belt: A term which is sometimes used synonymously with WIND-BREAK, but which is more usually now employed in those cases where protection against wind is provided by a belt of trees. Where the protection is afforded by shrubs or hedge, such a term as 'shelter-hedge' is often employed.

shimmer: The apparent distortion of terrestrial objects due to atmospheric inhomogeneities at low levels. It is also referred to as 'atmospheric boil'.

shock wave: A thin layer of a medium (in particular, the atmosphere) in which the temperature, pressure, density, and velocity suddenly jump to new values. Such an effect is produced, for example, by the sudden outward movement of air particles from the site of an explosion, or by the passage of an object through air at a supersonic speed. In the former case, the passage of the shock wave is marked by a jump to high values of air pressure and temperature. This is quickly followed (at places beyond a critical distance from the explosion) by a rather longer-lived period in which the pressure and temperature fall to values lower than those which prevailed before the arrival of the wave. These phases are termed the 'compression' and 'suction' phases, respectively.

In atmospheric flight at subsonic speeds, i.e. v (speed of flight) < V (speed of sound), pressure disturbances are propagated outwards through the atmosphere in all directions from the moving object as a series of non-overlapping spherical waves —see Figure 29(a). In contrast, the spherical disturbance waves emitted at successive time intervals by an object moving at a supersonic speed ($v > V$) intersect and are contained within a solid cone behind the object, the air in advance of the object being unaffected by the motion—see Figure 29(b). The semi-angle (α) of the cone ('Mach angle') is given by

$$\sin \alpha = \frac{Vt}{vt} = \frac{V}{v} = \frac{1}{M}$$

where t is time and M the MACH NUMBER. Lines AB and AC in Figure 29(b) are the two-dimensional representation of so-called 'Mach lines' within which the cone is contained. The position of the lines remains unchanged with respect to the moving body. Density of intersection of successive waves and, therefore, pressure disturbance are greatest along these lines. The waves propagated by the body advance in a direction normal to the surface of the cone with the speed of sound and are heard as a sharp report, coinciding with the arrival of the 'shock wave'.

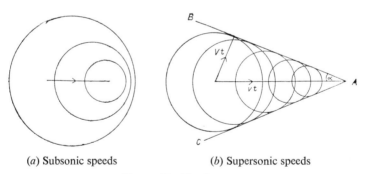

(a) Subsonic speeds   (b) Supersonic speeds

FIGURE 29—Shock waves.

short-wave radiation: In its common meteorological usage, solar radiation received near the earth's surface, of maximum intensity at about 0·5 micrometre (μm) and confined within the approximate limits 0·29 and 4μm. The term is used in contrast to the LONG-WAVE RADIATION emitted at terrestrial temperatures.

showalter index: An INSTABILITY INDEX (I) usually derived by assuming appropriate adiabatic ascent of an air parcel, originally at 850 mb, to the 500-mb level and subtracting the temperature (°C) so attained from the 500-mb environment temperature (T_{500}).

Thunderstorms become increasingly likely the further I decreases (algebraically) below a threshold value of about $+4$.

shower: In weather reports, solid or liquid precipitation from a CONVECTION cloud is designated a shower and is distinguished in such reports from the precipitation, intermittent or continuous, from layer clouds. Showers are often characterized by short duration and rapid fluctuations of intensity. Hail invariably implies a shower, while drizzle very seldom does.

For synoptic purposes, rain showers are classified as 'slight', 'moderate', 'heavy' or 'violent' for the rates of accumulation 0–2, 2–10, 10–50, or greater than 50 mm/h, respectively.

shred cloud: A term sometimes applied to the cloud species FLOCCUS.

Siberian anticyclone: A cold ANTICYCLONE which is a feature of the winter sea-level mean pressure distribution. In a January mean pressure chart, for example, it is represented by a centre over east central Eurasia, with central pressure over 1036 mb.

SID: An abbreviation for 'sudden ionospheric disturbance', which is a collective name for the effects of SCA, SEA, SPA, SWF and MAGNETIC CROCHET. These effects are due to the extra ionization produced suddenly in the low ionosphere at the time of a SOLAR FLARE. They are confined to the sunlit hemisphere, seldom last longer than an hour, and vary in intensity and duration approximately in relation to the associated flare intensity.

Similar effects in the high atmosphere are produced by an atmospheric nuclear explosion.

sidereal period: The sidereal period of a planet or the moon is the time required for the body to make a complete circuit of its orbit, with respect to the stars. See also DAY.

significance: In statistics, the property which distinguishes those results which should receive further attention from those which may be ignored.

The smaller the CHANCE EXPECTATION of a statistical result, the more confidently is significance attached to it. It is convenient to speak of a result as significant at the 5 per cent or the 1 per cent level, for example, when its chance expectation is less than 1/20 or 1/100, respectively. Results with a chance expectation greater than 1/20 are not described as significant, though they may be suggestive. The use of these standard significance levels is a matter of practical convenience, since it allows the use of published tables and facilitates the comparison of the results of different authors; other significance levels may however be quoted.

A significant result is not necessarily important since it may mean no more than that the data contain a copying mistake; a result whether significant or suggestive must be considered in relation to the background before its importance can be estimated. See also PROBABILITY.

significant-weather chart: A form of forecast weather chart in which only the main features of cloud, weather, icing and turbulence are presented. It has largely replaced the pictorial cross-section type of presentation. For fuller details reference should be made to the relevant ICAO document.*

---

* International Civil Aviation Organization. Procedures for air navigation services: Meteorology: Specification for meteorological services for international air navigation. Doc. 7605—Met/526/5, 5th edn. Montreal, 1970.

silver iodide: A substance used, in the form of a fine smoke which is scattered from the ground or from the air, as an ice-nucleating agent in CLOUD SEEDING experiments.

silver thaw: An expression of American origin. After a spell of severe frost, the sudden setting in of a warm, damp wind may lead to the formation of ice on objects which, being still at a low temperature, cause the moisture to freeze upon them and give rise to a 'silver thaw'. See GLAZE.

similarity theory of turbulence: A dimensional approach in which characteristic scales of velocity, length, etc. are formed from the physical quantities which are reasoned to determine the properties of flow. Vertical gradients of mean quantities (wind, temperature, etc.) and statistical properties of TURBULENCE (variances, spectral properties) are then expressed in dimensionally appropriate combinations of these scales and universal functions of non-dimensional parameters.

simoom: A hot, dry, suffocating wind or whirlwind which occurs in the deserts of Africa and Arabia. Most frequent in summer, it usually carries much sand and is short-lived (less than about 20 minutes).

single observer forecasting: Local weather forecasting which is based purely on observation of the weather elements for the same locality. The application of experience in synoptic meteorology, combined with physical reasoning, is capable of producing reasonably reliable local forecasts for a few hours ahead; in some situations reliability is possible for an appreciable time ahead. The observations chiefly used in such forecasting are those of pressure and pressure tendency, wind velocity at surface and higher levels, cloud, temperature.

singularity: An annual RECURRENCE TENDENCY in a meteorological element during a group of successive calendar dates. An example is a period of unseasonal warmth or cold, such as a BUCHAN SPELL.

The precise definition of singularities and their verification as reliable features of climate are often difficult and controversial. Their application in long-range weather forecasting appears to have some, though very limited, usefulness.

sink: See SOURCE.

sinusoidal pattern (of thickness): That synoptic pattern, approximating to a sine curve, which is formed by alternate thermal TROUGHS and thermal RIDGES of about equal amplitude (see Figure 30). The theory of thermal DEVELOPMENT implies the tendency for cyclogenesis (*C*) and anticyclogenesis (*A*) in the regions indicated. When the 'wave' amplitude is small, THERMAL STEERING predominates and surface

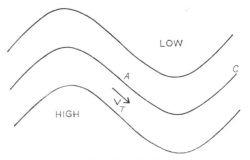

FIGURE 30—Sinusoidal pattern.

depressions move through the pattern without much deepening. With marked deepening and subsequent occluding of an associated depression, the pattern becomes distorted beyond recognition.

siphon barometer: A U-shaped mercury BAROMETER in which the areas of the upper and lower mercury surfaces are nearly equal. The instrument is adapted, as in the float barograph (see BAROGRAPH) to give a continuous record of pressure variation.

sirocco: See SCIROCCO.

site: In order to secure observations comparable with those at other stations, the site of a meteorological STATION has to be carefully selected in accordance with certain rules which are set out in the *Observer's handbook*.* A rain-gauge requires a certain amount of protection from the wind, but for other outdoor instruments the more open the site, the better. A compromise is usually effected. The latitude, longitude and height of the ground on which the rain-gauge stands are used to define the position of a station. See also EXPOSURE.

SI units: Abbreviation for 'Système International d'Unités'. This system has been adopted by the General Conference of Weights and Measures and has been endorsed by the International Organization for Standardization. It is expected that the system will become the generally accepted metric system throughout the world and that this acceptance will coincide with the adoption of the metric system by the United Kingdom.

SI is a coherent system based on six primary units, namely:

Quantity	Unit	Symbol
length	metre	m
mass	kilogramme	kg
time	second	s
electric current	ampere	A
temperature	kelvin	K†
luminous intensity	candela	cd

Derived units, related to the basic units by definition, have been given special names in some instances. For example, the unit of force in SI is called the NEWTON, this being the force which applied to a mass of 1 kg gives it an acceleration of 1 m/s².

Six's thermometer: A U-shaped thermometer, designed by J. Six, in which the positions of two iron indexes, subsequently reset by means of a magnet, indicate the maximum and minimum temperatures attained since the previous setting.

skewness: A measure of asymmetry in a FREQUENCY DISTRIBUTION. Skewness is regarded as positive when the longer tail is to the right of the MEAN and negative when it lies to the left. Thus, for example, most rainfall distributions are positively skew while distributions of minimum temperature tend to be negatively skew.

The coefficient of skewness γ_1 of a distribution is defined as $\gamma_1 = \mu_3/\sigma^3$, where μ_3 is the third moment about the mean and σ is the STANDARD DEVIATION.

---

* London, Meteorological Office. Observer's handbook, 3rd edn. London, HMSO, 1969, pp. 169–172.
† In October 1967 the thirteenth Conférence Générale des Poids et Mesures recommended that the kelvin, symbol K, be used both for thermodynamic temperature and for thermodynamic interval, and that the unit-symbols °K and deg be abandoned.
However, in this publication a temperature interval is described by the abbreviation 'deg' to distinguish it from a point in the temperature scale, e.g. degK, degC, degF as opposed to K, °C, °F.

skill score: A measure of the accuracy of a series of weather forecasts, usually in comparison with RANDOM FORECASTS or PERSISTENCE FORECASTS. For example, such an index may be given by

$$S = \frac{\text{number of forecasts correct} - \text{number correct by chance}}{\text{total number of forecasts} - \text{number correct by chance}}$$

skin-friction coefficient: An alternative for DRAG COEFFICIENT.

skip distance: The distance, measured along the earth's surface, between the point at which the ground signal from a radio-wave transmitting source is just undetectable by a receiver of normal sensitivity and that at which reception of the transmitted waves is first possible after reflection by the IONOSPHERE. This distance corresponds to a minimum angle of incidence of the waves on the ionosphere; if this angle is not attained, the energy passes on to outer space.

The 'zone of silence' corresponds to the skip distance in the case of sound waves —see AUDIBILITY.

sky light: An alternative for DIFFUSE RADIATION.

sky radiation: An alternative for DIFFUSE RADIATION.

sky, state of: Fraction of the sky obscured by cloud on a scale of 0 (cloudless) to 8 (sky entirely covered by cloud). In the BEAUFORT NOTATION letters, b is used for a total cloud amount 0–2/8, bc for a total cloud amount of 3/8–5/8 and c for a total cloud amount 6/8–8/8. For a uniform thick layer of cloud completely covering the sky o is used. The letter u, for an ugly, threatening sky, may be used with any of the preceding letters to indicate the general appearance of the sky. (See *Observer's handbook.**)

slant visibility: An alternative for OBLIQUE VISIBILITY.

slantwise convection: A type of motion, considered significant in the GENERAL CIRCULATION of the atmosphere, in which exchange of air between different levels is effected despite the absence of a superadiabatic lapse rate.

In a normal atmospheric state, isentropic (equal potential temperature) surfaces slope upwards from lower to higher latitudes. While, therefore, an air parcel may be prevented from rising vertically because of a subadiabatic lapse rate, poleward travel brings it to an environment more dense than itself, thus enabling it to rise. Air at higher levels and higher latitudes may similarly descend by equatorward movement. The effective slope along which the movement occurs is less than that of the original isentropic surfaces.

sleet: Precipitation of snow and rain (or drizzle) together, or of snow melting as it falls.

In American terminology, sleet is often used to signify ICE PELLETS.

slice method: The estimation of vertical STABILITY in the atmosphere by a method which takes some account of the compensating downward motion induced in the environment by upward rising parcels of air. Since the environment is assumed to warm at the dry ADIABATIC lapse rate during its descent, greater stability is in general deduced by this method, as compared with the PARCEL METHOD.

London, Meteorological Office; Observer's handbook, 3rd edn. London, HMSO, 1969, p. 68.

sling psychrometer: An alternative for WHIRLING PSYCHROMETER.

sling thermometer: A thermometer mounted on a frame pivoted about a handle so that it can be whirled in the hand, thus providing 'ventilation'. If the bulb is shielded from direct solar radiation, satisfactory readings of air temperature can thus be obtained in a simple and inexpensive manner. A pair of thermometers, dry- and wet-bulb, similarly used, constitute a 'sling' or 'whirling' psychrometer. See also PSYCHROMETER.

small circle: Any plane which does not pass through the centre of a sphere cuts the surface of the sphere in a 'small circle'.

Smith period: A simplified form of MILLS PERIOD in which a minimum relative humidity of 90 per cent, following rain, is substituted for leaf wetness.

smog: A term, being a contraction for 'smoke fog', which signifies a FOG in which smoke, or other form of atmospheric pollutant, besides playing an important part in causing the fog to form and to thicken (for example, by acting as condensation nuclei) has unpleasant or dangerous physiological effects. A noteworthy smog occurred in the London area in early December 1952.

Photochemical smog is a form of air pollution characterized by haze, eye irritation and plant damage. It is very marked in Los Angeles and is produced in stagnant or slowly moving air by the reaction, in the presence of sunlight, of certain hydrocarbons and oxides of nitrogen arising mainly from motor-car exhaust; OZONE and other oxidizing agents are produced.

smoke: The visible product of incomplete combustion; in Great Britain the main source is coal burning. Coal smoke comprises mainly carbon and hydrocarbon particles of very small size (about $0 \cdot 1$ μm) which remain in the air, on average, for 1 to 2 days.

Atmospheric smoke concentration is measured by a 'smoke filter' method in which the weight of smoke deposited on a white filter-paper by a known volume of air is inferred from the measured reflectance of the smoke stain. Typical annual mean values for 1969 are:
   (i) Country air—25 to 30 μg/m^3.
   (ii) City air—a wide range of values varying from 250 μg/m^3 for a few highly industrialized towns to, more typically, 75 to 100 μg/m^3 for most cities.

An over-all decrease in concentration has occurred during recent years. See ATMOSPHERIC POLLUTION, RINGELMANN SHADES.

smoothing: A process of performing space or time averaging of data to suppress local or short-period variations. The use of efficient filters or weighted averages enables the larger-scale or more lasting components to be preserved with little distortion, while the use of inefficient filters (which include unweighted averages) entails a risk of emphasizing insignificant oscillations compared with significant features. See FILTERING.

smudging: A method of FROST PROTECTION consisting of the production, by combustion or by chemical means, of a smoke pall over a confined area, e.g. vineyard. The resulting reduction in outgoing terrestrial radiation results, in suitably calm conditions, in a decrease in the rate of fall of temperature of the ground in the locality.

Snell's law: See REFRACTION.

snow: Solid PRECIPITATION which occurs in a variety of minute ICE CRYSTALS at temperatures well below 0°C but as larger SNOWFLAKES at temperatures near 0°C. 'Granular snow' consists of opaque grains, rather flattened in shape and generally less than 1 mm in diameter.

For synoptic purposes, snow (or a snow shower) is classed as 'slight', 'moderate', or 'heavy' for a rate of accumulation of snow (in the absence of drifting or melting) less than 0·5 cm/h, 0·5 to 4 cm/h, or greater than 4 cm/h, respectively.

See also SNOWFALL.

snow, day of: In British climatology, any period of 24 hours ending at midnight GMT upon which snow is observed to fall is regarded statistically as a 'day of snow'.

See also SNOWFALL.

snowdrift: When a strong wind blows, there is a strong tendency for falling snow, or fallen snow on the ground, to accumulate not in open places but in any sheltered place, as in the lee of natural or artificial obstacles, there forming 'drifts'.

The symbols ↮ and ↮ are employed, in synoptic chart plotting, to distinguish drifting snow at 'generally low' and 'generally high' levels, respectively.

snowfall: The depth of fresh snowfall is normally measured with a graduated ruler. Its measurement as RAINFALL (i.e. its water content) may be made in a suitable SNOW-GAUGE, or by melting the snow caught in a normal rain-gauge, or by collecting and melting samples of fresh snow which has fallen in the open. Thirty centimetres of freshly fallen snow has about the same water content as 25 mm of rainfall.

The amounts of snow which fall over the British Isles are measured as rain; separate statistics of snowfall amount are therefore not available. Separate records are, however, kept of the numbers of days of snowfall and of SNOW LYING, and also of snow depth.

The average annual number of days with snow falling on low ground up to about 60 m above mean sea level increases with increasing latitude and from west to east, and ranges from less than five days in southern Cornwall to over 35 days in north-east Scotland. At places above 60 m the average number of days increases by approximately one day for every 15 m of elevation; above 300 m the increase is greater than this, the corrections being about $+20$, $+52$, $+90$ and $+140$ days at 300, 600, 900 and 1200 m, respectively.

The average annual number of days with SNOW LYING is one of the most variable of meteorological elements over the British Isles. Factors which influence it include monthly mean temperature, frequency of snowfall, quantity of snowfall and the character of the station and its surroundings, such as its height, aspect and distance from sea. Values range from less than five days per year in southern and western coastal districts to over 100 days per year in the Grampians. In some winter months, such as February 1942 and February 1947, snow lying has been reported on every day of the month over large areas of Britain, whereas in February 1943 the majority of stations reported no single day of snow lying.

Falls of undrifted snow of depth exceeding 150 mm on level ground at low altitudes occur somewhere in England in one-third to one-half of winters; the average number of days in 10 years with snow lying at such depths ranges from less than 1 day in south-west coastal districts to between 30 and 40 days in eastern England.

snowflakes: Aggregates of ICE CRYSTALS which occur in an infinite variety of shape and form. At very low temperatures the flakes are small and their structure simple. At temperatures which are close to freezing-point, the individual flakes may be composed of a very large number of ice crystals (predominantly star-shaped) and the flake may then have a diameter up to several inches (1 in ≈ 25 mm).

snow-gauge: A device for the retention and measurement of SNOW. In the Hellmann–Fuess snow-gauge the snow is caught in a receiver supported on a balance, the displacement of which is continuously recorded, so that an autographic record of snowfall (and of the fall of rain and hail also) is obtained. Most snow-gauges are, however, merely rain-gauges fitted with jackets or other devices to make them suitable for collecting solid precipitation, and for melting it before taking the reading.

snow grains: Precipitation of very small white and opaque grains of ice. These grains are fairly flat or elongated; their diameter is generally less than 1 mm.

snow-line: The lower limit in altitude of the region of perpetual snow. In high polar latitudes the snow-line is at sea level; in northern Scandinavia it is at about 1200 m, in the Alps at about 2400 m, in the Himalayas at about 4500 m. These figures are only approximate, as the height of the snow-line varies on the north and south sides of a mountain and from one mountain to another in the same latitude or region. It has no direct relation to the mean annual temperature, depending more on the summer temperature, but many other factors exert an influence, such as amount of snow in winter, prevailing winds, exposure and steepness of the slopes, etc.

snow lying: This expression (international symbol ⊠) is used for occasions when one-half or more than one-half of the ground representative of the station is covered with snow. The ground representative of the station is defined as 'the flat land easily visible from the station and not differing from it in altitude by more than 30 m'. British statistics of snow lying refer only to occasions when this state of affairs exists at the hour of morning observation.
See also SNOWFALL.

snow pellets: Precipitation of white and opaque grains of ice (formerly 'soft hail'). These grains are spherical or sometimes conical; their diameter is about 2–5 mm.

snow rollers: Cylinders of snow, formed and rolled along by the wind.

snow survey of Great Britain: A survey started in 1937–38 by the British Glaciological Society, with an annual report covering each snow season published in the *Journal of Glaciology*. The survey lapsed during the Second World War but was restarted with the report for 1946–47 and continued by the Society until 1952–53.
Starting with the report for 1953–54 the work was taken over by the Meteorological Office; reports were first published in the *Meteorological Magazine* and later (from 1956–57 onwards) in *British Rainfall*.

sodium: A very minor constituent of the atmosphere, estimated to total about one tonne (1000 kg) in weight. The sodium D line (5893 ångströms) is observed in the NIGHTGLOW emission spectrum and is conspicuous in the TWILIGHTGLOW. The sodium is thought to have maximum concentration at about 85 km.
Experiments have been made in which a few pounds of sodium are injected by rocket into the high atmosphere to produce an artificial glow which, observed from the ground, is used as a measure of the high atmospheric winds.

soil moisture: The moisture content of soils is generally expressed as the percentage ratio of the mass of water to that of dry soil, but may be expressed also in terms of inches of water per given depth of soil. Soil moisture is of obvious importance in regard to the growing of plants and is of direct meteorological interest in affecting the thermal conductivity of soil and so the rate at which heat is conducted upwards to or downwards from the atmosphere, and also in affecting, in certain circumstances, the rate of EVAPORATION from the soil and of TRANSPIRATION from vegetation. Conversely, the rates of evaporation and transpiration are much affected by the meteorological variables temperature, sunshine, humidity and wind.

Among the measures of soil moisture which are employed are SATURATED SOIL, FIELD CAPACITY, and WILTING POINT. The varying force with which soil retains contained water is termed the CAPILLARY POTENTIAL.

soil moisture deficit: The amount of rainfall or irrigation required to restore soil to its FIELD CAPACITY.

soil temperature: See EARTH TEMPERATURE.

soil thermometer: See EARTH THERMOMETER.

solar activity: See SUN.

solar constant: The solar radiation flux at a surface normal to the sun's beam outside the earth's atmosphere at the earth's mean distance from the sun.

Measurement of the solar constant has been made at high-level observatories since 1902, mainly by the Smithsonian Institution. Direct solar radiation intensity is continuously measured and is related to an absolute standard; atmospheric attenuation is allowed for by measurement, at various solar zenith angles, of relative flux over a wide band of selected wavelengths and extrapolation to 'zero path length'. The generally accepted value of the solar constant is 139·6 mW/cm^2; this is equivalent to 2·00 IT cal/cm^2 min.

There has been much controversy as to whether measured day-to-day variations and longer time-average variations of the solar constant are real or spurious. As measuring technique has developed, the values have more nearly approached a constant value (for a given absolute standard). It is generally held that no significant variation of the solar constant has yet been experimentally shown and that the upper limit of such variation is about 0·1 per cent. This is in marked contrast to the large percentage solar-cycle variations known from geophysical evidence to occur in the far ends of the solar spectrum where, however, the amount of energy involved is very small.

solar corpuscular streams: Streams of charged corpuscles (particles) which are emitted at high speed from disturbed regions of the sun are affected by the earth's magnetic field in the event of their crossing the earth's path and, in turn, affect this field, causing geomagnetic and ionospheric storms. It is generally held that these particles do not penetrate the earth's atmosphere below about 70 km (above which height their interaction with the atmosphere gives rise to visible AURORA) and that no relation between the incidence of particle streams and lower atmosphere meteorological phenomena has yet been shown. See also GEOMAGNETISM.

solar cycle: The relative sunspot number varies in a quasi-periodic manner, with successive maxima separated by an average interval of about 11 years—the so-called 'solar cycle'. If reversal of sunspot magnetic-field polarity in a given hemisphere in successive 11-year periods is taken into account, the complete solar cycle may be considered to average some 22 years. See SUNSPOT.

solar day: See DAY.

solar flare: A solar explosion, unpredictable in nature and up to a few hours in duration, from a restricted region of the CHROMOSPHERE above certain types of SUNSPOT. Flares are classified on an ascending scale from 1 to 3+, on the visual basis of intensity of emitted light and solar area covered. In the sunlit hemisphere, short-lived 'sudden ionospheric disturbances' (SID) start almost simultaneously with the visual appearance of a great flare, and are attributed to the arrival in the high atmosphere of a flood of ionizing radiation released by the flare; the extra ionization is thought to occur mainly in the D-layer at about 70–90 km height.

An intense and world-wide magnetic storm follows a great flare, which is near the centre of the sun's disc, with a frequency which is much in excess of chance expectation. It is thus inferred that a SOLAR CORPUSCULAR STREAM is ejected almost radially from the flare region, the delay of about 20 hours corresponding to the slower speed of travel of the solar particles compared with that of the wave radiation.

solarimeter: An instrument for measuring the total solar radiation received on a horizontal surface. See PYRANOMETER.

solar radiation: See RADIATION.

solar-radiation thermometer: An alternative for BLACK-BULB THERMOMETER.

solar spectrum: See RADIATION and ELECTROMAGNETIC RADIATION.

solar-terrestrial relationships: The relationships between the (variable) physical state of the SUN ('solar activity') and the (variable) particle and wave radiations emitted by the sun on the one hand, and the resulting physical effects produced in the earth's atmosphere on the other. Meteorological events observed in the troposphere and stratosphere appear to depend very little, if at all, on observed solar variability. The latter is, however, very important in certain types of COSMIC RAY event, and in studies of the AURORA, GEOMAGNETISM and the IONOSPHERE.

solar wind: Term proposed for the motion of interplanetary gas outwards from the sun towards the earth near which it interacts with the earth's magnetic field. It is generally assumed that the strength of this 'wind' increases with increasing solar activity. See SUN.

solenoids: The intersection in a BAROCLINIC atmosphere of surfaces of constant pressure with surfaces of constant specific volume (isobaric and isosteric surfaces, respectively) forms three-dimensional 'isobaric–isosteric solenoids'; since isosteric (constant specific volume) surfaces are also isopycnic (constant density) the intersections may also be said to form 'isobaric–isopycnic solenoids'. 'Unit solenoids' are formed by the intersection of surfaces separated by one unit of pressure and specific volume, or of pressure and density.

The existence of such solenoids in the atmosphere tends to produce a so-called 'direct circulation'; in the absence of CORIOLIS FORCES the rate of production of circulation is proportional to the concentration of unit solenoids and is so directed as to cause the lighter air to rise and the denser air to subside.

solstice: The time of maximum or minimum DECLINATION of the sun when, for a few days, the altitude of the sun at noon shows no appreciable change from day to day. The summer solstice for the northern hemisphere (winter solstice for southern hemisphere) occurs on about 22 June, when the sun is farthest north of the equator; the winter solstice for the northern hemisphere (summer solstice for southern hemisphere) occurs on about 22 December, when the sun is farthest south of the equator.

sounding: A direct or indirect measurement of the vertical distribution of some physical or chemical property of the atmosphere. See also BALLOON SOUNDING, METEOROLOGICAL ROCKET.

sound waves: Sound passes through a medium by means of longitudinal waves whose velocity of propagation depends on the temperature and nature of the medium. The passage of sound waves at a point in the atmosphere is associated with air pressure fluctuations about a mean value. Where these fluctuations are small relative to the mean pressure, the velocity of the waves is given by

$$V = \sqrt{(\gamma RT)}$$

where γ is the ratio (c_p/c_v) of the specific heats of air, R is the specific gas constant for air, and T is absolute temperature.

V in air at 0°C is about 332 m/s or 760 mile/h. See also WAVE MOTION, MACH NUMBER.

source: In hydrodynamics, a 'source' is a point, line, area or volume within a fluid at which fluid is continuously created and from which it moves equally in all directions; the source 'strength' is measured by the rate of production of fluid. The converse is a hydrodynamic 'sink' to which fluid converges and at which it disappears.

In an analogous way a 'heat source' in a thermodynamic system is that part of the system in which heat is continuously generated and from which it is transferred to be continuously dissipated at a 'heat sink'. The large-scale atmospheric heat source is at low levels in low latitudes, the heat sinks are at low levels in high latitudes and at high levels in all latitudes.

source region (air mass): See AIR MASS.

southerly buster: A name given in south and south-east Australia to a sudden change of wind, usually from a north-westerly direction to a southerly direction, which is accompanied by a sudden fall in temperature. This change of direction occurs behind a cold front, and if the rise of pressure is considerable the southerly wind is violent. The arrival of the southerly wind is usually marked by a long crescent-shaped roll of cloud. The temperature sometimes falls as much as 20 degC in half an hour. These storms are sometimes accompanied by thunder and lightning. They are similar to the PAMPEROS of South America and the LINE-SQUALLS of middle latitudes. They are most prevalent from October to March.

southern oscillation: The best known of three indices derived empirically by G. T. Walker and associated by him with a suggested period of 2·33 years. The index was based on the observed seasonal distribution of pressure, and to a lesser extent temperature and rainfall, over a large and predominantly oceanic region of lower latitudes.

A 'North Atlantic oscillation' and a 'North Pacific oscillation', derived in a similar manner for these regions, were also employed by Walker.

SPA: An abbreviation for 'sudden phase anomaly'. When a SOLAR FLARE occurs, the extra ionization of the D-layer in the sunlit hemisphere causes the layer to lower as a reflecting medium with respect to long waves reaching it from the earth's surface; an associated sudden change of phase is observed between the direct ground wave and reflected sky wave, by a receiver at some distance from a long-wave transmitter. See also SID.

space charge: In ATMOSPHERIC ELECTRICITY an excess, within any specified portion of the atmosphere, of positive over negative IONS, or vice versa—positive or negative space charge, respectively.

The downward movement of positive electric charge, and upward movement of negative charge, in response to the existing POTENTIAL GRADIENT, implies a positive space charge in fair-weather regions where the field is directed downwards. This space charge is greatest at low levels where the field is greatest. Large space charges of either sign are, however, measured in association with precipitation elements.

specific heat: The specific heat of a substance is the heat required to raise the temperature of unit mass of it by one degree; it has normally been expressed in the unit 'calories/gramme degree kelvin'. In future in SI UNITS it will be expressed in 'joules/kilogramme degree kelvin'. The dimensions are $L^2T^{-2}\theta^{-1}$.

The specific heat of a substance is to some extent dependent on the temperature (see CALORIE). The specific heat of water is 4187 J/kg degK (1·00 IT cal/g degK); the specific heat of ice is 1450 J/kg degK at $-90°C$, 1884 J/kg degK at $-30°C$ and 2106 J/kg degK at 0°C (0·346, 0·450 and 0·503 IT cal/g degK, respectively).

The specific heat of dry air at constant pressure (c_p) is 1005 J/kg degK (0·240 IT cal/g degK), and at constant volume (c_v) 716 J/kg degK (0·171 IT cal/g degK). The ratio γ of the specific heats (i.e. c_p/c_v) is 1·40. The specific heats of water vapour at constant pressure (c_{pv}) and at constant volume (c_{vv}) are, respectively, 1846 J/kg degK (0·441 IT cal/g degK) and 1386 J/kg degK (0·331 IT cal/g degK). Admixture of water vapour increases the specific heat of air, to a degree negligible for most purposes, as follows:

$$c_{pm} \text{ (moist air)} \approx c_p (1 + 0·9r)$$
$$c_{vm} \text{ (moist air)} \approx c_v (1 + 1·0r)$$

where r (HUMIDITY MIXING RATIO) is expressed in kilogrammes/kilogramme.

1 IT calorie = 4·1868 JOULES.

specific humidity: The specific humidity (q)—also termed the 'mass concentration' or 'moisture content'—of moist air is the ratio of the mass (m_v) of water vapour to the mass $(m_v + m_a)$ of moist air in which m_v is contained, m_a being the mass of dry air, i.e.

$$q = \frac{m_v}{m_v + m_a}.$$

Since m_v is much smaller than m_a, specific humidity for a given sample is almost identical with HUMIDITY MIXING RATIO, which is now generally preferred.

specific volume: The volume occupied by unit mass of a substance, at a specified temperature and pressure. It is the inverse of DENSITY.

spectrobolometer: See BOLOMETER.

spectrophotometer: An instrument which measures the intensity of radiation of a given wavelength. In meteorology, such an instrument is used mainly in the measurement of ozone—see DOBSON SPECTROPHOTOMETER.

spissatus (spi): A CLOUD SPECIES (Latin, *spissatus* thickened).

'CIRRUS of sufficient optical thickness to appear greyish when viewed towards the sun.'* See also CLOUD CLASSIFICATION.

splintering: The splintering or fragmentation of ICE CRYSTALS, more especially of the delicate branched form of crystal, is considered to be an important source of the multiplication of ice crystals within a cloud. The electric charge separation associ-

---

* Geneva, World Meteorological Organization. International cloud atlas. Vol. 1, Geneva, WMO, 1956, p. 12.

ated with splintering has been proposed as a factor in the rapid rate of charge separation which occurs within a THUNDERSTORM cloud.

spontaneous nucleation: An alternative for HOMOGENEOUS NUCLEATION.

spring: See SEASONS.

squall: A strong wind that rises suddenly, generally lasts for some minutes, and dies comparatively suddenly away. It is distinguished from a GUST by its longer duration.

The term is often used in such a sense as to include the precipitation, thunderstorm, etc., which are a common accompaniment of the sudden increase of wind. The following definition of 'squall' was adopted in April 1962 by the Third Session of the Commission for Synoptic Meteorology of the WMO:

'A sudden increase of wind speed by at least 8 m/s (16 knots), the speed rising to 11 m/s (22 knots) or more and lasting for at least one minute. Note: When Beaufort scale is used for estimating wind speed, the following criteria should be used for the reporting of squalls: a sudden increase of wind speed by at least 3 stages of the Beaufort scale, the speed rising to Force 6 or more and lasting for at least one minute.'

squall line: The name originally given to what is now known as a COLD FRONT.

The use of the term is now generally confined to violent convective phenomena extending along a line or belt which is non-frontal in nature—see INSTABILITY LINE.

stability: A system which is subjected to a small disturbing impulse is said to be in stable, neutral (or indifferent), or unstable equilibrium, according to whether it returns to its original position, remains in its disturbed position, or moves farther from its original position, respectively, when the disturbing impulse is removed.

Investigation of the 'static stability' of the atmosphere is made most simply by the 'parcel method', in which an assessment is made of changes of kinetic energy of a test parcel of air, displaced vertically and adiabatically with respect to its environment as represented by an ascent curve on an AEROLOGICAL DIAGRAM. The environment is termed stable, neutral, or unstable (at defined points or in defined layers) according as the kinetic energy of the parcel decreases, remains constant, or increases, respectively.

The following rules apply to the most general case of a moist but unsaturated test parcel which is subject to a smaller or larger vertical ascent through a moist but unsaturated environment:

    (i) 'absolute stability' exists if the existing LAPSE of temperature (γ) is less than the saturated ADIABATIC lapse rate (Γ_s), i.e. if $\gamma < \Gamma_s$.
    (ii) 'absolute instability' exists if $\gamma > \Gamma_d$ (dry adiabatic lapse rate).
    (iii) 'conditional instability' exists if $\Gamma_d > \gamma > \Gamma_s$.

Case (iii) is subdivided into two classes defined by the vertical distribution of humidity in the environment curve, as follows:

    (*a*) The case is one of stability if none of the pseudo wet-bulb potential temperature (θ_{sw}) lines (with sufficient accuracy, the WET-BULB POTENTIAL TEMPERATURE (θ_w) lines) corresponding to possible test parcels intersects the environment curve—see PSEUDO WET-BULB TEMPERATURE.
    (*b*) The case is one of 'latent instability' if one or more θ_w lines intersects the environment curve. The latent instability is termed 'real latent' if the 'negative area' (that lying between the ascent curve of the test parcel and the environment curve, and to left of environment curve) is less than the 'positive area' (to right of environment curve); the latent instability is termed 'pseudo-latent' if the converse is true.

The term 'convective instability', or 'potential instability', is applied to the case in which a layer of air will become unstable on being lifted bodily (as over high ground) till it is saturated; the criterion for this case is that θ_{sw} (with sufficient accuracy θ_w) decreases with increasing height through the layer.

Estimation of the static stability by means of the parcel method and, in particular, calculations by this method of available kinetic energy in conditions of instability with associated convection, cloud formation and precipitation, are liable to serious error by the neglect of such factors as the mixing of rising parcels with the environment, the compensating downward motion induced in the environment, and the additional energy released by the cooling of the environment by the evaporation of precipitation into it. The method is, nevertheless, capable of producing useful results. See also DYNAMIC STABILITY.

standard: A prescribed measure or scale of any kind, such as a unit or scale of reference. The legal magnitude of a unit of measure or weight.

A 'primary standard' instrument (ABSOLUTE INSTRUMENT) is often used in the CALIBRATION of some 'sub-standard' or 'secondary standard' instruments which, in turn, may be used to calibrate many instruments for field use.

standard atmosphere: Hypothetical atmosphere, corresponding approximately to the average state of the real atmosphere, in which the pressure and temperature are defined at all heights. Such an atmosphere is adopted internationally as the basis for the calibration of altimeters, evaluation of aircraft performance, etc.

The principal standard atmospheres now in use are:

(i) The ICAO STANDARD ATMOSPHERE to 20 km and its extension to 32 km.
(ii) The COSPAR CIRA 1965 Reference Atmospheres 30–700 km which also give reference values plus winds seasonally and latitudinally in the STRATOSPHERE and MESOSPHERE and variations with the solar intensity in the upper THERMOSPHERE. The COSPAR Atmosphere is likely to be revised when new values are accepted.
(iii) The Russian 1964 GOST Atmosphere.
(iv) The U.S. Standard Atmosphere, 1962,* which extends the ICAO Atmosphere upwards from 32 km; values are given up to 700 km. The U.S. Standard Atmosphere Supplements, 1966, gives seasonal and latitudinal supplementary atmospheres for the stratosphere and mesosphere, linking at 120 km with thermosphere models which vary with solar activity. These values are widely used in the western hemisphere. Supplementary data on other atmospheric parameters are included.

The International Standards Organization is working towards the universal acceptance, as soon as possible, of an International Standard Atmosphere up to at least 32 km and probably to 50–60 km. Reference atmospheres plus variability data and information on other atmospheric parameters may be added later in consultation with WMO. See ATMOSPHERE, ICAO STANDARD ATMOSPHERE.

standard atmospheric pressure: A concept used in some physical definitions. It is the pressure exerted by a column of mercury 760 mm high, of density 13 595·1 kg/m^3, subject to gravitational acceleration 980·665 cm/s^2 (9·80665 m/s^2), and equals 101 325 N/m^2 or 1013·25 mb.

See STANDARD DENSITY, STANDARD GRAVITY.

standard ballistic atmosphere: See BALLISTICS.

---

* Washington, United States Committee on Extension to the Standard Atmosphere (COESA). U.S. Standard Atmosphere, 1962.

standard density: A conventional value of the density of mercury, adopted for the sake of uniformity in the conversion of pressure readings from units of pressure to units of height (mm or in), or vice versa. The value adopted by the World Meteorological Organization is the density at 0°C, equal to 13 595·1 kg/m³. See also STANDARD GRAVITY, BAR.

standard deviation: The standard deviation of a group of measurements is the root-mean-square departure of the observations from their mean. It is the square root of the VARIANCE and is the most generally useful measure of their dispersion. It is denoted by the Greek letter σ (sigma) and is defined by

$$\sigma_x = (V(x))^{\frac{1}{2}} = \left(\frac{1}{n} \sum_{j=1}^{n} (x_j - \bar{x})^2 \right)^{\frac{1}{2}}$$

where $x_1, x_2, \ldots x_n$ are measurements of the same kind.

When n is small, the standard deviation of the sample is an underestimate of that of the POPULATION from which it is derived; a better estimate of the population standard deviation, sometimes termed the 'standard error', is given by

$$\left(\frac{1}{n-1} \sum_{j=1}^{n} (x_j - \bar{x})^2 \right)^{\frac{1}{2}}.$$

If the variable x is normally distributed, then the range $(\bar{x} - 2\sigma_n, \bar{x} + 2\sigma_n)$ includes roughly 95 per cent of the values, so that a deviation from the mean exceeding twice the standard deviation is usually SIGNIFICANT in the statistical sense.

The standard deviation of the sum σ_s of n independent values drawn from different populations of standard deviation σ_1, σ_2, etc. is

$$\sigma_s = (\sigma_1^2 + \sigma_2^2 + \ldots + \sigma_n^2)^{\frac{1}{2}},$$

which reduces to $(n \sigma^2)^{\frac{1}{2}}$ when the sums are drawn from the same population of standard deviation σ.

The standard deviation of the difference (σ_d) of two independent means, of standard deviations σ_1 and σ_2, is

$$\sigma_d = (\sigma_1^2 + \sigma_2^2)^{\frac{1}{2}},$$

which reduces to $\sigma\sqrt{2}$ when the means are drawn from the same population of standard deviation σ.

See also NORMAL FREQUENCY DISTRIBUTION, RANGE.

standard gravity: A conventional value of the gravitational acceleration, adopted for the sake of uniformity. The value adopted by the World Meteorological Organization is 980·665 cm/s² (9·80665 m/s²). It is to this value, not to the value of 980·616 cm/s² which is the best-determined value of gravity at sea level in 45° latitude, that pressure data in height units (millimetres or inches) refer. (The value 980·665 cm/s² is a previous best-determined value.)

See also GRAVITY, BAR, STANDARD DENSITY.

standard temperature (of barometer): That temperature at which, under STANDARD GRAVITY, the indicated reading of a mercury barometer is correct. At any temperature of an ATTACHED THERMOMETER other than the standard temperature, a 'temperature correction' must be applied to the barometer reading to take account of differences between the density of the mercury and dimensions of the metal scale of the barometer and those values assumed in the CALIBRATION of the instrument.

WMO resolved that, with effect from 1 January 1955, the standard temperature for all mercury barometers should be 0°C.

standard time: Time referred to the mean time of a specified meridian. The meridian of Greenwich is the standard for western Europe. The standard meridian for other countries is generally so chosen as to differ from Greenwich by an exact number of hours or half-hours. See also ZONE TIME.

standard vector deviation: The standard vector deviation of a group of vectors is a measure of the scatter of the vector end points about the point which represents the end of the vector mean, all the vectors emanating from a common origin. The standard vector deviation (σ) is defined as the root-mean-square vector deviation; it is most readily calculated from the equation

$$\sigma^2 = \Sigma |\mathbf{V}_i|^2/n - |\mathbf{V}_m|^2$$

where $|\mathbf{V}_i|$ is the module (magnitude) of the individual vectors and $|\mathbf{V}_m|$ the module of the vector mean.

σ for the wind vector is found to increase with height in the troposphere in accordance with the approximate relation $\sigma\rho =$ constant, where ρ is air density. The average value of σ at a pressure level of 500 mb over the British Isles is about 30 knots in summer, 40–45 knots in winter.

Some 50 per cent of vector observations are contained within a circle centred on the end point of the mean vector and of radius $0\cdot83\sigma$, 95 per cent within radius $1\cdot73\sigma$, 99 per cent within $2\cdot15\sigma$.

See also CONSTANCY OF WINDS.

standing wave: In meteorology, an air wave which is (almost) stationary with respect to the earth's surface. Such wave or waves are commonly associated with airflow over mountains. See MOUNTAIN WAVE, LEE WAVES.

Stanton number: The reciprocal of the PRANDTL NUMBER.

starshine recorder: An alternative for NIGHT-SKY RECORDER.

state, equation of: An alternative for GAS EQUATION.

state of ground: See GROUND, STATE OF.

static: See ATMOSPHERICS.

static stability: See STABILITY.

static stability, equation of: An alternative for HYDROSTATIC EQUATION.

station: In meteorology, a location at which regular weather observations are made. Among the classes of station are the SYNOPTIC STATION, CLIMATOLOGICAL STATION, CROP WEATHER STATION, HEALTH RESORT STATION, RAINFALL STATION, OCEAN WEATHER STATION and AUTOMATIC WEATHER STATION.

station index number: A group of three figures used in synoptic messages to signify the particular station, within a given block area the boundaries of which coincide in most cases with national frontiers, at which the observation was made.

statistical inference: The process by which one proceeds from factors derived from samples of measurements which are available for analysis to reach conclusions about the statistical POPULATIONS of which these measurements form part, or about other samples of the same kind. Conclusions drawn by statistical inference are never absolute, but are always regarded as valid with a greater or less degree of PROBABILITY, depending on the nature and statistical SIGNIFICANCE of the evidence.

statistical theory of diffusion: Treatment of DIFFUSION involving the statistical properties of turbulence (VARIANCE, CORRELATION and ENERGY SPECTRUM) based on the velocity changes experienced by a moving particle, in contrast to the velocity changes evident at a fixed point in the flow.

statistics: The science dealing with the problems arising when numerical data are considered in groups or in the mass instead of as individuals. The science includes the definition of the statistics (plural) which sum up the essential features of many data in comparatively few numbers, and the establishment of relationships between variables occurring in groups or sets. There is a considerable statistical vocabulary of which the essential terms require more elaborate discussion than is possible in the *Glossary*. See, however, PROBABILITY, RANDOM NUMBERS, FREQUENCY, POPULATION, SAMPLE, NORMAL DISTRIBUTION, AVERAGE, STANDARD DEVIATION, CORRELATION, SIGNIFICANCE.

steam fog: An alternative for ARCTIC SEA SMOKE.

steering: In synoptic meteorology, controlling factor(s) in the direction and speed of movement of pressure systems, sometimes also of precipitation belts, thunderstorms, etc.

The forecasting problem relating to the movement of pressure systems is one of DEVELOPMENT combined with steering. The principle of THERMAL STEERING operates well when development is not very pronounced. Attempts to find by more empirical means an appropriate 'steering current', that is wind velocity at a particular 'steering level' or mean wind velocity in a particular 'steering layer', have had only limited success.

The movement of thundery precipitation belts is often found to accord well with the wind velocity at a level of 700 mb.

Stefan–Boltzmann law: The law, discovered empirically by Stefan and later shown theoretically by Boltzmann, that the total radiation in all directions from an element of a perfect radiator is proportional to the fourth power of its absolute temperature. See RADIATION.

steppe: A name given to the grassy, treeless plains in Russia and Siberia. The word is sometimes extended to mean similar plains and regions of semi-ARID CLIMATE elsewhere.

steradian: The unit of solid angle, being the solid angle subtended at the centre of a sphere of unit radius by a cap of unit area on the spherical surface. The whole sphere subtends a solid angle of 4π steradians at the centre of the sphere.

stereographic projection: See PROJECTION.

Stevenson screen: A standard housing for meteorological thermometers. It is similar to the 'large thermometer screen' but is smaller.

The Stevenson screen was designed by Thomas Stevenson, a civil engineer and father of Robert Louis Stevenson, the well-known author. See THERMOMETER SCREEN.

Stokes's law: A sphere of radius r m, moving with velocity v m/s through a fluid of viscosity η N s/m², experiences a viscous drag of F newtons, tending to oppose its motion, given by
$$F = 6\pi\eta\, rv \text{ newtons.}$$
This formula holds only for a small sphere moving with low velocity in conditions of laminar flow in a wide expanse of fluid. See also TERMINAL VELOCITY.

storage half-time: For a specified variable constituent of the atmosphere, the time required for one-half of the constituent to be removed from the atmosphere. See, for example, RADIOACTIVE FALLOUT.

storm: The term 'storm' is commonly used for any violent atmospheric phenomenon, such as a gale, thunderstorm, line-squall, rainstorm, duststorm, and snowstorm. In synoptic meteorology, the term is applied to an active centre of low pressure with which are associated gales, precipitation, etc.

storm cone: See GALE WARNING.

storm, eye of: See EYE OF STORM.

storm surge: A deviation, positive or negative, of the observed tide from the computed astronomical tide at the corresponding time and place. The storm surge is an essentially dynamical phenomenon; wind velocity is the main cause, with static pressure a minor contributory factor. A notable storm surge occurred in the North Sea on 31 January and 1 February, 1953 and caused widespread flooding in adjacent land areas.

s.t.p.: An alternative for N.T.P.

stratiformis (str): A CLOUD SPECIES (Latin, *stratus* flattened and *forma* appearance).
'Clouds spread out in an extensive horizontal sheet or layer. This term applies to ALTOCUMULUS, STRATOCUMULUS and, occasionally, to CIRROCUMULUS.'* See also CLOUD CLASSIFICATION.

stratocumulus (Sc): One of the CLOUD GENERA. (Latin, *stratus* flattened and *cumulus* heap).
'Grey or whitish, or both grey and whitish, patch, sheet or layer of cloud which almost always has dark parts, composed of tessellations, rounded masses, rolls, etc., which are non-fibrous (except for virga) and which may or may not be merged; most of the regularly arranged small elements have an apparent width of more than five degrees.'† See Plates 11 and 17. See also CLOUD CLASSIFICATION.

stratopause: The atmospheric boundary between the stratosphere and MESOSPHERE. See STRATOSPHERE.

stratosphere: That region of the ATMOSPHERE, lying above the TROPOSPHERE and below the MESOSPHERE, in which, in contrast to these regions, temperature does not decrease with increasing height. The stratosphere therefore extends from the TROPOPAUSE to a height of about 50 km, where the temperature reaches a maximum. Subdivision of the region is sometimes made into the 'lower', 'middle' and 'upper'

---

* Geneva, World Meteorological Organization. International cloud atlas, Vol. 1. Geneva, WMO, 1956, p. 12.
† Geneva, World Meteorological Organization. International cloud atlas, Vol. 1. Geneva, WMO, 1956, p. 11.

stratosphere, with approximate limits from tropopause to 20 km, 20 to 30 km, and 30 to 50 km, respectively.

An alternative definition of the stratosphere as that region from the tropopause to about 20 km in which temperature changes little with height, is not now favoured.

The stratosphere is a region which is characterized by relatively large amounts of ozone but by amounts of water vapour which are lower (mixing ratio of the order 10^{-2} g/kg or less) than in the high troposphere. These constituent gases, together with carbon dioxide, largely determine the radiation balance which, in general, controls the vertical temperature distribution of this region. Despite the absence of convective motion, the (lower) stratospheric region has vigorous circulations which are often clearly related to low-level pressure systems.

stratospheric oscillation: An alternative for QUASI-BIENNIAL OSCILLATION.

stratospheric warming: See SUDDEN WARMING.

stratus (St): One of the CLOUD GENERA (Latin, *stratus* flattened).

'Generally grey cloud layer with a fairly uniform base, which may give drizzle, ice prisms or snow grains. When the sun is visible through the cloud, its outline is clearly discernible. Stratus does not produce halo phenomena except, possibly, at very low temperatures.

Sometimes stratus appear in the form of ragged patches.'* See Plate 12. See also CLOUD CLASSIFICATION.

streak lightning: LIGHTNING discharge which has a distinct main channel, often tortuous and branching; the discharge may be from cloud to ground or from cloud to air.

stream function: At a level of non-DIVERGENCE in a horizontal air current a stream function (ψ) may be defined such that

$$\mathbf{V} = \mathbf{k} \wedge \nabla \psi$$

where $\mathbf{V}$ is the wind velocity vector, $\mathbf{k}$ the vertical unit vector and $\nabla \psi$ the GRADIENT of the stream function.

The wind velocity vector is normal to, and to the left of, $\nabla \psi$, that is the wind blows along the isopleths of ψ (and with low values to the left). The isopleths of ψ are therefore streamlines—hence the term 'stream function'. ψ is a scalar quantity with dimensions L^2T^{-1}.

streamline: A curve which is parallel to the instantaneous direction of the wind vector at all points along it. Isobars are streamlines only in strict GEOSTROPHIC FLOW.

Student's *t*-test: A statistical test which is used to indicate whether a given sample is derived from a given normal distribution, or to indicate whether two samples are derived from the same normal distribution. It is recommended when dealing with small samples and its use is described in statistical textbooks.

Stüve diagram: An AEROLOGICAL DIAGRAM with rectangular co-ordinates of T (temperature) and p^κ (p = pressure, $\kappa = (c_p - c_v)/c_p$ = constant (c_p and c_v are specific heats of dry air at constant pressure and volume, respectively).

---

* Geneva, World Meteorological Organization. International cloud atlas, Vol. 1. Geneva, WMO, 1956, p. 12.

subcooling: A seldom-used alternative for SUPERCOOLING.

subjective forecast: A FORECAST in which, in contrast to an OBJECTIVE FORECAST, the personal judgement of the forecaster plays a significant part. A forecast in SYNOPTIC METEOROLOGY, though based on physical and dynamical principles, is in some degree subjective.

sublimation: In chemistry, the conversion of a solid to a vapour, without melting. In meteorology, the term is applied with respect to water both in the above sense (direct evaporation from an ice surface) and in the converse sense (direct deposition of ice from water vapour).

sublimation nucleus: A type of NUCLEUS, the existence of which has not been definitely confirmed in the atmosphere, on which direct SUBLIMATION of water vapour to ice crystals may occur.

subsidence: The word used to denote the slow downward motion of the air over a large area which accompanies DIVERGENCE in the horizontal motion of the lower layers of the atmosphere. The greatest divergence is from regions of rapidly rising pressure and the subsidence is probably of the order of 30 to 60 m per hour in many cases. In stationary unchanging anticyclones the subsidence is due to the outward airflow at the earth's surface only, and is then very much slower. The subsiding air is warmed dynamically (see ADIABATIC) and its relative humidity therefore becomes low, occasionally falling below 10 per cent at about 1 to 2 km after prolonged subsidence. The downward movement, and consequent warming, increase with height, up to 3 km or perhaps more, so that the LAPSE rate of temperature is decreased, and INVERSIONS are often developed. The vertical velocity is zero at the horizontal ground, but turbulence often mixes up the lower layers and brings some of the warm dry air to the ground. Subsidence normally results in fine dry weather, but fog, stratus or stratocumulus clouds may occur in certain conditions.

substantial change: An alternative for LAGRANGIAN CHANGE.

subtropical high: One of the cells of high atmospheric pressure (ANTICYCLONES) which compose the quasi-permanent belt of high pressure of the 'subtropics' (i.e. that part of the earth's surface between the TROPICS and the 'temperate regions' whose equatorial boundaries are about 40°N and S). See GENERAL CIRCULATION.

sudden ionospheric disturbance: See SID.

sudden warming: A term applied to a relatively sudden temperature rise which occurs on some occasions in the stratosphere at higher latitudes, generally in late winter. The main warming, typically of about 50 degC in one or two weeks but sometimes much more rapid, occurs at levels of 25 km or above; modified effects occur, however, at lower levels. The warming is thought to be associated with downward motion of air at the levels concerned since, with the lapse rate which prevails in the stratosphere, relatively slow subsidence is able to produce appreciable warming. The phenomenon is also termed 'explosive warming'.

sulphur dioxide: Gas, of chemical formula SO_2, which occurs in minute and variable concentration in the atmosphere and is of industrial and volcanic origin. In populated regions sulphur dioxide is formed by oxidation of much of the sulphur content of coal or coke or of heavy fuel oils, on combustion. The gas is estimated to amount to about 3 per cent by weight of fuel burned. It dissolves readily in water to form sulphurous acid, and oxidizes photochemically in sunlight to sulphur trioxide which similarly becomes sulphuric acid. These acids cause damage by corrosion.

Sulphur dioxide concentration of air is measured either by finding the acidity of a sulphuric acid solution formed by the reaction between known volumes of air and hydrogen peroxide, or by exposing a prepared surface of lead peroxide to the air for a considerable time (usually a month) and measuring the yield of lead sulphate; in either case the sulphur-dioxide concentration of the air may be inferred. Typical annual mean values in Great Britain are, for country air 0·01, and for city air 0·08, part per million parts by volume. See ATMOSPHERIC POLLUTION.

sultriness: In meteorology, a combination of high atmospheric temperature and humidity.

sumatra: A SQUALL which occurs in the Malacca Strait, blowing from between south-west and north-west. There is a sudden change of wind from a southerly direction and a rise in speed is accompanied by a characteristic cloud formation—a heavy bank of cumulonimbus which rises to a great height. These squalls usually occur at night, and are most frequent between April and November. They are generally accompanied by thunder and lightning and torrential rain. There is a sudden fall of temperature at the moment the squall arrives.

summer: See SEASONS.

sun: A luminous gaseous sphere round which the earth moves in a slightly elliptical orbit at an average distance of $1·4953 \times 10^8$ km. The sun's diameter is $1·3914 \times 10^6$ km, its apparent diameter at the earth's mean distance 0·533 degree, its mass $1·9866 \times 10^{30}$ kg, its mean density $1·41 \times 10^3$ kg/m³. The RADIATION emitted from the sun's luminous disc (photosphere) corresponds to a black-body radiation temperature of about 5800 K, the internal gases being at a temperature of many millions of degrees. The gaseous regions above the photosphere, visible during solar eclipses, comprise the REVERSING LAYER, the CHROMOSPHERE, and finally the solar CORONA which extends outwards to a distance of several solar diameters. The Fraunhofer spectral lines due to absorption by gases in these regions show the presence of terrestrial elements. The SYNODIC PERIOD of rotation of the sun, as judged by sunspot movement is 26·9 days in latitude 0° and 28·3 days in latitudes $\pm 30°$.

The energy output of the sun, both particles and waves, varies with time. This so-called 'solar activity' is associated with disturbances which are observed in the photosphere and solar atmosphere and which are in large measure interrelated. Chief among the solar disturbances are SUNSPOTS and SOLAR FLARES but they also include, for example, faculae, flocculi, prominences and outbursts of solar 'radio noise'. The sun is described as 'quiet' or 'disturbed' if the disturbances are relatively few and weak or numerous and active, respectively. The relationships between solar activity and various geophysical phenomena to which they give rise are termed SOLAR–TERRESTRIAL RELATIONSHIPS.

sun dog: A popular alternative for mock sun or PARHELION.

sun drawing water: See CREPUSCULAR RAYS.

sun pillar: A vertical column of light above (and sometimes below) the sun, most often observed at sunrise or sunset. The colour may be white, pale yellow, orange or pink. The phenomenon, which is due to the reflection of sunlight from ice crystals, may be seen over a wide area. See Plate 25.

PLATE 25   Sun pillar: Jersey.
Part of the upper tangent arc to the 22° halo (not itself visible) is also seen and has, at this low solar elevation, a marked V-shape.

PLATE 26 Swell proceeding ahead of a tropical storm (actually approaching Nassau, Bahamas, from West Indian hurricane). Note that the swell waves are long, low and long-crested. The wind is blowing almost at right angles to the direction of motion of the swell (i.e. parallel to the crests), as is made clear by the breaking of sea waves in the foreground.

sunrise and sunset colours: The general explanation of the variety of colours that are to be seen in the sky about the time of sunrise or sunset is as follows. White light such as that from the sun may be regarded as composite, the constituents being light of all the colours of the spectrum. When the light waves meet obstacles in their course, such obstacles as the molecules of the atmospheric gases or larger obstacles such as particles of dust, the waves are broken and secondary waves proceed in all directions from the obstacles. The direct light is therefore reduced in strength and the farther the light goes through an atmosphere of such obstacles the more the strength is reduced, the energy being used up in producing scattered light. This effect is more pronounced with blue light, for which the wavelength is short, than with red light for which the wavelength is longer. Accordingly a beam of white light passing through air loses the constituents of shorter wavelength and becomes yellow, then orange, and finally red.

This accounts for the changing colour of the sun as it nears the horizon. The SCATTERING by air alone merely makes the setting sun yellow, but if there is dust in the air or even the nuclei on which water vapour is condensed then the sun becomes orange or red before it sets.

Clouds which are illuminated by the light from the setting sun are also red, whilst other clouds which are illuminated by scattered light in which the blue constituents are present are white or grey; higher clouds illuminated by light which has only passed through less dense and cleaner air may also appear white.

The colours of the sky itself are to be explained in the same way. When sunlight has already travelled a great distance through the lower atmosphere it has lost the constituents of short wavelength, and in the light which remains to be scattered the longer wavelengths predominate. In the further passage of scattered light to the eye of the observer the longer wavelengths again have the preference.

When we look at the sky in a particular direction we receive light which has been scattered by the atmosphere at all heights. Blue may predominate, whilst that coming from the lower levels may be red. The combination of light from both ends of the spectrum gives us purple. On the other hand in other parts of the sky the middle wavelengths may predominate and the resulting colours are green or yellow.

sunrise, sunset: The times at which the sun appears to rise and set, in consequence of the rotation of the earth on its axis. Owing to the effect of atmospheric REFRACTION, which increases the apparent angular altitude of the sun when near the horizon by about 34′, sunrise is earlier and sunset later than geometrical theory indicates. There is a further uncertainty caused by the fact that the sun has an appreciable diameter (32′) so that time elapses between the first and last contacts with the horizon.

For meteorological purposes allowance is made for normal refraction of 34′ and it is assumed that, for an observer at sea level with a clear horizon, sunrise and sunset occur when the sun's upper limb contacts the apparent horizon; at such a moment the true centre of the sun's disc is 50′ below the horizon.

The times of sunrise and sunset vary with latitude and with the declination of the sun. Diagrams illustrating the variations, so far as the British Isles are concerned, are given in the *Observer's handbook*.*

sunshine: Direct, as opposed to diffuse, solar RADIATION.

The routine measurements of the duration of sunshine which are made for climatological purposes refer in the British Isles, as in most other countries, to so-called 'bright' sunshine. Since different instruments differ in their response characteristics to the radiation, this term has lacked precise definition—but see

---

* London, Meteorological Office. Observer's handbook, 3rd edn. London, HMSO, 1969, pp. 184–186.

SUNSHINE RECORDER. Climatological means for places in the British Isles for the period 1931–60 are contained in *Averages of bright sunshine for Great Britain and Northern Ireland*, 1931–60.†

sunshine recorder: An instrument for recording the duration of bright SUNSHINE. Such instruments depend either on the heating action of the sun or on the chemical action produced by the sun's rays. The Campbell–Stokes recorder, an instrument of the former class, is in general use in the British Isles. A spherical glass lens focuses the solar image on a graduated card held in a frame of special design. The duration of sunshine is indicated by the length of the burnt track of the image. The World Meteorological Organization decided in 1962 to adopt the Campbell–Stokes recorder, used with record cards to specifications of the French Meteorological Office, as a standard of reference to which values of sunshine duration should be reduced in future. See also EXPOSURE.

sunspot: A relatively dark region on the disc of the SUN, with an inner 'umbra' of effective radiation temperature about 4500 K and an outer 'penumbra' of somewhat higher temperature.

Sunspot duration varies from a few hours to many solar rotation periods. Their frequency is quasi-periodic, with an average 'period' of about 11 years. In the typical sunspot 'cycle' there are at first few spots in about solar latitude $\pm 30°$, maximum spots in about $\pm 15°$ after some $4\frac{1}{2}$ years, and again few spots in about $\pm 8°$ after a further $6\frac{1}{2}$ years, these last spots overlapping in time the first high-latitude spots of the following cycle. There are, however, some large departures from these average figures. The mean time taken for a spot in solar latitude $0°$ to return to the central meridian, as seen from the earth, is 26·9 days; in latitude $\pm 30°$ the time taken is 28·3 days.

Sunspots are vortex-like disturbances with large associated magnetic fields. There is yet no accepted theory of their formation or quasi-periodic nature. There are well-established relationships of a general nature between sunspottedness and effects measured in the ionosphere, geomagnetism, etc., but not as yet in meteorology.

Sunspottedness is represented by the 'relative sunspot number' (R), introduced by R. Wolf of Zürich Observatory and expressed by the formula

$$R = k(10g + f)$$

in which g is the number of groups of spots plus single spots, f is the total number of spots counted in the groups and spots combined, and k is a factor whose value depends on the viewing instrument. The value of R for each day is compiled at Zürich from the results obtained at many observatories. Investigation has shown that the sunspot numbers derived in this arbitrary way are a reasonable index of 'spotted area' of the sun. $R = 100$ corresponds to about 1/500 of the sun's visible disc covered by spots.

The list of mean annual relative sunspot numbers since 1750 is contained in Table XII.

superadiabatic lapse rate: A 'lapse rate' (rate of fall of temperature with height) greater than the dry ADIABATIC lapse rate of 1 degC/100 m. Such a LAPSE rate does not occur within the free atmosphere, but the dry adiabatic rate is often exceeded by a factor of several times near a land surface which is strongly heated by solar radiation.

---

† London, Meteorological Office. Averages of bright sunshine for Great Britain and Northern Ireland, 1931–60. London, HMSO, 1963.

TABLE XII—*Sunspot numbers 1750–1969*

	0	1	2	3	4	5	6	7	8	9
1750	83	48	48	31	12	10	10	32	48	54
1760	63	86	61	45	36	21	11	38	70	106
1770	101	82	66	35	31	7	20	92	154	126
1780	85	68	38	23	10	24	83	132	131	118
1790	90	67	60	47	41	21	16	6	4	7
1800	14	34	45	43	48	42	28	10	8	2
1810	0	1	5	12	14	35	46	41	30	24
1820	16	7	4	2	8	17	36	50	64	67
1830	71	48	28	8	13	57	122	138	103	86
1840	63	37	24	11	15	40	62	98	124	96
1850	66	64	54	39	21	7	4	23	55	94
1860	96	77	59	44	47	30	16	7	37	74
1870	139	111	102	66	45	17	11	12	3	6
1880	32	54	60	64	64	52	25	13	7	6
1890	7	36	73	85	78	64	42	26	27	12
1900	10	3	5	24	42	64	54	62	48	44
1910	19	6	4	1	10	47	57	104	81	64
1920	38	26	14	6	17	44	64	69	78	65
1930	36	21	11	6	9	36	80	114	110	89
1940	68	48	31	16	10	33	93	152	136	135
1950	84	69	31	14	4	38	142	190	185	159
1960	112	54	38	29	10	15	47	94	106	106

supercooling: Supercooling of a liquid (sometimes termed 'subcooling' or 'undercooling') signifies the existence of a substance in the liquid state at a temperature below the normal freezing-point.

Although the supercooled state is regarded as unstable in the sense that its achievement in the laboratory requires very careful cooling of the liquid, supercooling of cloud droplets is common in the atmosphere. All clouds which extend above the 0°C isotherm contain supercooled droplets at some stage in their history; in particular, altocumulus is predominantly a water-droplet cloud though generally at a temperature well below 0°C and it is necessary to attain cirrus level to find clouds almost invariably in the form of ice crystals. The larger raindrops do not undergo a marked degree of supercooling.

Supercooling is fundamental in the ice-crystal process of PRECIPITATION.

superior air: A term sometimes used, in synoptic meteorology, in respect of air at higher levels which has been made very dry by the process of SUBSIDENCE.

superposed-epoch method: A method of statistical analysis (also called the '*n*-method') which is used, for example, to investigate the possibility of a RECURRENCE TENDENCY in a given TIME SERIES, or the relationship between two synchronous time series.

Where a single series is involved, average values are calculated of the *n* terms $(+1, +2, \ldots +n)$ of the series which follow various '0' terms selected on the basis of an objective criterion, for example the peak value in successive equal blocks of terms; statistical SIGNIFICANCE may be looked for in the departures of the computed average values from the over-all mean of the series, taking into account the standard deviation of the series.

Where the relationship between two series is involved, average values of terms

$+1$, $+2$, ... $+n$ are computed in each series, the epoch of the '0' term being selected on the basis of a criterion applied to one of the series; the significance of the distribution of the means may be assessed by the CHI-SQUARE TEST, for example. Corroboration of a suggested relationship between the series may be sought by defining the '0' epoch on the basis of a criterion applied to the second series and repeating the calculations on the data arranged on this basis.

superrefraction: See ANOMALOUS RADIO PROPAGATION.

supersaturation: See SATURATION.

surface chart: In synoptic meteorology, an alternative for WEATHER MAP.

surface free energy: An alternative for SURFACE TENSION.

surface inversion: An INVERSION of temperature through an atmospheric layer extending upwards from the earth's surface. This is frequently a RADIATION INVERSION, but it may form also as the result of a drift of air over a surface colder than itself.

surface temperature: Unless otherwise specified, the air temperature measured in the shade at a height of between 1·25 and 2 metres, as in the THERMOMETER SCREEN.

surface tension: Any surface of a liquid is subject to a tension, expressed in newtons per metre in SI UNITS, due to forces of attraction between the liquid molecules which act in such a way as to minimize the surface area.

The phenomenon, which is also termed 'surface free energy' (J/m^2), is important, for example, in NUCLEATION processes and in SOIL MOISTURE. The surface tension of pure water increases with decrease of temperature; values at $+20°C$, $0°C$ and $-20°C$ are, for example, $7·27 \times 10^{-2}$, $7·57 \times 10^{-2}$ and $7·91 \times 10^{-2} N/m$, respectively.

surface wind: Generally, the WIND velocity at a height of 10 metres in an unobstructed area.

surge: A term in synoptic meteorology which is sometimes used (first by Abercromby) to denote a substantial and general rise of atmospheric pressure over an area, the rise being greater than that attributable to the movement of depressions or anticyclones in the vicinity.

In tropical meteorology the term is used to signify a marked and sudden increase of strength of a monsoon or trade-wind current.

The term is also used of water disturbances—see STORM SURGE.

swell: Swell is wave motion in the ocean caused by a disturbance which may be at some distance away; the swell may persist after the originating cause of the wave motion has ceased or passed away. It often so continues for a considerable time with unchanged direction, as long as the waves travel in deep water. The height of the waves rapidly diminishes but the length and velocity remain the same, so that the long low regular undulations, characteristic of swell, are formed. Swell is often observed to have a wavelength greatly in excess of that of waves seen during a storm; the probable explanation is that the longer waves are then masked by the shorter and steeper storm waves. Swell observations are useful in denoting the direction in which sea disturbance due to tropical cyclones or other storms has taken place. See Plate 26. See OCEAN WAVES.

SWF: An abbreviation for 'short-wave (radio) fadeout'. When a SOLAR FLARE occurs, short radio-waves, which are normally received after reflection by the F-layer, suffer extra absorption in the sunlit hemisphere D-layer, with an associated sudden fadeout of reception of the waves. See also SID.

symbols, international meteorological: See WEATHER MAP.

symmetry point: In a graph of successive daily values of a meteorological element (particularly pressure) against time, a point about which the pattern displays a high degree of reflection over a period lasting perhaps several weeks; the reflection may be direct or reversed.

It has been suggested that, if such a symmetry point were recognized soon after its occurrence, the implied foreknowledge of subsequent variations of the element would be useful in long-range weather forecasting. Apart from the obvious difficulty of recognizing a symmetry point at a sufficiently early stage, the physical significance of the phenomenon appears very doubtful in the light of the statistical finding that, when the day-to-day PERSISTENCE of the element is taken into account, symmetrical patterns are no more remarkable or frequent than may be accounted for by chance.

synodic period: The synodic period of a planet or the moon is the interval of time between successive CONJUNCTIONS of the body and the sun, as viewed from the earth. See also DAY.

synoptic chart: A chart or map on which is represented the distribution of selected meteorological elements over a large area at a specified instant of time. 'Surface synoptic chart' is an alternative for WEATHER MAP.

synoptic meteorology: That branch of meteorology which is concerned with a description of current weather as represented on geographical charts and applied especially to the prediction of its future development.

The starting point for this branch of meteorology is the 'synoptic report' or 'synoptic message' containing a coded summary of the current weather at each of a large number of 'synoptic stations'. Many such reports, transmitted by land-line or wireless are, on receipt at selected centres, plotted symbolically on a 'synoptic chart' which thus provides a representation of the weather at a particular time, and generally over a large geographical area, in a synoptic (summary or condensed) form which is suitable for the purposes of synoptic meteorology, as defined above.

synoptic station: A STATION at which meteorological observations are made for the purposes of SYNOPTIC METEOROLOGY. The observations are made at the 'major synoptic hours' 00, 06, 12, 18 GMT and normally also at the 'minor synoptic hours' 03, 09, 15, 21 GMT, the observed elements being plotted symbolically on the WEATHER MAP.

T

tablecloth: The term applied to the orographic stratus cloud which often occurs on the windward side of Table Mountain at Capetown, South Africa.

tail wind: See EQUIVALENT HEAD WIND.

Technical Commission: The Technical Commissions of the WORLD METEOROLOGICAL ORGANIZATION are each composed of experts in the various meteorological fields, Atmospheric Sciences, Aeronautical Meteorology, Agricultural Meteorology, Climatology, Hydrometeorology, Instruments and Methods of Observation, Maritime Meteorology, and Synoptic Meteorology. Meetings of the Commissions take place at least once every four years.

telluric currents: An alternative for EARTH CURRENTS.

temperature: The condition which determines the flow of heat from one substance to another. Temperature has no recognized DIMENSIONS, the normal convention being to allot to it the special dimension θ.

temperature scales: The scales in common use are the Celsius (or centigrade), the Fahrenheit and the kelvin (or absolute) scales.

On the Celsius (centigrade) scale the freezing- and boiling-points of water at standard pressure are respectively 0°C and 100°C; on the Fahrenheit scale these points are 32°F and 212°F, the zero on this scale being the temperature of a mixture of common salt and ice. Alternative conversion formulae °C to °F and vice versa are:

(1) $\quad °C = (°F - 32) \times \dfrac{5}{9} \quad$ and $\quad °F = \left(°C \times \dfrac{9}{5}\right) + 32,$

(2) $\quad °C = (°F + 40) \times \dfrac{5}{9} - 40 \quad$ and $\quad °F = (°C + 40) \times \dfrac{9}{5} - 40.$

The use of the gas thermometer is based on the gas equation, combining BOYLE'S and CHARLES'S LAWS in the form

$$\frac{p_1 v_1}{p_2 v_2} = \frac{1 + \alpha t_1}{1 + \alpha t_2}$$

where α is the volume coefficient of expansion of the gas, found experimentally to be nearly the same for all pure gases and to be identical on extrapolation to zero pressure, the state of 'perfect gas', with the value 0·0036609 per degC. The equation shows that zero pressure occurs at the temperature $t = -1/\alpha$, i.e. $1/\alpha = 273\cdot15$ Celsius degrees below the ice point (0°C) and defined as the 'absolute zero'. Kelvin showed from thermodynamic reasoning that the perfect-gas scale, measured from absolute zero in Celsius units, is 'absolute' in the sense that it is independent of the thermometric substance. The symbol K is now recommended in preference to °A for temperatures measured on this kelvin (absolute) scale. It is the appropriate scale to use in basic physical equations which involve temperature. Its use has also the practical advantage of avoiding negative values. In meteorological practice $K = 273 + °C$ with sufficient accuracy.

tendency: A term used in synoptic meteorology to signify local time rate of change of an element, for example of surface pressure or of GEOPOTENTIAL (height) at a fixed pressure level. See BAROMETRIC TENDENCY.

tendency equation: The equation relating the change of pressure with time at a point at height h in the atmosphere (pressure TENDENCY) to the change in the weight of the air above is

$$\left(\frac{\partial p}{\partial t}\right)_h = -g \int_h^\infty \rho \left(\frac{\partial u}{\partial x} + \frac{\partial v}{\partial y}\right) dz - g \int_h^\infty \left(u \frac{\partial \rho}{\partial x} + v \frac{\partial \rho}{\partial y}\right) dz + (g\rho w)_h$$

where g is the gravitational acceleration (assumed here constant with height), ρ air density, and u, v, w the components of the wind velocity in the x, y, z directions, respectively (z the vertical co-ordinate).

The equation shows that the local pressure tendency at height h has contributions from three processes (the three terms, taken in order):
  (i) horizontal DIVERGENCE at heights greater than h,
  (ii) horizontal ADVECTION of air of different density at heights greater than h, and
  (iii) vertical motion of air at height h.

The third process does not normally operate in the surface pressure tendency $(\partial p/\partial t)_0$ because of the boundary condition $w = 0$ at the earth's (level) surface.

tenuity factor: In BALLISTICS, the ratio of the density of the air having the observed pressure at the surface and temperature equal to the BALLISTIC TEMPERATURE, to the density of air at pressure 30 in of mercury and temperature 60°F (1015 mb and 16°C).

tephigram: An AEROLOGICAL DIAGRAM with Cartesian co-ordinates T (temperature) and $\log \theta$ (potential temperature). See Figure 2.

tercentesimal scale: The name for the approximate absolute scale, sufficiently accurate for meteorological purposes, obtained by adding 273° to the Celsius temperature. See TEMPERATURE SCALES.

tercile: See PERCENTILE.

terminal velocity: The velocity of a body, falling through a fluid, reaches a 'terminal' value when the weight of the body is balanced by the combined upthrust and drag due to the fluid.

For very small particles, viscous drag predominates, STOKES'S LAW is obeyed, and the terminal velocity v of a sphere of radius r m and density σ kg/m³ is given by:

$$v = \frac{2}{9} g \frac{r^2(\sigma - \rho)}{\eta} \text{ m/s,}$$

where ρ is the density of the fluid in kilogrammes/metre³, η its viscosity in newton second/metre² and g the gravitational acceleration in metres/second². The expression is accurate to 1 per cent for water-drops up to 30 µm in radius. For larger drops, inertial drag becomes more important and the empirical relation $v = 0·085\,r - 0·15$ m/s (r = radius in µm) fits well for drops from 30 to 500 µm in radius. Drops with radius larger than 500 µm become non-spherical when falling and the expression

$v = 6 \cdot 5 \sqrt{r} - 0 \cdot 7 (1 \cdot 3 - r)^2$ m/s (r = radius in mm) is accurate to within 3 per cent for drops from 0·5 mm radius up to the largest raindrops (3 mm radius) which have a limiting terminal velocity of about 9 m/s. Terminal velocities at heights of 3 and 6 km exceed those at ground level by about 10 and 30 per cent respectively because of changes in air temperature and pressure.

Approximate terminal velocities are: for a single ice crystal 0·5 m/s, for a snowflake 1 m/s and for a rimed snowflake 2 m/s.

The terminal velocity of hailstones is a function of density, shape and surface roughness as well as radius and no formula will fit all stones. However, the terminal velocity is approximately proportional to $\sqrt{r}$ and the largest hailstones (about 50 mm radius) have terminal velocities around 45 m/s.

terrestrial magnetism: See GEOMAGNESTISM.

terrestrial radiation: RADIATION (also termed 'long-wave radiation' or, loosely, 'infra-red radiation'), which is emitted by the earth and atmosphere in the approximate temperature range 200–300 K. The radiation is confined within the wavelengths of about 3 and 100 μm and has maximum intensity at about 10 μm. Since the earth is nearly a perfect radiator, the radiation from its surface varies in close accord with the Stefan–Boltzmann law, i.e. directly as the fourth power of the absolute temperature of the surface.

tesla: Unit (T) of magnetic field strength in SI UNITS. It is that magnetic field intensity which results in a torque of 1 newton metre on a coil of magnetic moment 1 ampere metre placed with its axis perpendicular to the magnetic field. The tesla is 1 WEBER/metre^2.

The tesla is related to the GAUSS, the corresponding (obsolescent) unit in the C.G.S. SYSTEM, by the relationship

$$1 \text{ tesla} = 10^4 \text{ gauss}.$$

See also GAMMA.

thaw: The transition by melting from snow or ice to water. The term is especially used to indicate the end of a spell of FROST, which in the British Isles in winter is generally associated with the displacement of a stagnant or continental air mass by one of maritime origin.

theodolite, pilot-balloon: An instrument consisting of a telescope mounted to permit of rotation in elevation and azimuth, and fitted with a right-angled prism so that the observer continues to look horizontally into the eyepiece no matter what the elevation of the balloon.

thermal: A volume of air which possesses BUOYANCY on account of low density relative to its environment and so rises through the environment.

Thermals are produced in conditions of intense solar heating over land, with a resulting SUPERADIABATIC LAPSE RATE at low levels. Strong thermals tend to occur over regions where the earth's surface is warmer than the surrounding area; sunfacing slopes and towns are among the good thermal sources recognized by glider pilots. Cumulus clouds show the presence of thermals which, however, often exist without such evidence.

Observations have shown that the typical pattern of air motion within a thermal is one in which maximum vertical velocity occurs within the central core, with circulatory motions ('vortex rings') towards either edge of the thermal. A thermal grows in size by mixing with surrounding air at its upward-moving head and by the mixing of air into its wake. The temperature excess and buoyancy of the thermal relative to its environment are thus progressively reduced.

thermal capacity: The thermal capacity, also called heat capacity, of a body is the product of its mass and its SPECIFIC HEAT.

thermal conductivity: See CONDUCTIVITY, THERMAL.

thermal diffusivity: See CONDUCTIVITY, THERMAL.

thermal equator: The latitude of highest mean air temperature. Because of the non-uniform distribution of land and sea, this does not coincide, over a year, with the geographic equator. In northern summer the thermal equator is at about latitude 20°N, and in northern winter it is at about 0°, averaging about 10°N for the year as a whole.

thermal high, low: A closed centre of high (low) values of THICKNESS on a thickness chart. The centre is so called because it represents, to a close approximation a centre of high (low) mean temperature in the isobaric layer concerned. But see also THERMAL LOW.

thermal low: A surface depression whose formation is the direct result of differential solar heating of neighbouring land and sea areas; in summer, for example, air density and therefore also surface pressure over land tend to be reduced relative to the values over sea, because of the higher surface temperatures reached.

A thermal low may form over the British Isles during afternoon and evening in summer if the general pressure gradient is very slight; such a depression is weak, has little vertical extent and has no pronounced associated weather characteristics. Another example is the tendency for formation of shallow depressions in winter over the relatively warm Mediterranean. The Asiatic 'monsoon low' is an example of a thermal depression on a much larger scale. See also THERMAL HIGH, LOW.

thermal precipitator: An apparatus used for sampling small particles, especially Aitken and condensation nuclei—see NUCLEUS.

Air is drawn past a heated wire at a slow, controlled speed. The particles are unable to penetrate a dust-free barrier surrounding the wire and form line deposits on, for example, cold glass surfaces located at a convenient distance from the wire; the deposits are examined and counted microscopically.

thermal steering: The principle of thermal steering states that surface patterns of the vertical component of vorticity (corresponding closely to surface depressions and anticyclones) are 'steered' in the direction of the THERMAL WIND in the troposphere, with a speed proportional to that of the thermal wind. The process is one of forward development, rather than translation, of the vorticity patterns.

Thermal steering is always present but is often masked by other processes of development. Application of the steering principle is most successful in the type of situation illustrated in Figure 31, i.e. one in which almost straight thickness lines intersect a well-marked pattern of surface vorticity. See DEVELOPMENT.

thermal wind: The thermal wind ($\mathbf{V}_T$) in a specified atmospheric layer, at a given time and place, is the vertical geostrophic WIND SHEAR in the layer concerned. Where $\mathbf{V}_1$ and $\mathbf{V}_0$ are the geostrophic winds at the top and bottom of the layer, respectively, $\mathbf{V}_T$ is defined by the relationship

$$\mathbf{V}_T = \mathbf{V}_1 - \mathbf{V}_0.$$

The term 'thermal wind' was adopted because wind shear is determined by the distribution of mean temperature in the layer concerned. The direction of the wind is such that it blows parallel to the mean isotherms (or THICKNESS lines) of the layer and in the sense such as to keep low mean temperature (or low thickness) on the left in the northern hemisphere and on the right in the southern hemisphere. Its magnitude is given by the expression

$$V_T = \left| \frac{g}{f} \frac{\triangle z'}{\triangle n} \right|$$

where g is the gravitational acceleration, f the CORIOLIS PARAMETER, and $\triangle z'/\triangle n$ the gradient of thickness of the layer (expressed in geometrical units). See also VECTOR, HODOGRAPH ANALYSIS, POLAR VORTEX.

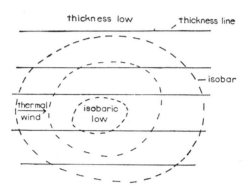

FIGURE 31—Thermal steering.

thermocline: An oceanic layer in which the rate of decrease of temperature with increasing depth is a maximum.

A thermocline is a stable region in which vertical mixing of water is strongly inhibited. A permanent thermocline exists at a depth of some hundreds of metres in low and middle latitudes. In summer, a 'seasonal thermocline' is formed at a shallower depth, especially in middle latitudes, because of the heat which is then absorbed in the surface layers and distributed by wave motion throughout a limited depth.

thermocouple: An instrument for measuring temperature. It consists basically of two wires of different metals joined at each end. One junction is kept at a fixed (known) temperature, the other put at the point where the temperature is to be measured. A thermoelectric e.m.f. is generated, of magnitude proportional to the temperature difference (for two given metals); this e.m.f., or the resulting electric current in the circuit, may be used as a measure of the temperature difference between the junctions.

thermodynamic diagram: A diagram on which may be represented graphically the states of air samples (or the varying state of a single air sample) in terms of pressure, temperature, and humidity or of other functions related to these. It is also termed an 'adiabatic diagram' or, in meteorology, an AEROLOGICAL DIAGRAM, many forms of which exist.

The term 'thermodynamic diagram' is restricted by many authors to those types of adiabatic diagram in which the area enclosed on the diagram by a curve which represents a cyclic thermodynamic process is proportional in all parts of the diagram to the amount of work performed in the process.

thermodynamics: That part of the science of heat which deals with the transformation of heat into other forms of energy, and vice versa.

thermodynamic temperatures: Definitions of thermodynamic dew-point, frost-point, ice-bulb, and wet-bulb temperatures are identical with those of DEW-POINT, FROST-POINT, ICE-BULB TEMPERATURE, and WET-BULB TEMPERATURE. The distinction in nomenclature is made in recognition of the fact that the measured values of these elements may differ slightly from the defined values because of instrumental limitations or procedure.

thermogram: The continuous record of temperature yielded by a THERMOGRAPH.

thermograph: A recording THERMOMETER. Many patterns are in use, the sensitive member being, for example, a bimetallic spiral, a Bourdon tube, a resistance element, or a steel bulb filled with mercury.

thermohygrograph: An alternative for HYGROTHERMOGRAPH.

thermometer: An instrument for measuring temperature, from one or other of the physical changes produced in matter by heat, e.g. expansion of solids, liquids or gases, changes in electrical resistance, production of electromotive force at the junction of two different metals, etc.

In normal meteorological practice, mercury-in-glass thermometers are used. At temperatures below $-38.9°C$ (freezing-point of mercury), and also in measuring minimum temperatures, alcohol (freezing-point $-114.4°C$) is used.

See BOURDON TUBE, EARTH THERMOMETER, MAXIMUM THERMOMETER, MINIMUM THERMOMETER.

thermometer screen: The standard housing for meteorological thermometers. It consists of a wooden cupboard, with hinged door, mounted on a steel stand, so that the wet-bulb and dry-bulb thermometers are 1·25 m (4 ft) above the ground. The whole screen is painted white. Indirect ventilation is provided through the bottom, double roof and louvered sides. Thermometers are placed within it to give a close approximation to the true air temperature, undisturbed by the effects of direct solar or terrestrial radiation.

The 'small thermometer screen' accommodates the dry-bulb, wet-bulb, maximum and minimum thermometers. The 'large thermometer screen' provides additional accommodation for a THERMOGRAPH and HYGROGRAPH. It also has another door at the back. See also STEVENSON SCREEN.

thermopile: An instrument for measuring RADIATION. It consists essentially of a number of THERMOCOUPLES, either connected in series (if e.m.f. is measured) or in parallel (if electric current is measured), as a means of increasing the sensitivity beyond that possible with a single thermocouple.

thermosphere: That part of the ATMOSPHERE, extending from the top of the MESOSPHERE at about 80 km to the atmosphere's outermost fringe, in which the temperature increases with increasing height.

thickness: The GEOPOTENTIAL height difference at a given place between specified pressure levels. Thickness values relating to selected standard pressure levels are obtained from radiosonde observations and are plotted on geographical 'thickness charts'. Contours are drawn, at an appropriate thickness interval, joining places of equal thickness and are termed 'thickness lines'. The analysis of such charts is termed 'thickness analysis' and has an important role in synoptic meteorology—see, for example, DEVELOPMENT.

thoron: Gas, of atomic mass 220 and atomic number 86, which is a radioactive isotope of RADON. It occurs in minute concentration in the atmosphere and plays a small part in the IONIZATION of the air at low levels.

thunder: The noise which follows a flash of LIGHTNING, caused by a sudden heating and expansion of air along the path of the lightning. The distance of a lightning flash may be roughly estimated from the interval between seeing the flash and hearing the thunder, counting one kilometre for every three seconds. The long duration of thunder compared with the associated lightning flash is explained by the different distances travelled by the sound from different parts of the flash and by echoing from neighbouring hills, Echoing causes intensity variations, which, however, also arise from the multiple and tortuous nature of many lightning strokes.

Thunder is seldom heard at distances greater than 20 km though distances up to about 65 km have been reported on occasion. Owing to refraction of sound waves in the lower atmosphere, thunder is sometimes inaudible at distances much less than 20 km, especially when the initiating lightning flash is not to ground.

thunderbolt: Popular term for a LIGHTNING discharge from cloud to ground.

thundercloud: Popular expression for CUMULONIMBUS, the cloud associated with thunderstorms.

thunderstorm: 'One or more sudden electrical discharges, manifested by a flash of light (lightning) and a sharp or rumbling sound (thunder.)*'

Necessary conditions for a thunderstorm are generally stated to be a CUMULONIMBUS cloud base lower than the 0°C isotherm, such vertical depth of cloud as to ensure a cloud top of temperature less than about $-20°C$ probably implying glaciation of the cloud top), and the occurrence of precipitation. There are, however, well substantiated observations, mainly in low latitudes, of the occurrence of lightning in clouds no parts of which were at a temperature below 0°C. The mature thundercloud often has a cellular structure and there are both updraught and downdraught regions; in earlier stages of development only updraught occurs, in later stages of dissipation only downdraught. Vertical updraught averages about 6 m/s and may reach 30 m/s; downdraught velocities are about half those of updraught.

Charge separation in thunderclouds is such as to produce a positive charge in the upper part of the cloud and negative charge in the lower part; small regions of positive charge near the base have also been observed. Initial charge separation has been attributed by various workers to mechanisms such as the selective capture of ions by water drops or ice particles, the frictional rupture of large raindrops, and surface and volume interactions of water, in its various phases, contained in the cloud. Experimental support has been obtained for most of the theories but the relative importance of the various mechanisms is yet uncertain.

Rainfall in a thunderstorm at a particular place typically reaches peak intensity more quickly than it dies away; behaviour is, however, sometimes complicated by the effects of two or more cells. Other well-marked surface effects are attributable to the arrival of the downdraught air which, cooled by evaporation of water drops, is accelerated earthwards. Surface wind suddenly becomes strong and gusty; temperature falls sharply, sometimes by 10 degC or more; relative humidity rises

---

* Geneva, World Meteorological Organization. International cloud atlas, Vol. 1. Geneva, WMO, 1956, p. 76.

unsteadily to nearly 100 per cent; and pressure rises sharply, reversing a previous fall. Thunderstorms are fairly frequently accompanied by hail. In Great Britain, snow in thunderstorms is mainly confined to exposed west and north coasts.

An adequate supply of moisture and a lapse rate of temperature in excess of the saturated adiabatic through a range of height not less than 3000 metres above cloud base are required for the development of a thunderstorm; high surface temperature is a common but not an essential condition. An initial 'trigger' action is often provided by orographic uplift or, especially, by horizontal convergence of surface air. This latter most frequently exists in a shallow depression, trough of low pressure (as at a cold front), or col. The line of convergence which often marks the farthest inland penetration of the sea-breeze may provide the necessary impulse.

In Great Britain thunder is a very variable element, the highest and lowest annual totals at many individual stations ranging from less than 5 in a quiet year to 20 or more in an active one. One consequence of this is that published maps showing the average frequency of days of thunder differ considerably in detail according to the period of records used. They agree, however, in showing that the average annual frequency is less than 5 days in western coastal districts and over most of central and northern Scotland and 15–20 days over the east Midlands and parts of south-east England. There is relatively little seasonal variation on the western seaboard but elsewhere summer is the most thundery season.

The most thundery part of the earth is the island of Java where the annual frequency of thunderstorms is about 220. C. E. P. Brooks has estimated that the earth has a total of about 44 000 thunderstorms per day and a total of about 100 lightning discharges each second.

thunderstorm day: A local calendar day on which THUNDER is heard.

tidal wave: A popular term for a destructive type of wave motion in seas and oceans, associated either with strong winds or with under-water earthquakes. In technical terms they are classified as STORM SURGE and TSUNAMI, respectively.

tide: The periodic rise and fall of the earth's oceans due to combined gravitational forces applied by the moon and sun. Similar, though more complex, effects occur in the earth's atmosphere—see ATMOSPHERIC TIDES.

tilt of a trough: The angular departure of a TROUGH line from a meridian. A trough whose axis is rotated anticlockwise from a meridian is said to have a positive tilt.

time: When the centre of the sun is due south of an observer, the time is called 12h or noon, local apparent time (LAT). The sun is said to 'transit' at this time. The interval between two consecutive transits of the sun is divided into 24 equal parts, and the times where the lines of division fall are numbered 13h, 14h ... 23h, 24h (or midnight), 1h 2h ... 12h. Local apparent time is recorded by sunshine recorders and sundials.

The interval of time between successive transits of the sun is not quite constant, but goes through a cycle of changes during the year. This is due to the two facts that the orbit of the earth is elliptical, and that the earth's axis is not at right angles to its orbit. As it would be very inconvenient in daily life if the length of the day varied in this way, astronomers have invented a 'mean sun' whose apparent motion round the earth is uniform throughout the year. The apparent positions of the real and mean suns are always very close, and coincide on the average of a year. The moment of transit of the mean sun is 12h (or noon) local mean time (LMT). The interval between successive transits is called a DAY and each day is divided into 24 equal parts. All hours, and all days, are equal in duration. Local mean time is obtained from local apparent time by adding (or subtracting, if sign is minus) the

'equation of time' which is given very accurately in *The astronomical ephemeris**
and elsewhere, and sufficiently accurately for meteorological purposes in the
Observer's handbook.† The equation of time varies from about $+14\frac{1}{2}$ minutes in
mid-February to about $-16\frac{1}{2}$ minutes in early November.

Local mean time at Greenwich is called Greenwich mean time (GMT). Differences in local mean time between different places are determined solely by longitude differences on the scale 1 hour per 15° of longitude. Thus, LMT for any place is derived from GMT by adding to GMT, or subtracting from GMT, a correction on this scale (4 minutes of time per degree of longitude), according as the place is to the east or west of the Greenwich meridian.

The inconvenient use of a clock time which varies continuously with longitude is avoided by the use of ZONE TIME. In the British Isles BRITISH STANDARD TIME, like BRITISH SUMMER TIME, is one hour in advance of GMT. However, GMT is adhered to throughout the year for meteorological observation purposes, other than evening reports for the Press from selected stations, which are based on clock time.

time series: In statistics, a series of values which are arranged in order of occurrence and which refer, in general, to equally spaced time intervals. Such series are common in geophysics. The values may be those which obtain at discrete intervals of time or which are averaged over successive periods of time.

Some degree of persistence normally exists between successive values in a time series, provided the time interval of separation is not too long. The presence of persistence implies that the number of statistically independent values in the series may be much smaller than the total number of values, with important implications in questions of statistical SIGNIFICANCE. See PERSISTENCE.

tipping-bucket switch: A switch, suitably fitted to the delivery tube of a RAIN-GAUGE, consisting of a delicately balanced bucket divided into two equal compartments each of which is adjusted to collect a precise quantity of rain. The compartments are presented to the delivery tube in the manner of a see-saw so that as one overbalances when the required amount of rain is collected, the other assumes the collecting position. Each tip causes a magnet to actuate a reed switch which, in turn, actuates an impulse counter or magnetic-tape event recorder.

tonne: Metric ton, equal to 10^3 kg (0·9842 ton).

topography: A term used to signify the configuration of contours of a surface. It is used both of natural contours which delineate higher and lower levels of the earth's surface, and, in synoptic meteorology, of contours of an isobaric surface.

tornado: A violent whirl, generally cyclonic in sense, averaging about 100 m in diameter and with an intense vertical current at the centre capable of lifting heavy objects into the air. Uprooting of trees and the explosive destruction of buildings, due to local pressure differences that occur in the intense horizontal pressure gradient near the tornado centre, mark the paths of tornadoes. The paths vary in length from a few hundred metres to some hundreds of kilometres: associated winds in extreme cases are estimated to attain speeds of about 200 knots. Heavy rain, and generally thunder and lightning, occur with the tornado.

---

* London, Her Majesty's Nautical Almanac Office. The astronomical ephemeris. London.

† London, Meteorological Office. Observer's handbook, 3rd edn. London, HMSO, 1969, pp. 224–225.

While the conditions required for the formation of a tornado are similar to those required for a severe THUNDERSTORM, namely great instability, high humidity, and horizontal convergence of winds at low levels, the precise conditions which cause tornadoes (rather than merely thunderstorms) are not yet known. Tornadoes are most frequent and intense in the United States, east of the Rocky Mountains, especially in the central plains of the Mississippi region where they form in unstable air of tropical origin and move towards north or north-east.

Destructive tornadoes have been reported in the British Isles (mainly south and central) with a frequency somewhat greater than once in two years. Because of their very local nature, however, some may be unrecorded and their actual frequency of occurrence is probably appreciably greater than this.

The term 'tornado' has also been used for thunderstorm squalls in West Africa.

torr: A little-used name for a unit of pressure equal to 1 mm mercury under conditions of STANDARD DENSITY and STANDARD GRAVITY.

Torricelli, Evangelista: The inventor of the barometer, born at Piancaldoli in 1608. In 1641 he became amanuensis to Galileo for three months until Galileo's death. Subsequently he was professor to the Florentine Academy until his death in 1647. Torricelli deduced from the fact that water would rise only about 10 m in a suction pipe, that the air had weight, and could, therefore, exert a definite pressure equivalent to a height of 10 m of water or 76 cm of mercury. This conclusion was confirmed by the well-known Torricellian experiment.

torrid zone: The torrid (or equatorial) zone is the region of the earth which lies between the Tropics of Cancer (23° 27' N) and Capricorn (23° 27' S).

tower of the winds: An octagonal tower, built in Athens in about the second century B.C. The tower carries on its sides the names of the winds associated with the eight compass points and also symbolic figures which represent the character of the winds. A description of the figures is given by Theophrastus in his treatise 'On winds and on weather signs' and is quoted by Sir Napier Shaw in *Manual of meteorology*, Volume 1,* together with an estimate of the corresponding weather.

trace of rain: In rainfall measurement, a 'trace' is recorded if either, (i) the measured fall is less than 0·05 mm (or 0·005 in) and the observer knows that the water in the gauge is not the result of water draining from the sides of the can after the previous measurement; or (ii) the gauge contains no water but the observer knows from his own observation that some rain or other precipitation has fallen since the previous observation.

tracer: A property, or a substance, which 'labels' a particular mass of air and so makes it possible to infer the three-dimensional flow of the mass over a period of time.

Wet-bulb potential temperature, ozone, dust and radioactive material are examples of tracers. A well-defined source of the property or substance, leading to large differences in its concentration throughout the atmosphere, is normally required. See, for example, BLUE MOON, OZONE, RADIOACTIVE FALLOUT.

trade winds: The trade winds (or 'tropical easterlies') are the winds which diverge from the subtropical high-pressure belts, centred at 30°–40° N and S, towards the

---

* SHAW, SIR NAPIER; Manual of meteorology, Vol. 1. Cambridge, University Press, 1942, p. 80.

equator, from north-east in the northern hemisphere and south-east in the southern hemisphere.

The characteristics of the trade-wind belt vary considerably with both latitude and longitude. Marked steadiness of the winds is a feature only in latitude belts some 10°–15° wide, centred on about 15° N and S (but varying somewhat with season), mainly in the eastern half of the tropical oceans. Fine weather prevails in the poleward and eastern sections of the belts, owing to the marked anticyclonic subsidence undergone by the trade-wind air. Towards the equator and the western oceanic regions of the belt, the stability of the air is decreased by added moisture; more cloudy, showery weather prevails in these regions, accentuated at times by horizontal convergence of air and the development of cyclones.

The name originated in the nautical phrase 'to blow trade', meaning to blow in a regular course or constantly in the same direction, afterwards shortened to 'trade'. The word is allied to the words 'track' and 'tread'; its use in the sense of commerce was a later development.

See also ANTI-TRADES.

trajectory: A curve drawn to represent the actual path of an air particle over a finite time interval. Such a path is in general three-dimensional; normally, only the horizontal projection of the path is drawn because of insufficient knowledge of the relatively small vertical component of motion.

tramontana: A local name in the Mediterranean for a northerly wind. It is usually dry and cold.

transit: Transit of a heavenly body is said to occur when the body is on the MERIDIAN of the observer. The body is at its maximum elevation at the moment of transit.

translucidus (tr): One of the CLOUD VARIETIES (Latin, *translucidus* transparent).

'Clouds in an extensive patch, sheet or layer, the greater part of which is sufficiently translucent to reveal the position of the sun or moon.

The term applies to ALTOCUMULUS, ALTOSTRATUS, STRATOCUMULUS and STRATUS.'*
See also CLOUD CLASSIFICATION.

transmission coefficient: A quantity τ, also called the 'transmissivity', which is the fraction of the radiation intensity incident on a medium which remains in the beam after passing through unit thickness of the medium. It is related to the ATTENUATION coefficient σ by the relation

$$\tau = e^{-\sigma}.$$

transmissivity: An alternative for TRANSMISSION COEFFICIENT.

transmissometer: Automatic visibility-measuring equipment which employs a light source and a photoelectric cell to measure atmospheric opacity or transmissivity; certain assumptions are then made to convert the measurements to visibility. The alternative term 'visibility recorder' is also used.

Equipment currently in use in the Meteorological Office employs a selenium barrier layer photocell at the focus of a condensing lens to measure the decrease in brightness of a beam from a small projector at a distance of 180 m. The equipment is designed to minimize the possibility of extraneous light reaching the photocell, and periodic adjustments are made to allow for the slow reduction in performance of the light source in the projector and of the photocell in the receiver.

---

* Geneva, World Meteorological Organization. International cloud atlas, Vol. 1. Geneva, WMO, 1956, p. 16.

Visibilities in the range 120–3600 m may be deduced from the strip-chart record provided.

transmittancy: The transmittancy T of a medium of thickness r is the pure number defined by
$$T = \tau^r$$
where τ is the TRANSMISSION COEFFICIENT of the medium.

transparency: The capacity of a medium for allowing RADIATION to pass. Of fundamental importance in meteorology are the different transparencies of the various atmospheric constituents in respect of radiation of a given wavelength, and of individual constituents in respect of radiation of different wavelengths. See also ABSORPTION.

transpiration: The process by which the liquid water contained in soil is extracted by plant roots, passed upwards through the plant and discharged as water vapour to the atmosphere. The process is one necessary for the health and growth of the plant. The rate of transpiration during the day is about the same as that from an open water surface in the same meteorological conditions but is almost zero during night hours.

If the root system of a plant is never short of water the process corresponds to so-called 'potential transpiration'. The actual rate of transpiration falls significantly below the potential value only in a long spell of dry weather, a difference between the rates appearing sooner with the shallower-rooted plants. Values of potential transpiration may be calculated in a similar way to values of potential EVAPORATION. See also EVAPOTRANSPIRATION.

tree line: A term used to signify, on the hemispherical scale, the latitudinal limit of tree growth; on the regional scale, the higher altitude limit of tree growth within a region.

In the northern hemisphere the tree line divides the TUNDRA regions from those of BOREAL CLIMATE. Within the British Isles the tree line is generally at a height of 450 to 600 m above sea level. The tree line depends mainly on summer temperature, an approximate criterion for growth being a mean temperature in excess of 10°C (50°F) in at least two months. Such physical factors as exposure and drainage are also important.

tree ring: See DENDROCHRONOLOGY.

triple point: That point on a pressure–temperature diagram which is the common meeting point of the liquid–vapour, solid–liquid, and solid–vapour lines for a given substance. These lines sharply define the conditions of pressure and temperature at which the changes of state from liquid to vapour, etc. occur. The triple point thus represents the pressure–temperature conditions, unique for a given substance, at which the substance may be solid, liquid or gas. The triple point of water substance has the co-ordinates $p = 4.58$ mm of mercury (6.11 mb), $T = +0.0075°C$.

tritium: A radioactive isotope of HYDROGEN which is continuously formed in the high atmosphere by the action of COSMIC RADIATION, and has also been injected into the atmosphere by thermonuclear explosions. Tritium is used as a TRACER of atmospheric motions.

tropical air: An AIR MASS originating in low latitudes, normally in the SUBTROPICAL HIGHS at around 30°–35° N and S.

Tropical air which reaches the British Isles generally originates in the Atlantic ('maritime tropical air'). It is characterized on and near the south and west coasts by stability at low levels, low stratus clouds, hill fog, poor visibility, and frequent drizzle. Inland and in the east, the clouds are more broken and have a higher base, especially in summer. Occasionally, tropical air reaching the British Isles is of continental (African) origin ('continental tropical air') and is then generally associated with fine, mild (in summer, very warm) weather.

tropical climate: A type of CLIMATE which obtains in most equatorial and tropical parts of the earth and is characterized by high temperatures and high humidity throughout the year and frequent rain throughout most of the year.

tropical continental air: See TROPICAL AIR.

tropical cyclone: A CYCLONE of tropical latitudes. WMO nomenclature is as follows: 'tropical depression' when winds on the Beaufort scale are up to and including force 7; 'moderate tropical storm' for Beaufort forces 8 and 9; 'severe tropical storm' for forces 10 and 11; and 'hurricane' for force 12.

The more intense tropical cyclones are confined to fairly specific regions and seasons which, broadly, are the western sides of the great tropical oceans, beyond 5° from the equator, towards the end of the hot season or seasons. More specifically, the main oceanic regions and times are: North Atlantic (West Indies), North Pacific off the west coast of Mexico, and North Pacific westwards of 170°E (China Seas), July to October; South Indian (Madagascar to 90°E and near north-west Australia) and South Pacific (150°E eastwards to 140°W), December to March; Bay of Bengal and Arabian Sea, April to June and September to December.

The mean annual cyclone frequencies for the North Atlantic (1887–1948) are 7·3 (all intensities) and 3·5 (hurricanes); the respective mean monthly frequencies for the month of maximum frequency (September) are 2·4 and 1·3. Months of maximum frequency of cyclones of hurricane intensity are not always, though usually, those of maximum frequency of cyclones of all intensities; for example, depressions are most common in the Bay of Bengal in July but development to hurricane intensity does not occur then.

Central pressure of the more intense tropical cyclones is often about 960 mb and pressure at the periphery about 1020 mb. These values are comparable with those of a mid-latitude depression but the tropical storm diameter is much smaller (some 800 km compared with 2400 km) and pressure gradients and winds are correspondingly greater. Very low surface pressures are sometimes attained; the lowest known value reduced to mean sea level is about 877 mb, recorded about 1000 kilometres east of Guam on 24 September 1958. Pressure tendencies are very large near the centre of an intense tropical cyclone; Figure 32 is a reproduction of a barogram during the passage of such a cyclone.

A tropical cyclone generally moves initially towards west or north-west in the northern hemisphere and towards west or south-west in the southern hemisphere; the speed is generally about 10 knots. 'Recurvature' of the cyclone, that is change of path direction towards north-east in the northern hemisphere and towards south-east in the southern hemisphere, sometimes occurs at about latitudes 20°–30°. Much more complex tracks, however, are not uncommon. After recurvature, the cyclone tends to assume the characteristics of a mid-latitude depression.

Apart from a central EYE region, usually between 20 and 40 km in diameter, heavy and continuous rain and multi-layer cloud occupy the central regions of the cyclone, with more showery precipitation towards the edges. Decay of a cyclone is usually rapid after passage inland.

A sufficient supply of both real and latent heat (sea surface temperature at least 27°C or 80°F) at a distance from the equator (at least some 5°) sufficient for the CORIOLIS FORCE to be active are necessary conditions for the formation of a tropical storm. The precise mechanisms which cause shallow tropical cyclones to form, or having formed, to intensify to a tropical storm or hurricane are as yet uncertain. An essentially dynamical explanation, rather than a frontal or convective explanation, is now favoured.

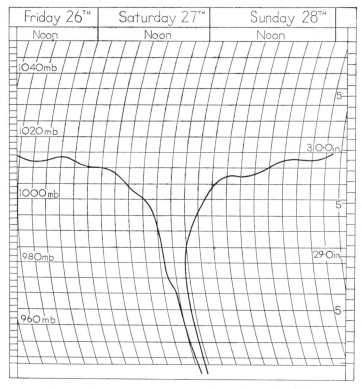

FIGURE 32—Barogram for Cocos Island, November 1909.

tropical maritime air: See TROPICAL AIR.

tropics: That region of the earth's surface lying between the Tropics of Cancer and Capricorn at 23° 27′ N and S, respectively.

tropopause: The atmospheric boundary between the TROPOSPHERE and the STRATOSPHERE.
  (i) The 'first tropopause' is the lowest level at which the lapse rate decreases to 2 degC/km or less, provided also that the average lapse rate between this level and all higher levels within 2 km does not exceed 2 degC/km.
  (ii) When, above the first tropopause, the average lapse rate between any level and all higher levels within 1 km exceeds 3 degC/km, then a 'second tropopause' can occur and is defined by the criteria of (i) above. This tropopause can either be above or within the 1-km layer.
  (iii) Further tropopauses may be defined similarly.

Day-to-day changes of tropopause height occur and there are also appreciable systematic seasonal changes (height greater in summer than in winter).

Detailed synoptic studies have revealed a complex, sometimes discontinuous, tropopause structure, usually in association with deep depressions, jet streams and fronts. Terms used to describe such complexities include 'tropopause funnel' (a bowl-shaped lowering of the tropopause to an unusually low level), 'folded' tropopause and 'multiple' tropopause (also termed 'laminated' or 'foliated' tropopause).

Synoptic studies by J. S. Sawyer* led him to the conclusions that the tropopause generally moves as a material surface embedded in the airstream and that short-period height changes are due in approximately equal measure to horizontal advection and vertical motion.

troposphere: The lower layers of the atmosphere, extending to about 16 km near the equator, 11 km in latitude 50°, 9 km near the poles, and with upper limit at the TROPOPAUSE.

The troposphere is characterized in general by a positive LAPSE rate of temperature and is the region to which precipitation and clouds (apart from certain rather rare types) are confined. (Greek, *tropos* turn.)

trough: A trough (of low pressure) is a pressure feature of the SYNOPTIC CHART; it is characterized by a system of isobars which are concave towards a DEPRESSION and have maximum curvature along the axis of the trough, or 'trough line' (see Figure 33). The trough is said to be 'deep', or 'shallow', according as the maximum curvature of the isobars along the trough line is great, or small, respectively; the former corresponds to the V shape referred to in the obsolete term 'V-SHAPED DEPRESSION'. If the isobars of a depression are circular the trough line is generally taken to be the line through the centre perpendicular to the line of advance of the centre.

A FRONT is necessarily marked by a trough but the converse is not true. Those troughs which are not frontal in character are, however, also generally marked by cloudy, showery weather.

The term 'trough' is also used in meteorology to signify an elongated region of low values of any specified element, e.g. 'thickness trough', 'temperature trough'.

trowal: A term used, mainly in Canada, to signify the projection on the earth's surface of a tongue of warm air aloft, such as may be formed during the OCCLUSION process of a depression. This feature is often found to mark a line of discontinuity of surface weather, cloud type and pressure tendency.

truncation error: If a quantity is defined by a mathematical series it may be approximated by cutting off or truncating the series after a chosen number of terms. The error which arises by using such an approximation is termed a truncation error.

tsunami: A 'tidal wave' generated by an under-water upheaval of the earth's crust. Such a wave moves out in all directions from the point of origin and is capable of causing great destruction on arrival at a coast.

***t*-test:** See STUDENT'S *t*-TEST.

tuba (tub): A supplementary cloud feature (Latin, *tuba* trumpet).

'Cloud column or inverted cloud cone, protruding from a cloud base; it constitutes the cloudy manifestation of a more or less intense vortex.

This supplementary feature occurs with CUMULONIMBUS and, less often, with

---

* SAWYER, J. S.; Day-to-day variations in the tropopause. *Geophys Mem, London*, **11**, No. 92, 1954.

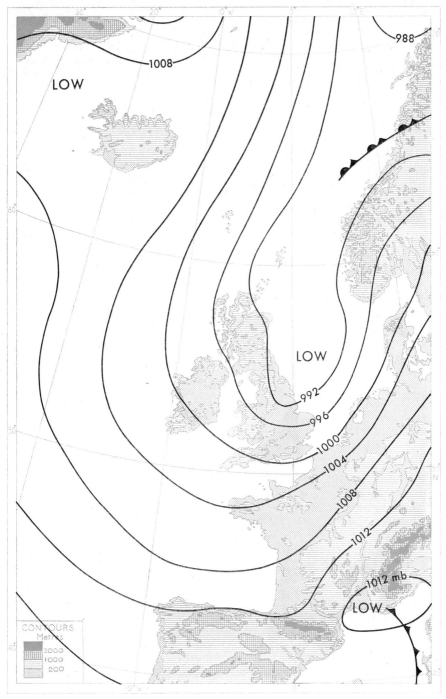

FIGURE 33—Non-frontal trough extending over the southern British Isles from the north-east, 06 GMT, 28 October 1961.

CUMULUS.'* It is commonly known as a 'funnel cloud'. See also CLOUD CLASSIFICATION.

tundra: Treeless lands of northern Canada and Eurasia which lie mainly just inside or just outside the Arctic Circle. In these regions mean monthly temperature rises above freezing-point in some 2 or 3 months in summer but remains below freezing-point throughout the year at a depth of about 30 cm.

turbidity: That property of a cloudless atmosphere which produces ATTENUATION of solar RADIATION. Measurements of atmospheric turbidity (turbidity factor) are generally concerned with the attenuation which is additional to that associated with molecular SCATTERING, the particles responsible being DUST, SMOKE, etc.

turbulence: Turbulent motion is defined by O. G. Sutton as one which contains random oscillations of finite size, leading to irregularities in the path of a particle of scale comparable with lengths which determine the kinematics of the mean motion, such as the shape of the boundary. While there is no precise mathematical definition of turbulence, it is generally taken to comprise the complex spectrum of fluctuating motion which is superimposed on a 'mean flow'. The theory of atmospheric turbulence has been mainly developed in relation to processes which are very restricted in both space and time. The phenomenon is, however, of fundamental importance in meteorological processes of even the largest space and time dimensions.

Small-scale atmospheric turbulence is evident in the fluctuations of wind speed and direction recorded by an anemograph. Such fluctuations are greatest within the 'atmospheric BOUNDARY LAYER' and are much influenced by the nature of the surface over which the air flows and by other factors such as the degree of static stability of the air.

A fundamental property of turbulence is the vertical interchange of mass, momentum, heat and vapour which is effected by eddies of a variety of shapes and sizes. The effect of such eddies is to increase the effective atmospheric DIFFUSIVITY and VISCOSITY far beyond the values which would be appropriate to purely molecular action.

See, for example, LAMINAR FLOW, AERODYNAMIC ROUGHNESS, AERODYNAMIC SMOOTHNESS, EDDY, EXCHANGE COEFFICIENT, K-THEORY, MIXING LENGTH, REYNOLDS NUMBER, SIMILARITY THEORY OF TURBULENCE, STATISTICAL THEORY OF DIFFUSION, LAGRANGIAN SIMILARITY.

turbulence spectrum: An alternative for EDDY SPECTRUM.

turbulent boundary layer: A very shallow layer of air, adjacent to a fixed boundary, in which the air velocity increases from zero at the boundary to a value which then changes little throughout the layer. The REYNOLDS STRESS in such a layer greatly exceeds the molecular viscous stress.

turbulent diffusion: An alternative for EDDY DIFFUSION.

turbulent flux: An alternative for EDDY FLUX.

twilight: The pre-SUNRISE or post-sunset period of partial daylight.
Twilight is caused by the reflection and scattering of sunlight towards an observer on the earth by the upper atmosphere, when the sun is below the observer's horizon.

---

* Geneva, World Meteorological Organization. International cloud atlas, Vol. 1. Geneva, WMO, 1956, p. 18.

The amount of light received progressively diminishes after sunset as the sun sinks farther below the horizon and as the sunlight is scattered by progressively higher and less dense air and is subject more and more to multiple scattering before reaching the observer.

Various stages of twilight are recognized. Thus, in the evening, 'astronomical twilight' (A.T.) ends when the sun's centre is 18° below the horizon, corresponding to the last trace of daylight. 'Civil twilight' (C.T.) ends when the sun's centre is 6° below the horizon, corresponding to the lower limit of sufficiency of daylight for outdoor activity. Intermediate is 'nautical twilight' when the sun's centre is 12° below the horizon.

While the intensity of indirect illumination from sunlight (assuming no cloud or haze) is fixed by the angular depression of the sun below the horizon, the duration of twilight (morning or evening) varies with latitude and season as shown in Table XIII.

TABLE XIII—*Duration of astronomical twilight (A.T.) and civil twilight (C.T.) in different latitudes and seasons*

	Equator		50°		60°	
	A.T.	C.T.	A.T.	C.T.	A.T.	C.T.
	h m	h m	h m	h m	h m	h m
Winter solstice	1 15	0 26	2 1	0 45	2 48	1 9
Equinox	1 10	0 24	1 52	0 37	2 31	0 48
Summer solstice	1 15	0 26	—	0 51	—	1 59

At midsummer, between the Arctic Circle and latitude $48\frac{1}{2}°$N, there is a belt with no true night, twilight extending from sunset to sunrise.

The following figures are quoted from Kimball and Thiessen for the intensity of illumination, in foot-candles (1 foot-candle = 10·7639 lumens/m²), of a horizontal surface in cloudless conditions: sun in zenith, 9600; sun on horizon, 33; sun 6° below horizon, 0·4; sun 18° below horizon, 0·0001; full moon in zenith, 0·02.

twilight arch: The 'primary twilight arch' appears after the sun has set, as a bright, but not very sharply defined segment of reddish or yellowish light resting on the western horizon. The 'secondary twilight arch' is the slightly luminous segment near the western horizon in the last stages of TWILIGHT.

twilight flash: An alternative for TWILIGHT GLOW.

twilight glow: A marked intensification (also termed 'twilight flash') of the brightness of certain lines, notably the SODIUM D line at 5893 ångströms, in the AIRGLOW emission spectrum near the times of sunrise and sunset. Observation of the variation of intensity of the emission with solar zenith angle near these times has enabled the height distribution of sodium in the high atmosphere to be inferred.

twinkling (of stars): See SCINTILLATION.

type: Different distributions of atmospheric pressure are characterized by more or less definite kinds of weather. Accordingly, when a certain form of pressure distribution is seen on a chart the weather is described as being of a given type. The types are therefore defined by the shape or general trend of the isobars. Thus an 'anticyclonic' or a 'cyclonic' type denotes that an ANTICYCLONE or a DEPRESSION is the main feature of the pressure distribution; on the other hand a 'westerly' type indicates that the isobars run in more or less parallel lines over a considerable distance from west to east, having the lowest pressure to the north; a 'northerly' type will have isobars running north and south with the low pressure to the east, etc.

The weather associated with each type varies with season but members of the same type have nearly always something in common; thus, an anticyclonic type usually has rainless weather, the cyclonic, wet weather; the southerly type in the northern hemisphere will in general be relatively warm and the northerly type cold. The westerly type is very persistent and often gives rise to long periods of rather unsettled weather. The easterly type gives in winter suitable conditions for severe frosts, while in summer, in at least the southern part of the British Isles, the weather is usually very warm.

typhoon: A name of Chinese origin (meaning 'great wind') applied to the intense TROPICAL CYCLONES which occur in the western Pacific Ocean. They are of essentially the same type as the Atlantic 'hurricane' and Bay of Bengal 'cyclone'.

U

Ulloa's circle (or **ring**): A white rainbow or fogbow. See RAINBOW and BOUGUER'S HALO.

ultra-violet radiation: ELECTROMAGNETIC RADIATION in the approximate wavelength range from 10 to 4000 ångströms, i.e. in the wavelength region below visible radiation. See VISIBLE SPECTRUM.

The relatively small fraction of the total energy contained in solar RADIATION in the 'far ultra-violet' is strongly absorbed in the high atmosphere, resulting in various photo-chemical reactions including that of OZONE formation and a sharp cut-off, at about 2900 Å, of the solar spectrum observed at the earth's surface. The latter is, therefore, in large measure protected from the strongly actinic and biological effects which are produced by ultra-violet radiation.

Umkehr effect: An effect which is used to infer the vertical distribution of OZONE from surface measurements.

A series of measurements of the relative intensities (I), in light scattered from the zenith sky, of two selected wavelengths, one (A) strongly absorbed by ozone, the other (B) less strongly absorbed, is made when the sun is near the horizon. The ratio I_A/I_B decreases with increasing zenith angle (Z) of the sun, owing to increasing path length through the ozone, up to the point at which the 'effective' scattering height for both wavelengths (increasing with increase of Z) lies below or within the OZONE LAYER. A critical point is reached (e.g. $Z > 85°$) when most of wavelength A, but not of B, reaches the observer after being scattered from above the ozone layer: during a further increase of Z by a few degrees, the ratio I_A/I_B increases, constituting a reversal (German '*Umkehr*') of the previous trend. The vertical distribution of ozone may on certain assumptions be inferred from the precise variation of I_A/I_B with Z.

uncinus (unc): A CLOUD SPECIES (Latin, *uncinus* hooked).

'CIRRUS often shaped like a comma, terminating at the top in a hook, or in a tuft the upper part of which is not in the form of a rounded protuberance.'* See also CLOUD CLASSIFICATION.

undercooling: A seldom-used alternative for SUPERCOOLING.

undersun: A HALO phenomenon produced by reflection of sunlight on ice crystals in clouds. 'It appears vertically below the sun in the form of a brilliant white spot, similar to the image of the sun on a calm water surface. It is necessary to look downward to see the undersun; the phenomenon is therefore only observed from aircraft or from mountains.'†

---

* Geneva, World Meteorological Organization. International cloud atlas, Vol. 1. Geneva, WMO, 1956, p. 12.

† Geneva, World Meteorological Organization. International cloud atlas, Vol. 1. Geneva, WMO, 1956, p. 72.

undulatus (un): One of the CLOUD VARIETIES (Latin, *undulatus* waved).
'Clouds in patches, sheets or layers, showing undulations. These undulations may be observed in fairly uniform cloud layers or in clouds composed of elements, separate or merged. Sometimes a double system of undulations is in evidence.'*
See also CLOUD CLASSIFICATION.

Universal Decimal Classification (UDC): A method of classifying the subject matter of books and documents. It divides all knowledge into 10 classes, numbered from 0 to 9, which can each be further divided and subdivided in steps of 10 according to a decimal notation. The scheme is internationally agreed and is recommended by WMO for use in meteorological libraries. Meteorology is allotted the number 551·5 and is subdivided into main divisions as follows: 551·50 Practical meteorology (methods, data, instruments, forecasts and other applications); 551·51 Structure, mechanics and thermodynamics of the atmosphere in general; 551·52 Radiation and temperature; 551·54 Atmospheric pressure; 551·55 Wind; 551·57 Aqueous vapour and hydrometeors; 551·58 Climatology; 551·59 Various phenomena and influences.

universal time (UT): TIME determined by the average rate of the apparent diurnal motion of the sun relative to the MERIDIAN of Greenwich. In geophysics, the term signifies a common reference time of day (GMT) in all longitudes. While systematic diurnal effects of most phenomena progress entirely according to local solar time (LT), a systematic UT effect is found in certain cases, for example, potential gradient measurements in ATMOSPHERIC ELECTRICITY.

upbank thaw: The precedence of a THAW in a valley situation, sometimes by many hours, by a thaw or marked rise of temperature at mountain level in the same vicinity. The phenomenon is usually caused by the arrival at higher levels of the warm air in advance of a surface warm front; it may also be caused by the subsidence and dynamical heating of air at the higher level.
The associated inversion of the normal temperature lapse rate is a contributory cause of GLAZE.

upper-level trough: In synoptic meteorology, a line along which there exists in the upper air a TROUGH of low pressure (or the analogous contour trough on an isobaric chart), with an associated change of wind direction.
By implication, such a feature is not associated with a trough or front on the surface chart. It may, however, especially in summer, be associated with a line of convective phenomena—see INSTABILITY LINE.

upslope fog: FOG which is formed on the windward slopes of high ground by the forced uplift of stable, moist air till saturation is reached by adiabatic expansion.

upwelling: The term applied to the movement of cold water from moderate depths up to the surface, as near a coast on occasions when the warmer surface water is driven from the coast by the wind.

---

* Geneva, World Meteorological Organization. International cloud atlas, Vol. 1. Geneva, WMO, 1956, p. 14.

V

valley wind: Usually, the ANABATIC WIND which blows up a valley during the day in quiet, clear conditions. The counterpart at night is the KATABATIC 'mountain wind'.
The term 'valley wind' is sometimes also used in the same sense as RAVINE WIND.

Van Allen radiation belts: Two regions above the earth, more particularly in and near the equatorial plane, in which the flux of penetrating radiation due to particles of high energy reaches a maximum. The regions were discovered in satellite measurements made in 1958 under the leadership of J. A. Van Allen.
The inner belt is confined to lower latitudes and is at a height of about 3000 km. The outer belt, markedly horn-shaped, is at about 16 000 km near the equator but is narrower and at much lower levels in high (magnetic) latitudes. The charged particles concerned, protons and electrons, perform a spiral motion along magnetic lines of force across the equatorial plane to high latitudes where their motion is reversed. While thus essentially trapped by the magnetic field the particles suffer some continuous loss by collisions with atmospheric particles at lower levels and are also continuously replenished—it is thought by cosmic rays and streams of solar particles in the lower and higher belts, respectively.

vane: See WIND VANE.

vapour: A gas which is at a temperature below its 'critical temperature', i.e. at a temperature at which it can be liquefied by pressure alone. Water vapour is the main example of such a gas in the earth's atmosphere. Carbon dioxide and sulphur dioxide are also technically vapours at atmospheric temperatures, but in their low concentrations are not liquefied in the ranges of pressure and temperature that exist in the atmosphere.

vapour concentration: The density of water vapour (d_v) in a mixture of water vapour and dry air, being defined as the ratio of the mass of water vapour (m_v) to the volume (V) occupied by the mixture, i.e.
$$d_v = m_v/V.$$
This quantity is of the order 10 g/m³.
The alternative terms 'absolute humidity' and 'vapour density' applied to this quantity are not now favoured.

vapour density: An alternative for VAPOUR CONCENTRATION.

vapour pressure: In meteorology, that part of the total atmospheric pressure which is exerted by WATER VAPOUR. The vapour pressure e' of water vapour in moist air at total pressure p and with MIXING RATIO r is defined by
$$e' = \frac{r}{0.62197 + r} p.$$
Vapour pressure is measured indirectly from dry- and wet-bulb temperature readings, with the aid of humidity slide-rule or tables (see PSYCHROMETER).
If r_w, r_i denote SATURATION mixing ratio of moist air with respect to a plane

surface of pure water and ice, respectively, then the 'saturation vapour pressure' with respect to water (e_w') and that with respect to ice (e_i'), of moist air at pressure p and temperature T are, respectively, defined by

$$e_w' = \frac{r_w}{0{\cdot}62197 + r_w} p,$$

$$e_i' = \frac{r_i}{0{\cdot}62197 + r_i} p.$$

The implied pressure dependence of e_w' and e_i' is in practice negligible; to a close approximation the saturation vapour pressure of moist air depends only on temperature, as is strictly true of pure water vapour in equilibrium with a plane (or ice) surface, in accordance with the values shown in Table XIV.

In the table the values of e_w' at temperatures 0°C and below are those which obtain with respect to supercooled water. At such temperatures, the excess of e_w' over e_i' (which has a maximum of 0·27 mb at about -12°C) is important in the formation of PRECIPITATION.

For equilibrium conditions other than those at a plane surface of pure water or ice, the values of saturation vapour pressure are changed, relative to those in the table, to a degree which is significant in the CONDENSATION process, as follows:

(i) Kelvin showed that at a given absolute temperature (T) the equilibrium vapour pressure (e_r') over a drop of pure water of radius r is greater than the corresponding value (e_w') appropriate to a plane surface, i.e.

$$\rho RT \log_e \frac{e_r'}{e_w'} = \frac{2\sigma}{r}$$

where ρ is the water density, R the gas constant for water vapour, and σ is the surface tension of the water drop.

(ii) The saturation vapour pressure over water which contains dissolved substance is reduced, by an amount $\triangle e$, in accordance with Raoult's law, i.e.

$$\frac{\triangle e}{e_w'} = -\frac{in'}{n + in'}$$

where n and n' are the numbers of moles of water and solute, respectively, and i is a factor which varies with the concentration of the solution.

TABLE XIV—*Variation of e_i' and e_w' with temperature*

T (°C)	-40	-30	-20	-10	0	$+10$	$+20$	$+30$
e_i' (mb)	0·13	0·38	1·03	2·60	6·11	—	—	—
e_w' (mb)	0·19	0·51	1·25	2·86	6·11	12·27	23·37	42·43

vardarac (or vardar): A cold northerly wind which blows through the Morava–Vardar gap in the rear of a depression and affects the Thessaloniki region; it is a type of RAVINE WIND.

variance: A statistical measure of variability, equal to the square of the STANDARD DEVIATION. The success of a forecasting technique is often measured by the percentage reduction in variance found by expressing the variance of the forecast errors as a percentage of that of the element being forecast.

variance-ratio test: See F-TEST.

vector: A quantity which requires both direction and magnitude for its complete specification. Meteorological examples are wind velocity, vorticity, pressure gradient, as opposed to such SCALAR quantities as temperature and pressure. The magnitude of a vector is termed the 'modulus' of the vector.

A vector quantity may be represented by a straight line drawn in a specific direction and of specific length. Graphical addition of two (or more) forces not acting in the same straight line or, alternatively, resolution of a single force into two (or more) 'components' is done by using the parallelogram (or polygon) law relating to vector quantities. Thus, for example, in Figure 34 $\overrightarrow{AC}$ is the vector sum or 'resultant' of $\overrightarrow{AB}$ and $\overrightarrow{BC}$, which are thus components of $\overrightarrow{AC}$. A meteorological illustration relating to this diagram is that the GEOSTROPHIC WIND at an upper isobaric level ($\overrightarrow{AC}$) may be regarded as the vector sum of the geostrophic wind at a lower isobaric level ($\overrightarrow{AB}$) and the THERMAL WIND in the isobaric layer concerned ($\overrightarrow{BC}$).

Vector mean is the vector sum divided by the number of observations. The vector sum may be obtained by graphical addition, or may be computed by resolution of each vector into north and east components, algebraic addition of the respective components, and recombination of the two sums into a single vector.

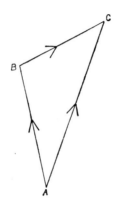

FIGURE 34—Vector addition.

veering: The changing of the wind in the direction of the motion of the hands of a clock, in either hemisphere. The opposite to BACKING.

velocity: A VECTOR quantity signifying rate of change of position with time in a specified direction.

In meteorology, this term is often loosely used, e.g. in relation to motion of air or pressure systems, as being synonymous with speed, which is a SCALAR quantity.

velocity potential: A scalar function (φ) which always exists in irrotational fluid motion and is defined by the equation:

$$\mathbf{V} = - \nabla \varphi$$

where $\mathbf{V}$ is the velocity vector. The equation implies that $\mathbf{V}$ is normal to the equipotential lines and is directed from high to low potential. See GRADIENT.

velum (vel): (Latin, *velum* sail of ship.)
'An accessory cloud veil of great horizontal extent, close above or attached to the upper part of one or several cumuliform clouds which often pierce it.
Velum occurs principally with CUMULUS and CUMULONIMBUS.'* See also CLOUD CLASSIFICATION.

vendavales: Strong, squally south-west winds in the Straits of Gibraltar and off the east coast of Spain. Associated with depressions mainly between September and March, they bring stormy weather and heavy rain.

venturi tube: A tube used in the measurement of fluid velocity, as in wind-tunnel experiments. The fluid velocity (V) is related to the measured pressure difference ($\triangle p$) at the tube entrance, relative to that at a constriction in the tube through which the fluid passes, by the equation

$$V = \sqrt{\frac{2\triangle p}{\rho(r^2 - 1)}}$$

where ρ is the fluid density and r the ratio of the tube cross-sections at entrance and constriction.

veranillo: The two or three weeks of fine weather which break the rainy season near midsummer in tropical America.

verano: The long, dry season near midwinter in tropical America.

verification of forecasts: The process of obtaining a measure of the success of FORECASTS by relating predicted weather to actual weather. While simple comparison may serve to reveal certain features, for example a systematic bias towards optimism or pessimism, the verification process generally consists of deriving an index by one or other of a variety of methods which depend on the nature of the forecast. Such an index may then be compared with various standard indices based, for example, on a RANDOM FORECAST, a PERSISTENCE FORECAST, or a forecast based on climatic normals.
Among the many purposes of forecast verification are the evaluation of forecasting techniques, the nature of forecasting errors, the short- or longer-period variation of accuracy of forecasts, the economic value of forecasts, and the skill of individual forecasters.

vernier: A contrivance for estimating fractions of a scale division when the reading to the nearest whole division is not sufficiently accurate. The vernier is a uniformly divided scale which is arranged to slide alongside the main scale of an instrument. An example of a vernier (on a barometer) and the method of reading are given in the *Observer's handbook*.†

vertebratus (ve): One of the CLOUD VARIETIES.
'Clouds, the elements of which are arranged in a manner suggestive of vertebrae, ribs, or a fish skeleton.
This term applies mainly to CIRRUS.'‡ See also CLOUD CLASSIFICATION.

---

* Geneva, World Meteorological Organization. International cloud atlas, Vol. 1. Geneva, WMO, 1956, p. 18.
† London, Meteorological Office. Observer's handbook, 3rd edn. London, HMSO, 1969, pp. 97, 98.
‡ Geneva, World Meteorological Organization. International cloud atlas, Vol. 1. Geneva, WMO, 1956, p. 14.

vertical visibility: The visual range of a dark object, of moderate angular size, viewed vertically upwards against a sky background in daylight.

This element, required in synoptic observations on occasions when the sky is obscured by fog, etc., may be measured by pilot balloon and theodolite, the vertical visibility being taken as $h \operatorname{cosec} E$ where h is the height of the balloon and E its angular elevation at the moment of its disappearance from view.

virga (vir): A supplementary cloud feature (Latin, *virga* rod).

'Vertical or inclined trails of precipitation (fallstreaks) attached to the under surface of a cloud, which do not reach the earth's surface.

This supplementary feature occurs mostly with CIRROCUMULUS, ALTOCUMULUS, ALTOSTRATUS, NIMBOSTRATUS, STRATOCUMULUS, CUMULUS and CUMULONIMBUS.'*
See also CLOUD CLASSIFICATION.

virtual height: In radio echo sounding of the IONOSPHERE, the equivalent height of reflection of the radio waves obtained from the time delay between emission and reception of the wave on the assumption that the wave travels with the speed of light.

virtual temperature: The virtual temperature of a sample of moist air is that temperature at which completely dry air of the same total pressure would have the same density as the given sample.

The following closely approximate relations hold between absolute virtual temperature (T_v) and absolute air temperature (T):

$$T_v \approx T / \left(1 - \frac{3}{8}\frac{e'}{p}\right).$$

'Virtual temperature increment' $\equiv T_v - T \approx 0.61qT \approx 0.61rT$, where e' is the vapour pressure, p the total pressure, q the SPECIFIC HUMIDITY (kg/kg), r the MIXING RATIO (kg/kg).

As a further, but generally sufficient, approximation,

$$T_v - T \approx q/6 \approx r/6$$

where q and r are expressed in grammes/kilogramme.

viscosity: That property of a fluid whereby it resists deformation. In a fluid in which different layers move with different velocities, molecular viscous forces operate so as to tend to make the velocities more uniform; for two layers a short distance apart, both parallel to the direction of flow, the viscous stress per unit area (τ) is proportional to the velocity gradient, the constant of proportionality being the coefficient of (dynamic) viscosity (μ), i.e. $\tau = \mu \partial u/\partial z$. The ratio of the dynamic viscosity to the density (ρ) of the fluid is termed the kinematic viscosity (ν). Air near the earth's surface has the approximate values $\mu = 1.8 \times 10^{-5}$ kg/m s and $\nu = 1.5 \times 10^{-5}$ m²/s.

In the atmosphere, turbulent eddies are very much more important in effecting mixing of momentum than is the molecular viscosity. By analogy with the definition of ν, the vertical 'eddy viscosity' (K_M), for example, is defined by the equation

$$\tau/\rho = K_M \, \partial \bar{u}/\partial z$$

where τ is the corresponding REYNOLDS STRESS. K_M varies with height and is generally of the order 1 m²/s, i.e. some 10^5 times greater than ν. See also DIFFUSIVITY.

visibility: Visibility is defined as the greatest distance at which an object of specified characteristics can be seen and identified with the unaided eye in any particular

---

* Geneva, World Meteorological Organization. International cloud atlas, Vol. 1. Geneva, WMO, 1956, p. 17.

circumstances, or, in the case of night observations, could be seen and identified if the general illumination were raised to the normal daylight level. Lower visibilities are expressed in metres or yards, higher visibilities in kilometres or miles. Reports generally refer to a visibility based on all directions; where there is marked variation with direction, the lowest visibility is recorded for synoptic purposes, with an appropriate entry in a 'special phenomena' group.

'Visibility objects' by day are ideally confined to black or nearly black objects which appear against the horizon sky. Night visibility objects comprise mainly unfocused lights of moderate and known intensity at known distances. Conversion of such night observations to daylight scales involves an assumption of the different values of CONTRAST THRESHOLD appropriate to the visibility objects by day and night. Various types of VISIBILITY METER are also used for observation by night.

Visibility, though to some extent dependent in its measurement on extraneous physiological and physical factors, is an element which is governed mainly by the atmospheric EXTINCTION COEFFICIENT associated with solid and liquid particles held in suspension in the atmosphere; the extinction is primarily caused by scattering, rather than by absorption, of the light. While visibility is an element which is characteristic, in a general way, of an air mass—it is, for example, broadly much better within air masses which originate in high latitudes and move equatorward than in those which originate in low latitudes and move poleward—local variations of visibility associated with precipitation, atmospheric pollution and other factors prevent its use as a reliable air-mass indicator. See also METEOROLOGICAL OPTICAL RANGE, CONTRAST THRESHOLD OF THE EYE, KOSCHMIEDER'S LAW, OBLIQUE VISIBILITY, VERTICAL VISIBILITY.

visibility meter: A class of instruments designed to measure VISIBILITY by the determination of either the EXTINCTION (e.g. the Gold visibility meter) or SCATTERING of light by the atmosphere. In an instrument measuring the latter of these properties the assumption is made that the reduction of visibility due to direct ABSORPTION is negligible.

Visibility meters are not ABSOLUTE INSTRUMENTS and require calibration in terms of the daylight visibility scale. Their use in practice tends to be limited to night observations in places where suitable night visibility points are not available.

visibility ratio: An alternative for LUMINOSITY.

visibility recorder: An alternative for TRANSMISSOMETER.

visible spectrum: That part of the ELECTROMAGNETIC RADIATION spectrum, between about 0·4 and 0·7 μm, to which the human eye is sensitive. Within the visible spectrum the wavelength increases through the range of colours violet, indigo, blue, green, yellow, orange and red. Of the total solar RADIATION intensity 41 per cent is contained within this part of the electromagnetic spectrum. See also LUMINOSITY.

volcanic dust: Volcanic dust is known to have spread in the stratosphere as a veil covering more than half the surface area of the globe in some instances and to have persisted in observable quantities for up to three years. The latitude zones which are sooner or later affected probably depend greatly on the latitude of injection. Such dust veils are associated with certain atmospheric optical effects (see, for example, BISHOP'S RING). It is also probable that significant effects on atmospheric circulation and world weather are caused by the scattering of solar radiation by widespread and persistent veils.

According to H. H. Lamb's estimate of the magnitude of eruptions dating later than 1600 (estimate based on evidence of dust veils and/or of quantities of solid material ejected), meteorological effects were probably significant in at least the following cases, listed in approximate order of magnitude: 1783, Skaftárjökull or Laki (Iceland), Eldeyjar (Iceland) and Asama (Japan), effects lasting until 1785; 1815, Tambora (8°S 118°E); 1883, Krakatoa (6°S 105°E), effects lasting until 1885–86; 1680, Krakatoa and Tonkoko (1°N 125°E); 1831, group of major eruptions including Pichincha (0°S 79°W), Mediterranean submarine eruption (37°N 12–13°E), Babuyan (19°N 122°E), Etna and Vesuvius; 1821–24, group of major eruptions including Kluchevskaya Sopka (55°N 161°E) in 1821, Eyafjallajökull (Iceland) from 1821, Vesuvius in 1822, Galunggung (7°S 108°E) in 1822, and Lanzarote (Canary Islands) in 1824; 1902–04, group of major eruptions including St Vincent (West Indies) from 1902 to 1903, Mont Pelée (15°N 61°W) in 1902, Santa Maria (Guatemala) from 1902 onwards, and Colima (Mexico) in 1903; 1835, Coseguina (Nicaragua) and several in Chile (mainly between 33°S and 44°S); 1755–56, Katla (Iceland).

In addition to the above cases, thick volcanic ash layers in various parts of the world supply evidence of former volcanic activity of a magnitude likely to have had significant climatic effects. In some cases they can be approximately dated. It appears that volcanic activity was particularly frequent from A.D. 1500 to 1900, around 500–0 B.C., around 3000 B.C., and around 7500 B.C. Both hemispheres were apparently affected by the three earlier waves of activity, but the A.D. 1500–1900 period seems to have affected mainly the northern hemisphere and the equatorial zone.

See also DUST.

vortex: A fluid flow which possesses VORTICITY. A 'vortex line' is one drawn from point to point of a fluid such that it coincides at all points with the instantaneous direction of the axis of rotation of the fluid. A 'vortex tube' is the surface which contains all the vortex lines which intersect a closed small curve within the fluid. A 'vortex filament' is the fluid contained within a vortex tube. A 'vortex sheet' is a surface of discontinuity of velocity which separates two adjacent streams of a fluid and on which the vorticity is infinite.

vorticity: The vorticity at a point in a fluid is a vector which is twice the local rate of rotation of a fluid element. The component of the vorticity in any direction is the CIRCULATION per unit area of the fluid in a plane normal to that direction. The dimensions are T^{-1}.

Vorticity is a three-dimensional property of the field of motion of a fluid. In large-scale motion in the atmosphere the vorticity component of chief significance is that which occurs in the horizontal plane (i.e. rotation about the vertical axis); the other components are, however, significant in some dynamical problems.

In vector notation the vorticity of a velocity vector **V** is written as curl **V** or rot **V** or $\nabla \wedge$ **V**. In Cartesian co-ordinates the vertical (z) component of 'relative vorticity' (i.e. rotation in a horizontal plane, evaluated from winds measured relative to the rotating earth) is

$$\text{vorticity}_z = \zeta = \left(\frac{\partial v}{\partial x} - \frac{\partial u}{\partial y}\right).$$

Similarly, $\text{vorticity}_x = \xi = \left(\dfrac{\partial w}{\partial y} - \dfrac{\partial v}{\partial z}\right)$

and $\text{vorticity}_y = \eta = \left(\dfrac{\partial u}{\partial z} - \dfrac{\partial w}{\partial x}\right).$

The expression for the vertical component of vorticity in terms of velocity V, radius of curvature of the streamlines r, and differentiation along the normal to the streamlines $\partial V/\partial n$ (see NATURAL CO-ORDINATES), namely

$$\zeta = \frac{V}{r} - \frac{\partial V}{\partial n},$$

shows that the vertical component of vorticity may be regarded as the sum of components due to curvature (V/r) of horizontal flow and to horizontal wind shear ($-\partial V/\partial n$). Thus, for example, the contribution to vorticity about the vertical axis made by the horizontal wind shear associated with a westerly jet stream in the northern hemisphere is strongly cyclonic poleward of the jet axis and anticyclonic equatorward of the axis.

In 'solid rotation' of angular velocity ω the vorticity is 2ω. In latitude φ, where the ANGULAR VELOCITY OF THE EARTH about the vertical axis is $\Omega \sin \varphi$, the earth has a vorticity about this axis of $2\Omega \sin \varphi$ which is cyclonic in sense. Air partakes of the vorticity of the earth appropriate to its latitude, in addition to any relative vorticity it may possess. Thus, in latitude φ, 'absolute vorticity$_z$' is given by

$$\zeta_a = \zeta + 2\Omega \sin \varphi.$$

Relative vorticity in a cyclonic sense is reckoned positive, in an anticyclonic sense negative. See also GEOSTROPHIC VORTICITY.

vorticity equation: The vorticity equation as used in meteororology relates the rate of change of the vertical component of VORTICITY to the horizontal DIVERGENCE. It is derived by eliminating geopotential (or pressure) from the equations of motion. In PRESSURE CO-ORDINATES the vorticity equation can be written

$$\underbrace{\frac{d}{dt}(\zeta + f)}_{} = \underbrace{- (\zeta + f) \operatorname{div}_p V}_{(a)} + \underbrace{\left(\frac{\partial \omega}{\partial y} \frac{\partial u}{\partial p} - \frac{\partial \omega}{\partial x} \frac{\partial v}{\partial p} \right)}_{(b)}.$$

In CARTESIAN CO-ORDINATES it is

$$\frac{d}{dt}(\zeta + f) = \underbrace{- (\zeta + f) \operatorname{div}_H V}_{(a)} + \underbrace{\left(\frac{\partial w}{\partial y} \frac{\partial u}{\partial z} - \frac{\partial w}{\partial x} \frac{\partial v}{\partial z} \right)}_{(b)} + \underbrace{\left(\frac{\partial p}{\partial x} \frac{\partial \alpha}{\partial y} - \frac{\partial p}{\partial y} \frac{\partial \alpha}{\partial x} \right)}_{(c)}.$$

Here ζ = vertical component of vorticity = $\dfrac{\partial v}{\partial x} - \dfrac{\partial u}{\partial y}$

f = Coriolis parameter
ω = dp/dt (equivalent of vertical velocity in pressure co-ordinates)
w = vertical velocity
α = specific volume
p = pressure.

The first term (a) on the right-hand side of the equation is the dominant one in large-scale atmospheric motion. The second term (b) is sometimes known as the 'twisting term' and represents the transformation of vorticity from the horizontal to the vertical component. It is believed to be important within smaller-scale motions as in fronts. The third term (c) in the second form represents the direct generation of vorticity by horizontal density and pressure gradients and is usually unimportant.

vorticity theorem: The vorticity theorem is derived from the circulation theorem of V. Bjerknes. It relates the local generation of VORTICITY to the local BAROCLINITY of the atmosphere and may be written

$$\operatorname{curl} (\rho \dot{\mathbf{V}} - \rho \mathbf{C}) = \nabla \rho \times \nabla (-\Phi)$$

where ρ = density
  $\dot{\mathbf{V}}$ = acceleration of the air
  $\mathbf{C}$ = vector representing apparent deviating force per unit volume due to earth's rotation
  Φ = geopotential.

Direct application of the vorticity theorem to large-scale motions is limited because the term arising from the deviating force is usually in approximate balance with the term arising from the density gradients.

V-shaped depression: An obsolete term for a sharply defined TROUGH of low pressure, with the isobars in the form of a V.

W

warm anticyclone: See ANTICYCLONE.

warm front: A FRONT whose movement is such that the warmer AIR MASS is replacing the colder.

As a warm front approaches, temperature and dew-point within the cold air gradually rise and pressure falls at an increasing rate. Precipitation usually occurs within a wide belt some 400 km in advance of the front. Passage of the front is usually marked by a steadying of the barometer, a discontinuous rise of temperature and dew-point, a veer of wind (in northern hemisphere), and by a cessation or near cessation of precipitation. Substantial lanes of clear air separating cloud layers are found in all except a small minority of warm fronts.

The average slope of a warm-frontal surface is about 1 in 150. A warm front moves, on average, at a speed some two-thirds of the component of the geostrophic wind normal to the front and measured at it.

warm-front wave: A secondary WAVE DEPRESSION which forms on an extended WARM FRONT at a point usually a considerable distance, some 1500 km, from the parent depression; after formation, it moves quickly east or south-east away from the parent depression. This type of depression, which is not common, seldom becomes deep but is responsible for a considerable spread and intensification of the warm-front precipitation. Formation may be aided either by a frontal distortion produced by a range of hills or by movement towards a col.

warm-occlusion depression: A SECONDARY DEPRESSION which forms at the point where a cold and warm front unite to form a warm OCCLUSION. Such a secondary generally moves quickly away from the primary depression and deepens, though seldom to a marked extent, at the expense of the primary.

warm pocket: A term applied in upper air analysis to a closed centre of high pressure on a FRONTAL CONTOUR CHART. Such a region on the chart indicates the isolation of warm air at low levels from the main body of warm air which is seen on the chart, usually at lower latitudes.

warm rain: The term 'warm rain' is sometimes applied to rain which falls from clouds whose tops do not reach the freezing level. Such rain is initiated by the coalescence process—see PRECIPITATION.

warm ridge: A pressure RIDGE (or ridge on an isobaric contour chart) in which temperature is generally higher than in adjacent areas.

warm sector: In the early stages of the life history of at least the majority of the DEPRESSIONS of temperate latitudes, and of the more important SECONDARIES, there is a surface sector of warm air, which disappears as the system deepens and the cold front catches up the warm front (see OCCLUSION). The warm sector is usually composed of tropical air, sometimes of maritime polar air.

washout: The removal of solid material from the air, and its deposition on the earth's surface, due to capture by falling PRECIPITATION elements. See also FALLOUT.

water: The oxide of hydrogen of chemical formula H_2O. Its maximum density of 999·97 kg/m^3 occurs at 4°C. Its thermal conductivity at 0°C and 20°C is, respectively, 0·5527 and 0·5987 W/m degC. Its physical properties are slightly modified by the small but variable amounts of impurities, due mainly to dissolved salts, which occur in natural water. It constitutes, as liquid or ice, 70·8 per cent of the earth's total surface—see EARTH.

Water plays a fundamental part in the energy balance of the earth–atmosphere system, notably because of the LATENT HEAT exchanges involved in its widespread changes of state. See also ICE, WATER VAPOUR, SALINITY.

water (-droplet) cloud: A cloud which is composed entirely of water droplets, either in the supercooled state or at temperatures above 0°C, as opposed to ice crystals. The CLOUD GENERA Ac, St, and Cu are normally water clouds.

watershed: In physical geography, the line separating the head streams which are tributaries to different river systems or basins, i.e. the line enclosing a CATCHMENT AREA.

water sky: Term applied, mainly in polar regions, to the dark appearance presented by the underside of a cloud layer which lies above a water-covered region relative, in particular, to that of a cloud layer above a snow- or ice-covered region (see ICEBLINK). Such an appearance is often useful in indicating the presence of open water which is not itself then visible.

water smoke: An alternative for ARCTIC SEA SMOKE.

waterspout: A funnel-shaped TORNADO cloud which extends from the surface of a sea or inland water to the base of a cumulonimbus cloud.

A cone-like point of cloud descends from the cumulonimbus base to the agitated sea below and assumes the appearance of a column of water, the diameter of which may vary between a few tens and a few hundreds of feet (10 ft $\approx$ 3 m). The duration of a waterspout ranges up to about half an hour, during which time the column may be appreciably bent by vertical wind shear. A circular and violent circulation of air is caused near the waterspout with an associated confused sea. The phenomenon is more common in the tropics and subtropics than in higher latitudes. See Plates 27 and 28.

water table: The depth at which the soil is persistently saturated with water. Such depth generally varies appreciably with the wetness of the season.

water vapour: Water substance in the vapour form is, meteorologically, the most important constituent of the atmosphere and is also the most variable in space and time.

Supplied to the atmosphere by EVAPORATION and SUBLIMATION at the earth's surface, water vapour has a concentration which decreases fairly steadily with height from a mass ratio to dry air of about 1×10^{-2} near the ground to about 2×10^{-6} in the lower stratosphere. There are some recent indications of an increase of concentration with increase of height between about 20 and 35 km. Partial dissociation of water vapour by ultra-violet radiation into HYDROGEN (H) atoms and HYDROXYL (OH) molecules may be effective above about 60 km.

The meteorological importance of water vapour derives from the part it plays in forming cloud and precipitation elements, in controlling the long-wave radiation balance of the atmosphere, in determining atmospheric stability, and in affecting

PLATE 27   Waterspout in the eastern Mediterranean, 27 October 1943, 0910 GMT. View from aircraft at about 150 m.

PLATE 28   Waterspout seen from Malta, 14 October 1930, 1645 GMT.

the heat balance conditions of the earth–atmosphere system by the powerful absorption of heat in the course of evaporation and sublimation from liquid water and ice, and by the eventual release of the stored latent, or 'hidden', heat which is involved in the reverse processes.

The amount of water vapour held in the atmosphere is specified by various 'humidity elements' which include VAPOUR PRESSURE, HUMIDITY MIXING RATIO, RELATIVE HUMIDITY, VAPOUR CONCENTRATION, DEW-POINT, FROST-POINT and WET-BULB TEMPERATURE. Different types of HYGROMETER are commonly used to measure the humidity at different atmospheric levels.

Water vapour is by far the most strongly absorbing constituent of the atmosphere and has a wide range of absorption bands over a range of wavelengths extending from the near infra-red upwards—see Figure 1, p. 4. A conspicuous feature of the absorption spectrum, of particular importance at terrestrial radiation temperatures, is the region between about 5·5 and 7 μm in which water vapour is almost opaque—see also ATMOSPHERIC WINDOW.

See also SPECIFIC HEAT and LATENT HEAT for values referring to water vapour.

watt: The unit of power in SI UNITS. It is the rate at which energy is transformed into heat in a lamp using 1 ampere at 1 volt.

1 watt $\qquad$ = 1 JOULE per second
$\qquad\qquad\qquad$ = 10^7 ERGS per second
1000 watts = 1 kilowatt = $1\tfrac{1}{3}$ horse-power.

Units derived from the watt are used by meteorologists for the measurement of the intensity of radiation, e.g. 1 milliwatt/cm².

wave clouds: Clouds which form in the crests of MOUNTAIN WAVES.

wave depression: A DEPRESSION which forms at the tip of a wave-like distortion of a FRONT. Most of the depressions of middle and high latitudes are of this type.

wave motion: An oscillatory movement of the particles of a medium as the result of which 'waves' are propagated through the medium. If the particle movement is perpendicular to the direction of wave propagation, the waves are 'transverse'; if the particle movement is a rhythmic advance and retreat along the direction of wave propagation, the waves are 'longitudinal'. Energy but not, in general, matter is propagated with the waves. In ELECTROMAGNETIC RADIATION, periodic disturbances of electric and magnetic fields, and not movement of particles, are, from the viewpoint of electromagnetic theory, involved; such waves can be propagated through space or through a medium.

The simplest type of wave is the 'simple harmonic wave', as represented by a sine curve. The main characteristics of waves are the amplitude, a, (half the distance between the extremes of the oscillations), the wavelength, λ, (distance between successive maxima), the period, τ, (time interval between successive crests passing the same point), and the frequency, f, (number of complete oscillations per second). In a TIME SERIES the wavelength and period are identical. The speed of propagation, v, of the wave pattern (termed the PHASE velocity) is related to τ, λ and f by $v = \lambda/\tau$ and $v = f\lambda$. The wave number, k, is alternatively defined as $k = 1/\lambda$ or $k = 2\pi/\lambda$ (i.e. number of waves per unit distance, or 2π times this quantity).

A large variety of types of wave motion, or of quasi-wave motion, occurs in the atmosphere. See, for example, SOUND WAVES, SHOCK WAVES, ATMOSPHERIC TIDES, GRAVITY WAVE, SHEAR WAVE, INERTIA WAVE, LONG WAVE, BAROTROPIC WAVE, BAROCLINIC WAVE, LEE WAVES, WAVE DEPRESSION.

wave number: See WAVE MOTION.

wave recorder: An instrument for recording OCEAN WAVES. The record obtained from such an instrument shows the variation of the height of the sea surface above a fixed point with time.

weakening: See INTENSIFICATION.

weather: The changing atmospheric conditions, more especially as they affect man, which, in synthesis, constitute the CLIMATE of a region.

Weather in its wider sense is the study pursued in SYNOPTIC METEOROLOGY. In this branch of meteorology, however, the term itself is used in a more limited sense to denote the state of the sky and the occurrence of precipitation or of mist or fog. Codes of 'present weather' and of 'past weather' are two of the codes used in synoptic meteorology.

A concise system of notation of weather introduced by Admiral Beaufort is described under BEAUFORT NOTATION. A concise, international method of recording weather and optical phenomena by means of symbols is given in the following tables and text taken from the *International cloud atlas*.*

Symbols of meteors

In order to facilitate the representation and entry of meteors in meteorological documents, symbols have been assigned to most of them.

The following table summarizes the classification of meteors and shows the basic symbols.

TABLE OF METEORS AND THEIR SYMBOLS

Hydrometeors

Designation of meteor	Symbol	Designation of meteor	Symbol
Rain	•	Drifting or blowing snow	+→
Freezing rain	∽	Drifting snow	+↓
Drizzle	❜		
Freezing drizzle	∽	Blowing snow	+↑
Snow	✶	Spray	⏋
Snow pellets	⨉	Dew	⌒
Snow grains	—⌂—		
Ice pellets	△	White dew	
Hail	▲	Hoar-frost	⌐⌙
Ice prisms	↔	Rime	V
Fog	≡		
Ice fog	⇄	Glaze	∽
Mist	═	Spout	)(

---

* Geneva, World Meteorological Organization. International cloud atlas, Vol. 1. Geneva, WMO, 1956, pp. 63–65.

TABLE OF METEORS AND THEIR SYMBOLS—(*continued*)

Lithometeors

Designation of meteor	Symbol	Designation of meteor	Symbol
Haze	∞	Dust storm or sandstorm	⚡
Dust haze	S	Wall of dust or sand ...	⚡
Smoke	⌒⌒	Dust whirl or sand whirl (dust devil)	ε
Drifting or blowing dust or sand	$		
Drifting dust or sand ...	$		
Blowing dust or sand ...	$		

Photometeors

Designation of meteor	Symbol	Designation of meteor	Symbol
Halo phenomena: solar ...	⊕	Bishop's ring	⊙
lunar ...	▽	Mirage	⋊⋉
Corona: solar	⊙	Shimmer	No symbols established
lunar	∪	Scintillation	
		Green flash	
		Twilight colours	
Irisation on clouds ...	⊘	Crepuscular rays ...	
Glory	⊙		
Rainbow	⌒		
Fog bow	⌒		

Electrometeors

Designation of meteor	Symbol	Designation of meteor	Symbol
Thunderstorm	℞	Saint Elmo's fire ...	⚡
Lightning	⟨	Polar aurora	⌒
Thunder	T		

It is possible to provide information concerning the character (intermittent or continuous) and intensity (slight, moderate or heavy) of precipitation by certain arrangements of the basic symbols. The following table, established for rain, illustrates various arrangements which may be used for this purpose.

Intensity	Character	
	intermittent	continuous
slight	•	••
moderate	• •	•.• •
heavy (dense)	• • •	•.• •.•

Combinations of two basic symbols of meteors may be used to indicate the occurrence of mixed precipitation or the occurrence of a thunderstorm accompanied by precipitation or duststorm or sandstorm. For example, the symbol ⁂ or ⁂ denotes a mixture of falling raindrops and snowflakes; the symbol ⚡ indicates thunderstorm with rain at the place of observation.

In addition to the basic symbols, several auxiliary symbols have been established to provide information concerning the showery character of precipitation and also the variation with time of various meteors and their location with respect to the station. These symbols are the following:

▽   shower, slight

▽̄   shower, moderate or heavy

|×   has increased (or formed) during the preceding hour

×| has decreased during the preceding hour

×]   during the preceding hour, but not at the moment of observation

(×)   not at the station, but within sight [estimated distance less than 5 km (3 miles)]

)×(within sight and at an estimated distance of more than 5 km (3 miles)

Useful supplementary information about meteors can thus be given by combining the above auxiliary symbols with one, or sometimes two, basic symbols. For example, the symbol ≡| denotes fog which has become thinner during the preceding hour; the symbol ▽̇] indicates slight shower(s) of rain during the preceding hour, but not at the time of observation.

Weather Centre: In the Meteorological Office, an office set up to provide a service to commerce and industry and to supply meteorological information and advice to the general public through radio, television, the Press, by telephone and personal inquiry.

weather lore: Empirical weather forecasting rules, world-wide in origin, many of which are expressed in rhyme. They include rules based on the influence of the moon and tides, the appearance of plants and trees, the behaviour of animals, the weather prevailing on specified key dates and the colour and appearance of the sky. Comparison of the various rules reveals many contradictions, while tests of statistical significance lend no support except to certain of the short-period forecasting rules based on local observation of sky and wind, etc. See *Weather lore* compiled by R. Inwards.*

weather map: A chart of a geographical area on which selected meteorological elements observed at a particular time at various points over the area are plotted in symbolic code; the positions of mean-sea-level isobars and surface fronts (also, on occasion, of other features, for example isallobars) are subsequently drawn.

The elements usually plotted on the weather map, which is also termed 'synoptic chart' or 'surface chart', are: atmospheric pressure, reduced to mean sea level; barometric characteristic and tendency; wind direction and force; air temperature; dew-point; visibility; 'present weather'; 'past weather'; type, amount and height of clouds. In the case of ship observations, sea temperature, the direction and amplitude of the swell and the direction and speed of movement of the ship are also plotted. See SYNOPTIC METEOROLOGY.

weather report: Statement of the values of meteorological elements observed at a specified place and time. The elements included depend upon the purpose for which the report is required.

It is a record of an observation, not a FORECAST.

weather routeing: The planning of a (ship's) route according to predicted weather, wind, waves and other factors in such a way as to achieve the most economic passage.

weber: The unit (Wb) of magnetic flux in SI UNITS. See also TESLA.

wedge: In synoptic meteorology, an alternative for RIDGE (of high pressure); the term is mainly applied to a relatively fast-moving ridge on a surface synoptic chart.

weighted (moving) averages: See FILTERING, WEIGHTS.

weights: A termed used for coefficients introduced when averaging estimates of the same kind to give more weight to the more reliable estimates. If $x_1, \ldots x_n$ are the estimates and $w_1, \ldots w_n$ the weights, then

$$\frac{w_1 x_1 + w_2 x_2, \ldots + w_n x_n}{w_1 + w_2 + , \ldots + w_n}$$ is the weighted average. In this application the weights are generally all positive, and are often adjusted for computational convenience so that their sum is unity.

wet adiabatic: An alternative for 'saturated adiabatic'. See ADIABATIC.

wet air: A term used to define the condition when objects become wet even when rain is not falling. It occurs when a warm, saturated or practically saturated air mass replaces a cold dry air mass and is denoted by the letter 'e' in the BEAUFORT NOTATION.

---

* INWARDS, R.; Weather lore, 4th edn. London, Rider and Company, 1950.

wet-bulb depression: In a wet- and dry-bulb PSYCHROMETER, the amount by which the wet-bulb reading is below that of the dry bulb.

wet-bulb potential temperature: The wet-bulb potential temperature (θ_w) at any level is obtained on an AEROLOGICAL DIAGRAM as that temperature at which the saturated ADIABATIC through the WET-BULB TEMPERATURE at the level concerned intersects the 1000-mb isobar.

θ_w is for practical purposes conservative for such processes as evaporation or condensation and both dry adiabatic and saturated adiabatic temperature changes; it is therefore a useful property in AIR-MASS ANALYSIS.

wet-bulb temperature: That temperature (T_w) at which pure water must be evaporated into a given sample of air, adiabatically and at constant pressure, in order to saturate the air at temperature T_w under steady-state conditions. The temperature recorded by the WET BULB of a psychrometer may not exactly accord with this definition. See also THERMODYNAMIC TEMPERATURES, PSYCHROMETER.

wet day: Defined for statistical purposes as a period of 24 hours, commencing normally at 09 GMT, on which 1·0 mm (0·04 in) or more of rainfall is recorded. See also RAIN DAY.

wetness recorder: An instrument used in plant pathology to detect and record the amount of moisture deposited on foliage.

wet season: An alternative for RAINY SEASON.

wet spell: See RAIN SPELL.

whirling psychrometer: A PSYCHROMETER in which the thermometers are mounted on a frame which is rapidly rotated by hand in order to provide the required ventilation of the bulbs. It is also termed a 'sling psychrometer'.

whirlwind: A small revolving storm of wind in which the air whirls round a core of low pressure. Whirlwinds sometimes extend upwards to a height of many hundreds of metres and cause DUST WHIRLS when formed over a desert.

whistlers: A type of disturbance heard on a suitable radio receiver. It comprises a succession of whistles which progressively become fainter and take longer to fall through the audio range of frequencies.

The disturbance, strongest at about 55° geomagnetic latitude, originates in a burst of electromagnetic waves in the audio-frequency range, emitted by a lightning discharge or produced artificially. The interval between successive whistles corresponds to the time taken by the waves to travel along the lines of force to the geomagnetically conjugate point in the opposite hemisphere, and, on being reflected there, to return along the same path to the radio receiver. The arrival of the waves produces a whistle because the higher frequencies travel faster through the IONOSPHERE than do the lower. The initiating discharge is close either to the receiver or to the conjugate point.

Detailed study of the dispersion of frequencies in whistlers has led to inferences concerning the geomagnetic field and state of ionization above the ionosphere. See also DAWN CHORUS, SFERICS FIX.

white-out: A term applied to that condition in which the contours and natural landmarks in a snow-covered region become indistinguishable. The associated meteorological conditions appear to be a uniform layer of relatively low cloud;

under such conditions the light which reaches the surface arrives in nearly equal measure from all directions, with a resulting absence of shadows.

Wien's (displacement) law: See RADIATION.

willy-willy: The name given in Western Australia to a severe TROPICAL CYCLONE.

wilting point: The point at which the soil contains so little water (measured as per cent of dry soil) that it is unable to supply it at a rate sufficient to prevent permanent wilting of plants. It varies with the type and structure of the soil, being about 3 per cent for light sand and 20 per cent for heavy clay, and corresponds to a CAPILLARY POTENTIAL of about 150 metres of water (pF = 4·2). See SOIL MOISTURE.

wind: The (horizontal) movement of air relative to the rotating surface of the earth; the vertical component of air movement, generally much the smaller, is identified as such, where appropriate.

In meteorology, the specified wind direction is that, relative to true (geographic) north, from which the wind blows. The converse practice has, however, been used with certain high-level winds, e.g. those inferred from radio measurement of METEOR trails; more commonly, doubt is now removed in such cases by reference, for example, to a 'westward' wind, i.e. an east wind in the normal meteorological sense. The wind direction is generally specified as a bearing in degrees clockwise from true north; the compass point direction (8, 16 or 32 points according to the accuracy required) is also used. By international agreement (1956) the KNOT is the meteorological unit of wind speed. The specifications of the Beaufort forces are given under BEAUFORT SCALE, together with the dynamic pressure exerted on a flat disc. The relationships between the knot and alternative speed units are:

$$1 \text{ knot} = 0·515 \text{ m/s} = 1·152 \text{ mile/h} = 1·853 \text{ km/h} = 1·689 \text{ ft/s}.$$

Surface wind velocity is normally measured by some form of ANEMOMETER or ANEMOGRAPH, and WIND VANE. Appropriate EXPOSURE of the instrument to ensure reasonable comparability of observations at different stations is especially difficult in wind measurement; 'surface wind' in synoptic reports refers to an 'equivalent' height of 10 metres. Upper-level winds are normally measured by the PILOT BALLOON, or by RADAR WIND techniques. They may be inferred from measurements of the angular velocity of clouds, excluding certain (wave) types of cloud which do not move with the wind.

Wind velocity is intimately related to the pressure distribution in extratropical regions. In large-scale motion the relationship gives rise to BUYS BALLOT'S LAW, the GEOSTROPHIC WIND and the GRADIENT WIND, while, for example, changing pressure distribution in space or time is related to the AGEOSTROPHIC WIND. Height changes of wind velocity are generally considered in terms of the THERMAL WIND. On the more local scale the pressure–wind relationship gives rise to such winds as LAND- and SEA-BREEZES, the KATABATIC WIND and the ANABATIC WIND.

Diurnal variation of atmospheric TURBULENCE is associated with variations of surface wind velocity due to stronger vertical mixing of air at lower levels by day than by night (appreciable increase of speed and very slight veer by day). Associated with the systematic diurnal variation of pressure is a variation of wind velocity of an amplitude which is too small at the surface to be revealed except by averaging over a long period but which increases with height to such an extent as to be a prominent feature of winds at very high levels—see ATMOSPHERIC TIDES.

wind-break: A term sometimes used in such a sense as to include both natural and artificial barriers to wind flow which provide shelter to animals, crops, etc. More usually the term is now restricted to artificial barriers (palings etc.). See also SHELTER-BELT.

The degree of shelter from wind afforded by a barrier depends, among other factors, on the height, lateral extent and permeability of the barrier. Nevertheless, it is now generally accepted that whatever the type of barrier employed, the shelter zone for a barrier of height h reaches to a distance of about $30h$ in the lee of the barrier, and that for significant wind-speed reductions of 20 per cent or more, only distances of up to $20h$ should be considered. Strongly eddying motion is a feature of the airflow to a distance of about $15h$ downwind from a dense barrier.

Other physical effects of wind-breaks include alterations of air temperature and humidity, soil temperature and moisture, and evaporation rate over the region affected by the presence of the barrier.

wind rose: One of a class of diagrams illustrating, for a particular place and extended period, the relative frequencies of wind direction and, in general also, of wind speed. A common form is illustrated in Figure 35; the length of each symbol is proportional to its frequency of occurrence and the radius of the circle is proportional to the frequency of calms which is also written within the circle.

A wind rose may be adapted to demonstrate the relationship between wind velocity and other meteorological variables.

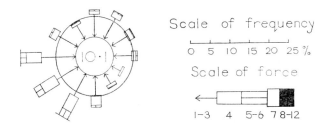

FIGURE 35—Wind rose for London/Heathrow Airport, 1949–58, showing annual frequency of wind direction and velocity.
The number in the inner circle gives the percentage frequency of calms. The other frequencies are measured from the inner circle and the outer circle indicates the 5 per cent frequency.

wind shear: The rate of change of the vector wind (**V**) with distance (n) in a specified direction normal to the wind direction, or $\partial \mathbf{V}/\partial n$. If the vector wind is horizontal then with the usual notation the 'horizontal shear' is the vector with components $\partial u/\partial y$, $\partial v/\partial x$, and the 'vertical shear' is the vector with components $\partial u/\partial z$, $\partial v/\partial z$; these reduce to $\partial V/\partial y$ and $\partial V/\partial z$ respectively when the wind vector is directed along the x-axis.

Shear is an important property of the wind field. For example, shear at a well-marked frontal surface may produce SHEARING INSTABILITY, while horizontal shear is closely associated with the vertical component of VORTICITY. The integral of wind shear through a vertical interval is the vector difference between the (horizontal) winds at the top and bottom of the interval and so is identical with the THERMAL WIND if the motion is geostrophic. See also JET STREAM.

wind vane: A device for indicating or recording the direction from which the wind is blowing. It usually consists of a horizontal arm carrying at one end a fin, either a vertical flat plate with its edge to the wind or an aerofoil, and at the other end a balance weight which also serves as a pointer; the arm is carried on a vertical spindle mounted on bearings which allow it to turn freely in the wind.

For indication only, the position of the vane can be estimated with reference to a fixed cross below the vane showing the four cardinal points. For more accurate indication, the spindle may be coupled by a suitable direction-transmitting rod to a pointer and dial situated at a convenient level below the vane. For remote indication and recording, the vane spindle is coupled to some form of electrical transmitter by means of which the movements of the vane are reproduced at a distance, and displayed as dial and pointer indications, or pen recordings on a chart.

windward: Windward of a point signifies the 'upwind' direction from the point, e.g. westward in the case of a west wind.

winter: See SEASONS.

WMO: Abbreviation for WORLD METEOROLOGICAL ORGANIZATION.

World Meteorological Organization: The World Meteorological Organization (WMO) is a specialized agency of the United Nations encompassing the field of meteorology. It replaced the IMO (International Meteorological Organization) in 1951. The WMO comprises over 130 States and Territories and has a permanent Secretariat in Geneva.

The purposes of the WMO are, *inter alia*, to facilitate world-wide co-operation in the establishment of networks of meteorological observation stations and to promote the development of centres charged with the provision of meteorological services; to promote the rapid exchange of weather information and the standardization of meteorological observations and their publication; to further the application of meteorology to human activities and to encourage research and training in meteorology.

World Weather Watch: A world-wide meteorological system composed of the national facilities and services provided by individual Members, co-ordinated and in some cases supported by WMO and other international organizations. Its primary purpose is to ensure that all Members obtain the meteorological information they require both for operational work and for research. Its essential elements are: the Global Observing System, the Global Data-Processing System, the Global Telecommunications System, and research and training programmes.

WWW: An abbreviation for WORLD WEATHER WATCH.

X, Y, Z

xenon: One of the INERT GASES, comprising $8 \cdot 0 \times 10^{-8}$ and $3 \cdot 6 \times 10^{-7}$ part per part of dry air by volume and weight, respectively. Its molecular weight is $131 \cdot 3$.

x-rays: ELECTROMAGNETIC RADIATION in the approximate band of wavelengths from $0 \cdot 1$ to 10 ångströms. The x-rays which are contained in the solar radiation incident on the high atmosphere are responsible for an appreciable part of the IONIZATION of the region.

year: The time taken by the earth to revolve once in its orbit round the sun. See CALENDAR.

zenith: The point of the sky in the vertical produced upwards from the observer. The word is now commonly used to denote a more-or-less extensive stretch of sky immediately overhead.

zenith distance: The zenith distance of a body is the angle between the body and the ZENITH, as observed at a particular point of observation.

zenith, magnetic: The direction indicated by the upper end of a suspended magnetic needle. In the north of the British Isles it is some 17°, and in the south some 23°, south-south-east of the geographical zenith. See DIP, MAGNETIC.

zephyr: A westerly breeze with pleasant warm weather supposed to prevail at the summer solstice.

zero: The point of origin in the graduation of an instrument; for example, the freezing-point of water on the CELSIUS SCALE of temperature is assigned the value of '0'. An error in the positioning of the entire scale of an instrument may be regarded as an incorrect location of the zero, and the term 'zero error' is commonly applied to it.

zero-temperature level: Term used, in aircrew briefing, for the lowest height above mean sea level of the 0°C isotherm. See FREEZING LEVEL.

zodiac: The series of constellations in which the sun is apparently placed in succession, on account of the revolution of the earth round the sun, are called the Signs of the Zodiac, and in older writings give their names and symbols to the months, thus:

Month	Symbol	Month	Symbol
March	Aries, the Ram	September	Libra, the Scales
April	Taurus, the Bull	October	Scorpio, the Scorpion
May	Gemini, the Heavenly Twins	November	Sagittarius, the Archer
June	Cancer, the Crab	December	Capricornus, the Goat
July	Leo, the Lion	January	Aquarius, the Watercarrier
August	Virgo, the Virgin	February	Pisces, the Fishes

Owing to precession, the position of the equator relative to the zodiacal constellations has altered a good deal since classical times. The sun now enters Aries late in April and reaches the other zodiacal constellations with the same retardation, but in textbooks of astronomy the point at which the sun crosses the equator at the spring equinox, 21 March, is still called the first point of Aries.

zodiacal band: A very faintly luminous band, a few degrees wide, joining the apices of the morning and evening ZODIACAL LIGHTS.

zodiacal light: A cone of faint white light in the night sky, extending along the ZODIAC from the western horizon after evening TWILIGHT and from the eastern horizon before morning twilight.

The phenomenon is caused by the scattering of sunlight from a cloud of particles lying in the ecliptic. The composition and origin of these particles—whether of dust or molecules or electrons, solar or terrestrial—is not yet certain. Molecular emission may also play a part.

zonal flow: West-to-east airflow. East-to-west airflow is generally reckoned as negative zonal flow.

zonal index: A measure of the strength of the ZONAL CIRCULATION, either at the surface or in the upper air, for a specified (large) area and period of time. Thus, for example, the mean surface pressure difference between the circles of latitude 35° and 55° is a convenient zonal index of surface airflow for mid-latitudes of the northern hemisphere.

zonal wave number: The number of complete wave forms, usually of the contours of a particular isobaric surface, around a selected circle of latitude. Such wave numbers (and the corresponding amplitudes) may be derived by the application of HARMONIC ANALYSIS to contour-height data.

The small wave numbers refer to large-scale atmospheric motion. Wave number one is a measure of the eccentricity of the flow relative to the pole; all other wave numbers represent those parts of the total flow which are symmetric about the pole, ridge opposing ridge in even-number flow and ridge opposing trough when the wave number is odd. See WAVE MOTION.

zone of silence: See AUDIBILITY.

zone time: A system of local TIME classification, differing from GMT in steps of 1 hour per 15° of longitude. The individual zones are distinguished by the letters A, B, C, etc. (omitting J) for areas centred on 15°E, 30°E, 45°E, etc., respectively; and by the letters N, O, P, etc. for areas centred on 15°W, 30°W, 45°W, etc., respectively. GMT is in this system designated Z time.